Virginia
1860 Agricultural Census
Volume 3

Linda L. Green

WILLOW BEND BOOKS
2011

WILLOW BEND BOOKS
AN IMPRINT OF HERITAGE BOOKS, INC.

Books, CDs, and more—Worldwide

For our listing of thousands of titles see our website
at
www.HeritageBooks.com

Published 2011 by
HERITAGE BOOKS, INC.
Publishing Division
100 Railroad Ave. #104
Westminster, Maryland 21157

International Standard Book Numbers
Paperbound: 978-0-7884-5308-3
Clothbound: 978-0-7884-8734-7

Introduction

This census names only the head of the household. Often times when an individual was missed on the regular U. S. Census, they would appear on this agricultural census. So you might try checking this census for your missing relatives. Unfortunately, many of the Agricultural Census records have not survived. But, they do yield unique information about how people lived. There are 48 columns of information. I chose to transcribe only six of the columns. The six are: Name of the Owner, Improved Acreage, Unimproved Acreage, Cash Value of the Farm, Value of Farm Implements and Machinery, and Value of Livestock. Below is a list of other types of information available on this census.

Linda L. Green
217 Sara Sista Circle
Harvest, AL 35749

Data Columns

Column/Title

1. Name
2. Acres of Land Improved
3. Acres of Land Unimproved
4. Cash Value of Farm
5. Value of Farm Implements and Machinery
6. Horses
7. Asses and Mules
8. Milch Cows
9. Working Oxen
10. Other Cattle
11. Sheep
12. Swine
13. Value of Livestock
14. Wheat, bushels of
15. Rye, bushels of
16. Indian Corn, bushels of
17. Oats, bushels of
18. Rice, lbs of
19. Tobacco, lbs of
20. Ginned cotton, bales of 400 lbs each
21. Wood, lbs of
22. Peas and beans, bushels of
23. Irish potatoes, bushels of
24. Sweet potatoes, bushels of
25. Barley, bushels of
26. Buckwheat, bushels of
27. Value of Orchard products in dollars
28. Wine, gallons of
29. Value of Products of Market Gardens
30. Butter, lbs of
31. Cheese, lbs of
32. Hay, tons of
33. Clover seed, bushels of
34. Other grass seeds, bushels of
35. Hops, lbs of
36. Dew Rotten Hemp, tons of
37. Water Rotted Hemp, tons of
38. Other Prepared Hemp
39. Flax, lbs of
40. Flaxseed, bushels of
41. Silk cocoons, lbs of
42. Maple sugar, lbs of

43. Cane Sugar, hunds of 1,000 lbs
44. Molasses, gallons of
45. Beeswax, lbs of
46. Honey, lbs of
47. Value of Home Made Manufactures
48. Value of Animals Slaughtered

Table of Contents

Hanover County, Virginia
1860 Agricultural Census

The Agricultural Census for Virginia 1860 was microfilmed by the University of North Carolina Library under a grant from the National Science Foundation from original records at the Virginia Department of Archives and History in 1963.

There are forty-eight columns of information on each individual. Only the head of household is addressed. I have chosen to use only six columns of information because I feel that this information best illustrates the wealth of individuals. The columns are:

1. Name
2. Improved Acres of Land
3. Unimproved Acres of Land
4. Cash Value of the Farm
5. Value of Farming Implements and Machinery
13. Value of Livestock

Jno. L. Thompson, 120, 400, 8000, 100, 1250
Jno. Pryor, 1 ¾, 5, 2000, 50, 300
Jno. S. Janes, 2 ½, -, 4000, 20, -
Jas. A. Harwood, ¾, -, 1200, -, 400
Chas. B. Stuart, ½, 5, 1100, 10, 210
David H. Smith, 1 ½, -, 2000, -, -
Saml. D. Leake, 40, 25, 8000, 100, 500
W. J. Leake, 1 ½, -, 200, -, -
Jno. K. MacMurdo, 10, 15, 4000, 50, 150
Isaac Cennon(Cannon), ¾, -, 1000, -, -
Jno. S. James, ¾, -, 2000, -, -
Hiram James, ¾, -, 1000, 10, 20
Nancy Cross, 150, 80, 2170, -, -
F. Waldrop, 50, 65, 1200, 150, -
Jesse Harlow, 125, 150, 2700, 200, -
M. M. Gray, 10, 60, 600, 15, -
W. G. Thomason, 30, 100, 1040, 100, -
Wm. A. Wren, 10, 38, 240, -, -
O. H. Harris, 75 176 ½, 125750, 50, -
B. J. Nash, 250, 148, 2500, 200, -

Susan Cauthan, 50, 100, 750, -, -
S. A. Ellis, -, -, -, 10, -
Jas. F. Bowles, 75, 75, 20000, 300, -
W. J. Hall, 7, 30, 228, -, -
Gemima Harris, 5, -, 100, -, -
W. J. Harris, 30, 25, 225, 30, -,
J. B. Bowles, -, -, -, 50, -
J. Freeman, 5, 30, 170, -, -
Nancy Ford, 72, 100, 500, -, -
J. B. Alley(Allen), -, -, -, 30, -
W. Chiles, 250, 320, 5660, 400, -
Chas. Toler, 15, 10, 300, 25, -
A. Thorp, -, -, -, 10, -
S. Hogg, 5, 5, 200, -, -
E. A. Hutchinson, 15, 100, 200, -, -
A. L. Deitrick, 30, 60, 1800, 50, -
Jno. O. Kelley, -, -, -, 50, -
S. T. Bowles, 100, 109, 1000, 100, -
Fleming Tatmon, 60, 60, 1300, 600, -
A. McDowell, 150, 75, 6480, 200, -
Wm. A. Dick, 500, 1000, 15000, 600, -
T. C. Woodson, 30 700, 800, 30, -
R. M. Carver, 120, 317, 3500, 100, -
Jas. T. Francis, 20, 20, 1200, 50, -

Wm. K. Winn, 300, 350, 9750, 500, -
C. Davis, 50, 150, 1000, 10, -
E. Gurthrow, 50, 100, 1500, 20, -
Jno. J. Davis, 80, 100, 1800, 100, -
Jno. D. Lively, 90, 60, 2000, 250, -
F. A. Woodson, 80, 100, 1800, 100, -
Benj. Thomas, 100, 240, 3000, 20, -
Geo. Kelley, 50, 100, 1500, 40, -
W. Q. Laurence, -, -, -, 30, -
T. C. Woodson, 30, 70, 800, 20, -
John Edwards, 40, 60, 750, 30, -
John Thacker, 10, 40, 500, 10, -
A. C. Davis, 100, 130, 2300, 200, -
J. _. Luck, -, -, -, 30, -
John Dulany, 60, 24, 850, 50, -
B. Stanley, 400, 412, 8700, 200, -
T. J. Woolridge, 257, 60, 12500, 700, -
R. Jenkins, 175, 625, 16000, 500, -
Wm. Anderson Jr., 450, 450, 9000, 1000, -
Jas. Stone, -, -, -, -, -
W. N. Waldrop, 200, 107, 3500, 150, -
John Christian, -, -, -, 30, -
John S. Jenkins, 100, 60, 1280, 50, -
John Snead, 100, 130, 1840, 200, -
R. A. Thornton, 100, 207, 6141, 100, -
H. Vass, -, -, -, 5, -
L. Bowles, 10, 40, 1200, -, -
W. B. Harris, -, -, -, 30, -
A. R. Chapman, 50, 50, 2000, 10, -
John M. Swift, 25, 25, 1000, 200, -
Lewis Collins, 10, 30, 400, -, -
E. B. Toler, 45, 65, 900, 50, -
Wm. H. Winston, 350, 96, 7000, 30, -
Jane D. Winston, 583, 398, 25000, 1000, -
Edmund Winston, -, -, -, 1000, -
N. B. Priddy, -, -, -, -, -
W. C. Wickham, -, 10, 200, -, -
G. W. Gilmore, -, -, -, 5, -
Thos. Bowen, 75, 65, 2800, 100,-
E. T. Brooks, -, -, 100, 30, -
Thos. M. Blackburn, -, -, -, 30, -
Wm. W. Toler, 90, 100, 1500, 50, -
E. Davis, 40, 10, 500, 30, -
N. Davis, 50, 46, 480, -, -
S. Redd, 100, 65, 1320, 300, -
L. A. Cunshaw, -, -, -, 30, -
P. Perrin, 40, 30, 1400, 50, -
W. T. Huffman, 6, 194, 1500, 200, -
Parke Perrin, -, -, -, -, -
P. H. Tignor, 100, 12, 3000, 100, -
R. Kersey, 50, 20, 700, 200, -
J. W. Tignor, -, -, -, 30, -
John R. Anderson, 100, 139, 5000, 300, -
S. R. Wingfield, 60, 35, 800, 30, -
H. Kersey, 50, 65, 1800, 100, -
B. Kersey, 60, 24, 1600, 50, -
P. C. Wingfield, -, -, -, 30, -
Wm. J. King, 30, 80, 1500, 200, -
P. H. Perrin, 40, 37, 700, 200, -
J. L. Patterson, 40, 56, 500, 200, -
J. D. King, 35, 70, 1050, 15, -
Tho. Eubank, 150, 450, 4800, 200, 100
Geo. C. Morgan, 75, 70, 1500, 150, 100
Jno. C. Brock, 400, 400, 8000, 200, 1000
R. A. Fox, 400, 100000, 50000, 1000, -
F. Carter, 200,123, 3330, 300, 300
J. M. Bazile, 50, 25, 1500, 150, 250
J. B. Jones, -, -, -, 100, 2000
Wm. Carter, 50, 50, 1000, 50, 100
H. M. Yarbrough, -, -, -, 50, 100
S. P. Wingfield, 100, 126, 2712, 200, 500
Geo. W. Peatross, 60, 30, 1000, 100, 100
J. F. Satterwhite, 200, 400, 7000, 100, 600
P. Jones, -, -, -, 30, 20
Robt. Smith, 75, 55, 1920, 75, 400
N. H. B. Campbell, 160, 140, 3600, 30, -

W. L. Wingfield, 50, 150, 1570, 30, 500

Jos. S. Wingfield, 300, 300, 12000, 600, 50

S. Ford, -, -, -, -, -

Jas. D. Bourne, 100, 167, 2740, 30, 150

Jas. T. Sutton, 700, 379, 21580, 1000, -

Benj. Smith, 6, 10, 500, 30, -

P. Slaughter, 100, 100, 2000, 30, 350

N. Yarbrough, 100, 100, 2000, 200, 300

Mary Bowe, 200, 220, 6000, 300, 1000

J. F. Cross, 200, 472, 7320, 200, 200

M. M. Green, 60, 46, 1378, 50, 200

Jas. E. Leitch, -, -, -, 200, 200

J. C. Browning, -, -, -, 30, 150

W. Wyatt, 120, 131, 5000, 3, 400

Mary Cross, 70, 20, 900, 500, 500

Henry Brice, 200, 204, 5000, 500, 1800

Wm. Priddy, -, -, -, 200, 250

C. Wyatt, 50, 80, 1570, 100, 175

A. B. Timberlake, 900, 385, 28700, 700, 1920

W. H. Vaughan, 300, 133 ½, 10337, 150, 600

W. H. Timberlake, 200, 75, 6870, 200, 700

D. A. Timberlake, -, -, -, 100, 800

N. A. Hazlegrove, 20, 11, 360, 30, 250

N. B. Clarke, 300, 254, 13850, 300, 1500

E. C. Clarke, 300, 100, 6000, 100, 500

E. W. Poindexter, -, -, -, 50, 350

W. F. Atkinson, 30, 40, 700, 20, 120

A. Short, 300, 408, 9000, 100, 400

R. E. Gardner, 90, 35, 1800, 40, 500

D. B. Moore, 75, 50, 3235, 30, -

Thos. W. Sydnor, 140, 58, 4950, 50, 600

N. Bowe, -, -, -, 150, 400

A. S. Jones, 125, 75, 6000, 50, 300

E. Poindexter, 150, 70, 6600, 200, 800

H. Teasrum, 20, 10, 900, 20, 400

Jane Timberlake, 200, 100, 4920, 30, 400

Geo. Thomas, 50, 50, 250, 10, 250

T. N. Green, 150, 50, 1630, 20, 300

S. F. McKenzie, 100, 90, 1700, 100, 400

P. Tinsley, 400, 300, 10500, 300, 2000

W. N. Gardner, 200, 185, 5770, 100, 600

S. N. Hill, 300, 100, 9100, 300, 900

E. P. Meridith, 85, 80, 3500, 100, 400

Thos. Gentry, 40, 26, 780, 20, -

Geo. Carter, 100, 88, 3760, 100 700

R. Gentry, 30, 30, 1800, 50, 600

Jas. D. Poindexter, 30, 20, 1400, 50, 300

C. R. Cullen, 100, 30, 5000, 200, 1000

F. _. Fleshman, 4, 13, 1000, -, 150

C. P. Courtney, 5, 20, 700, -, 100

P. Davis, 5, 20, 700, -, -

Jos. Grubbs, 30, 18, 2000, 25, 250

K. H. Hooper, 1 ½, -, 300, -, 175

Jas. E. Jones, 150, 70, 4220, 30, 500

Tho. H. Bowles, -, -, -, 50, 125

N. B. Ladd, 70, 50, 3600, 100, 400

Wm. C. Fleshman, -, -, -, 50, 220

R. E. Atkinson, 40, 20, 600 75, 120

J. G. Lumpkin, 350, 150, 23500, 500, 2400

Thos. B. Bates, 58, -, 1900, 30, 300

W. Catlin, 150, 25, 23000, 100, 800

L. C. Binford, 200, 30, 9200, 500, 600

J. C. Ellyson, 126, 50, 12000, 110, 600

W. S. Austin, 170, 5, 2134, 50, 500

W. B. Sydnor, 300, 500, 2000, 2000, 2500

L. M. Nunelley, 10, 50, 5500, 50, 300
V. O. Corby, 25, 25, 1110, 50, 200
S. Grubbs, 50, 41, 1820, 20, 300
G. W. Puller, 100, 20, 2000, 20, 300
E. Sydnor, 375, 155, 11500, 700, 1500
J. B. Bussick, 50, 50, 3000, 100, 400
H. Curtis, 250, 202 ½, 7000, 350, 1000
W. Hogan, 350, 107, 13500, 600, 1800
Jas. O. McGhee, -, -, -, 30, 150
Jno. R. Starke, 175, 40, 6450, 100, 400
Jas. White, -, -, -, -, -
Jas. Tucker, 3, -, 120, 20, -
W. Tucker, 4, 4, 80, -, -
Martha Talley, 100, 92, 3840, 20, 550
John A. Mosby, 200, 160, 10800, 600, 1000
H. M. Turner, 100, 92, 5700, 50, 500
F. W. Johnson, 300, 130, 10754, 300, 962
H. Hughes, 10, 20, 500, 10, 150
J. King, 125, 125, 3750, 100, 600
John Mills, 120, 30 3000, 50, 500
H. Basher, 72, 25, 850, 50, 100
J. Gibson, 100, 57, 3440, 100, 600
C. Hay, -, -, -, 10, 160
C. Martin, 10, 21, 600, 20, 100
M. A. Dustry, 6, -, 100, 50, 150
E. Dustry, 7, -, 102, 30, 120
John Basket, 40, 43, 2050, 30, 200
J. Jenkins, 25, 10, 396, 20, 120
B. Martin, 30, 40, 1460, 30, 175
H. B. Christian, 50, 59, 2180, 40, 200
M. L. Carlton, 100, 56, 4680, 50, 350
W. G. Glass, 25, 25, 1250, 10, 100
H. C. Richardson, 50, 50, 1200, 30, 150
T. Tucker, 14, 14, 360, 50, 150
E. J. Martin, 10, 16 ½, 312, 20, 80
E. Curby, 10, 20 ½, 360, -, 40

J. Matthews, -, -, -, 50, 260
J. T. Adams, 50, 74, 1250, 20, 250
C. Martin, 30, 26 ½, 560, 30, 150
A. G. Allison, 130, 30, 1900, 10, 420
David Woody, 150, 125, 3750, 60, 250
R. Burnett, 300, 400, 8400, 200, 900
W. Gibson, 300, 310, 9150, 160, 1000
W. Gilman, 10, 50, 3000, 10, 30
H. Dunn, 25, 25, 1500, 20, 100
J. H. Mantlo, 100, 110, 1680, 50, 450
E. Mantlo, 30, 12, 680, 30, 200
W. Tucker, 80 48 ½, 1270, 30, 400
H. Tucker, 100, 130, 3650, 50, 752
B. Talley, 69 ½, 100, 3480, 40, 400
Jesse Tucker, 100, 69, 4550, 40, 500
A. Martin, 60, 40, 1200, 30, 400
W. Allen, 200, 450, 9080, 140, 1100
E. L. Crenshaw, 150, 650, 39000, 1000, 2310
W. B. Hudson, 75, 26, 3000, 30, 500
R. Kent, 15, 30, 900, 10, 100
J. Crouch, 30, 40, 800, 20, 150
W. Warren, 25, 25, 1250, 10, 160
_. T. Shelton, 20, 134, 4920, 30, 500
W. T. Taliaferro, 134, 100, 4780, 50, 950
_. Shelton, 488, 400, 18860, 1000, 1484
T. W. Talley, 100, 465, 7725, 150, 650
H. C. Wingfield, 177, 32, 3600, 20, 370
N. H. Talley, 41, 72, 2800, 50, 290
A. Munder, 20, 20, 600, 20, 200
A. Tinsley, 50, 34, 1260, 20, 230
W. Jones, 600, 416, 25400, 1000, 3570
G. R. Hundley, 150, 131, 5270, 250, 800
Geo. W. Richardson, 170, 130, 9000, 150, 1600
E. C. Tuck, 50, 34, 2100, 40, 300
R. Martin, 30, 8, 950, 20, 300
Jas. Basher, 50, 53, 3500, 100, 400

B. Warren, 25, 25, 1000, 30, 200
J. E. Cosby, 50, 50, 2500, 20, 300
E. Collens(Colleus), 30, 40, 840, 10, 300
H. Pollard, 50, 25, 2250, 30, 300
W. Whitlock, 30, 40, 1400, 10, 300
F. T. Whitlock, 100, 30, 1300, 20, 400
Jesse Boaze, -, -, -, 10, 250
A. T. Arnold, 20, 10, 300, 10, 130
Wm. Gaines, 700, 398, 22320, 150, 2000
J. C. McGhee, 60, 24, 1008, 100, 370
W. F. Gaines, 1000, 800, 64000, 1800, 3135
Jos. Adams, 200, 160, 150, 2, 740
S. B. Wall, 250, 279, 10540, 150, 800
S. McGhee, 300, 234, 6408, 150, 500
T. Daverson, 40, 20, 762, 40, 120
R. Parsons, 30, 40, 600, 50, 100
A. D. Woody, 30, 36, 690, 30, 130
T. J. Wade, 40, 44, 880, 2, 100
T. Fauster, 100, 39, 1608, 100, 400
T. A. Boaze, 70, 54, 1240, 50, 150
C. Stuart, 20, -, 504, 30, 400
M. Gathright, 75, 45, 1440, 70, -
M. Tucker, 150, 210, 4352, 60, -
J. F. McGhee, 60, 24, 1600, 50, -
J. J. Williams, 100, 125, 4500, 150, -
Jesse Barker, 17, 200, 3600, 60, 320
C. Barker, 52, 10, 550, 30, -
W. E. Tyler, 133, 133, 2660, 100, 550
E. White, 84, -, 1008, -, -
J. M. White, 29, -, 300, 50, 200
W. E. Gauldin, 50, 48, 980, 30, -
N. Wickes, 175, 175, 3500, 50, 300
B. Gathright, 47 ½, 47 ½, 970, 30, 150
A. D. Wickes, 150, 75, 2250, 235, 900
W. H. McGhee, 75, 75, 1500, 60, 300
G. W. Livesay, 100, 77, 1880, 60, 200

J. W. Peace, 200, 35, 1350, 60, 300
J. C. Gauldin, 40, 20, 600, 15, 100
W. H. Tyler, 117, 117, 3460, 150, 700
G. Barker, 250, 250, 5000, 60, 700
W. Parsley, 118, 118, 2360, 60, 500
Ann Wade, 259, 129, 3380, 150, 900
S. Sigon, 199, 194, 3880, 150, 1000
E. D. Wade, 116 88, 2660, 160, 700
P. P. Moore, -, 222, 2220, -, -
Mary Alexander, 200, 100, 2400, 60, 600
John P. Parlsey, 229, 229, 4580, 150, 800
S. C. Via, 100, 70, 1360, 50, 300
Jos. Parsley, 200, 175, 6590, 150, 1000
W. A. Baker, 250, 450, 14000, 150, 800
N. T. Lipscomb, 140, 320, 6750, 150, 700
E. F. & E. White, 650, 650, 9600, 200, 900
M. F. Anderson, 200, 200, 4800, 150, 1400
S. West, 150, 160, 3840, 50, 250
M. W. Ellett, 400, 400, 1200, 500, 1400
C. H. Dabney, 200, 100, 6000, 550, 500
J. S. Cullen, 200, 117, 9500, 200, 1800
P. C. Clopton, 200, 70, 4050, 150, 770
J. H. Ernest, 150, 160, 4340, 150, 800
T. G. Turner, 556, 200, 25000, 300, 2000
John Beal, 200, 20, 8800, 60, 900
G. WS. Bassett, 2500, 2150, 139830, 1500, 5120
Wm. J. Alexander, 100, 100, 3000, 150, 700
Thos. & W. T. Otey, 100, 150, 2580, 60, 500
James Brown, 50, 200, 1700, -, -

Nancy Irons, 30, -, 1000, -, -
P. B. Jurden, 100, 145, 1690, 50, 300
P. D More, 50, 150, 2000, 50, -
Mary Brown, 50, 150, 1600, -, -
Margaret Brown, 50, 150, 1600, -, -
A. Laughter, 80, 100, 1800, 60, 300
Martha E. Wood, 150, 100, 3000, 50, 250
M. M. Peace, 120, 125, 25550, 60, 300
Martha Adams, 6 ½, 13, 120, -, -
Mary Adams, 5 ½, 15, 100, -, -
Mary Wade, 5 ½, -, 120, 30, 200
Thos. Adams, 30, -, 360, 20, 100
Thos. Terry, 30, 65, 950, 10, 200
Wm. H. Hughes, 30, 20, 800, 50, 250
Clocke & Oliver, 900, 300, 48000, 2000, 4245
J. T. Street, 132, 100, 3550, 300, 1500
M. G. Braxton, 1000, 310, 45000, 1500, 5400
Wm. P. Braxton, 180, 130, 6020, -, 460
J. G. Mantilow, 60, 27, 2055, 40, 400
R. M. Tomlin, 410, 125, 21000, 300, 700
Benjamin Tyler, 50, 50, 1000, 50, 180
J. G. Mantilow, 100, 80, 2055, 150, 500
M. L. Burton, 100, 40, 1200, 50, 250
J. W. Tomlin, 230, 198, 17120, 500, 1350
M. E. Tomlin, 300, 100, 14000, 500, 1500
John Wright, 100, 50, 1820, 60, 670
J. A. Lipscomb, 100, 72, 6000, 60, 600
Wm. Philips Jr., 130, 250, 7600, 150, 400
B. Wickes, 40, 40, 4000, 40, 400
Lewis Johnson, 143, 140, 5720, 150, 700
Wm. Says & Company, 500, 555, 58200, 2000, 2800

Michael Jas & J. A. Meridith, 800, 400, 36000, 100, 3500
E. W. Mills, 300, 150, 5400, 100, 650
Mary W. Brown, 30, -, 600, -, -
Spotswood Wickes, 30, 53, 2075, 60, 175
M. W. Pate, 200, 300, 7700, 500, 700
E. T. Tally, 400, 340, 11300, 200, 700
Wm. E. Norment, 300, 221, 7810, 250, 800
B. W. Tally, 250, 200, 6000, 250, 3300
Sarah B. Tally, 150, 112, 3000, 100, 200
R. R. Horne, 150, 83 ½, 3402, 150, 700
Gilson Via, 250, 317, 11355, 500, 800
P. R. Norment, 400, 239, 14000, 150, 1300
Wm. A. Tignor, 100, 60, 2400, 50, 500
Henry R. Kerby, 25, -, 500, 20, 300
B. Suthard, 20, 5, 500, 10, 150
James Via, 60, 45, 1200, 50, 400
E. Gaulden, 100, 70, 3400, 100, 400
Miss Slaughter, 100, 70, 2040, -, -
D. Melton & Sisters, 100, 66, 1772, -, -
J. L. Linney, 75, 30, 2100, 60, 300
C. Hughes, 60, 20, 1600, 70, 500
Wm. Bowles, -, -, -, 50, 300
Lewis King, 10, 11, 240, 10, 200
R. Tate, 40, 26, 1300, 30, 300
B. Smith, 40, 40, 1000, 50, 250
J. E. Burnet, 100, 70, 1770, 30, 250
M. Whitlock, 200, 100, 3600, 50, 450
T. Butler, 5 ½, -, 100, 5, 60
J. Bowles, 60, 40, 2000, 60, 400
T. Glass, 5, 100, 1000, 5, 100
B. Bowles, 60, 30, 1800, 50, 400
F. T. Tate, 60, 31, 1820, 50, 300

A. R. Timberlake, 100, 50, 3000, 20, 150

B. Suthard, 20, 5, 500, 10, 135

J. P. Songan, 100, 87, 3200, 80, 500

L. Overton, 100, 30, 2500, 60, 600

N. B. Tyler, 200, 244, 5328, 50, 400

H. Timberlake, 40, 424, 820, 20, 150

S. Hatt(Hott), 60, 40, 1000, 80, 150

L. L. Pollard, 260, 200, 6900, 200, 700

E. _. A. T. Archer, 100, 30, 1460, 30, 150

A. Q. Woody, 25, 25, 600, 40, 160

P. D. Woody, 65, 65, 1300, -, 150

J. J. Woody, 115, 115, 2772, 70, 300

A. C. Blake, 150, 80, 3060, 60, 354

Blake for G. Taylor, 800, 400, 24000, 500, 1400

John Thomas, 60, 20, 960, 40, 150

M. M. Hair, 120, 145, 5300, 60, 500

John Hair, 200, 120, 1172, 400, 700

C. P. Cross, 100, 60, 1920, 60, 300

G. W. Pollard, 800, 300, 22000, 1200, 1844

G. W. Elliot, 200, 77, 5540, 300, 650

T. R. Bickenbough, 430, 150, 20550, 600, 2117

Wm. B. Newton, 700, 300, 30000, 1000, 3840

Wm. R. Nelson, 400, 610, 22250, 500, 3000

Martha Hunley, 600, 400, 20000, 300, 2000

J. H. Taliaferro, 600, 270, 17500, 400, 2060

Laney Jones, 600, 475, 16124, 600, 2030

A. E. Pollard, 120, 69, 44265, 150, 410

T. J. Taliaferro, 200, 285, 7660, 200, 740

Wm. M. Wingfield, 70, 134, 3632, 10, 150

E. Hill, 125, 51, 3521, 30, 350

Wm. F. Dickhaus, 1500, 2000, 87000, 3000, 7036

Wm. D. Winston, 360, 355, 17875, 450, 2390

L. B. Price, 1200, 600, 45000, 1000, 4035

William Foster, -, -, -, 1000, 2000

J. T. Priddy, 100 75, 2000, 50, 600

B. L. Winston, 270, 185, 16000, 350, 2200

B. C. Norment, 100, 40, 5000, 60, 400

T. J. Blunt, -, -, - 20, 360

J. A. Wingfield, 300, 100, 12500, 100, 890

Charles Carter, 1950, 800, 39000, -, -

Wickes W. Ivers estate, 57, 18, 900, -, -

Braxton Gartick, 7, -, 3000, -, -

Doc. Wumley, 7, -, 3000, -, -

H. B. Tomlin, 100, 30, 1560, -, -

B. Smith, 100, 50, 3000, -, -

Wm. Sled, 500, 500, 12000, 100, 800

Wm. D. Hott, 10, 52, 722, 10, 250

P. H. Winston, 300, 328, 12560, 300, 1598

J. P. Tyler, 50, 25, 900, 20, 300

James Tyler, 30, 20, 500, 10, 200

L. Norment, 100, 30, 2600, -, -

C. White, 100, 100, 4000, -, -

John D. G. Brown, 300, 165, 6000, 200, 650

Michael Korb, 115, 48, 3260, 30, 400

Larkin B. Hancock, 180, 180, 5000, 200, 700

Cornelius Anthony, 60, 40, 1600, 5, 350

Edmund Taylor(Zaylor), 400, 200, 6000, 150, 400

James M. Pollard, 150, 55, 4000, 300, 928

Samuel Vaughan, 600, 663, 12630, 300, 4338

Judith Kimborugh, 200, 68, 2500, 155, 510

John Stanly, 275, 95, 3348, 200, 796

James W. Vaughan, 325, 75, 4000, 200, 1025

Washington B. Crenshaw, 275, 291, 4532, 125, 722

Hana Fleming, 130, 100, 2500, 50, 375

Henry H. Grubbs, 20, 30, 1000, 10, 105

Alixander M. Morris, 400, 200, 8000, 300, 2035

Thomas Tiller, 30, 20, 700, 12, 105

James M. Taylor, 350, 550, 5000, 155, 700

George W. Cosby, 400, 350, 12000, 200, 1003

William W. Mallory, 300, 150, 4900, 150, 500

James M. Terry, 30, -, 300, 30, 120

Theodore L. Garnett, 350, 255, 14000, 235, 1450

Cora DeJarnett, 100, 280, 4000, 75, 375

Thomas Doswell 800, 700, 30000, 1500, 30000

John V. Wyatt, 200, 156, 8900, 270, 850

William Miller, 30, 200, 7000, 300, 800

John W. Terry, 75, 155, 2000, 20, 170

Thomas H. Fox, 350, 150, 14000, 250, 1200

Reuben C. Johnson, 8, -, 300, 25, 86

James Holmes, 80, 76, 2340, 100, 246

John N. Mallory, 110, 105, 3440, 100, 410

John R. Taylor, 350, 700, 12000, 200, 1000

John Montgomery, 300, 237, 6000, 200, 320

James M. Doswell, 300, 237, 6000, 255, 800

John T. Anderson, 600, 860, 20000, 800, 2624

Edward Lowry, 150, 150, 2400, 40, 675

James B. Matthews, 175, 125, 3000, 100, 701

James Lowry, 250, 155, 4800, 150, 785

Edmund M. Anderson, 535, 283, 10000, 500, 1790

John R. Noel, 300, 326, 6000, 100, 800

James Fontain, 400, 245, 12000, 300, 2720

Thaddeus Brown, 200, 243, 5000, 400, 1200

Francis Wheat, 120, 74, 2500, 125, 304

Sanal Redd, 500, 430, 10000, 200, 1790

Isaac O. Butler, 500, 200, 11030, 390, 1766

Chas. Vest, 450, 250, 8500, 150, 866

E. C. Duke, 60, 36, 1000, 45, 230

Wm. Priddy, 350, 209, 10000, 350, 3700

Spot. James, 120, 50, 1360, 20, 280

Stephen V. Stone, 50 29, 474, 10, 115

S. V. Stone Jr., 30, 39, 928, 25, 240

Tho. A. Spicer, 75, 243, 3180, 60, 275

Jesse G. Hewlett, 150, 92, 2420, 25, 171

E. P. Perkins, 25, 25, 500, 30, 200

James L. Hanes, 35, 62, 700, 20, 80

Wm. L. Mallory, 12, 65, 450, 30, 245

Caroline M. Burriss, 250, 250, 8500, 350, 1049

John Martin, 300, 200, 7500, 100, 1200

_____ Anderson, 120, 155, 4125, 200, 1830

Rob. H. Nelson, 600, 500, 20000, 1500, 2000

Jno. C. Corker, 30, 5, 280, 5, 35

Edmund Fontain Jr., 350, 100, 9000, 200, 650

Henry R. Carter, 880, 750, 30000, 2000, 4000

Dudley Hall, 153, 62, 1500, 30, 160

Wm. Phillips, 30, 31, 1000, 35, 225

John G. Mitchell, 500, 477, 14655, 200, 4324

John W. Terrell, 225, 452, 4575, 40, 316

Earland N. Thompson, 250, 200, 4580, 150, 1010

J. W. Goodwin, 30, 33, 10000, 200, 1300

Alfred Duke, 300, 258, 10000, 253, 800

Chas. J. Terrell, 600, 550, 18000, 400, 2025

Wm. M. Thompson, 150, 135, 4290, 250, 530

Albert Spicer, 80, 125, 2040, 15, 330

Nancy N. Harris, 125, 75, 2400, 30, 400

Wm. Hatch, 300, 200, 15000, 1000, 4000

Joe Z. Terrell, 250, 450, 15000, 300, 2000

Wm. P. Yearman 150, 46, 1560, 50, 230

Richd. R. Fletcher, 30, 25, 330, 25, 85

James H. Phillips, 700, 300, 15000, 300, 1700

Jno. L. Johnson, 35, 40, 600, 25, 163

Jno. W. Farmer, 20, 10, 240, 10, 30

E. L. Farmer, 25, 10, 240, 5, 45

Henry P. Farmer, 44, 25, 828, 150, 165

J. B. Cason, 106, 105, 2120, 80, 286

Wm. H. Anthony, 50, 63, 2260, 150, 372

R. D. Rowzie, 200, 100, 3000, 200, 630

Abram R. Hall, 75, 28, 1000, 100, 268

Malmon White, 75, 8, 600, 10, 200

Silas White, 200, 97, 3000, 100, 300

J. E. Luck, 200, 200, 3000, 50, 275

Jno. Leay, 50, 53, 1000, 7, 180

Wm. G. Maddox, 235, 150, 6000, 173, 825

E. A. Rowzie, 430, 200, 6000, 300, 1370

Polly Thacker, 79, 8, 636, 100, 300

E. S. Wash, 100, 120, 2000, 75, 372

B. B. Dickinson, 30, 32, 2970, 150, 555

Jos. Hancock, 240, 230, 2820, 50, 575

W. S. Dickinson, 300, 150, 6000, 50, 200

C. C. Sims, 70, 60, 1300, 75, 135

J. B. Yeaman, 100, 90, 1140, 50, 320

J. S. Smith, 100, 138, 446, 100, 492

E. H. Hill, 90, 130, 4400, 80, 440

B. A. Smith, 160, 159, 4121, 150, 550

J. H. Johnson, 12, -, 240, 10, 130

Mary W. Goodwin, 222, 200, 3376, 300, 910

T. W. Gentry, 120, 66, 1116, 60, 300

C. B. Richardson, 180, 70, 2000, 120, 135

Pleasant Yeaman, 580, 100, 4800, 150, 890

S. A. Luck, 75, 25, 600, 30, 365

Wm. A. Wash, 100, 81, 2500, 75, 400

E. Fontaine, 800, 700, 22500, 600, 2400

Jos. C. Tinell, 200, 253, 3171, 50, 600

Eliz. Tinell, 80, 60, 1000, 40, 200

Jas. Wash, 100, 140, 1680, 25, 135

Wm. M. Strong, 200, 125, 1624, 10, 160

W. C. Winston, 450, 285, 7342, 300, 1014

Philip Nelson, 300, 200, 5000, 300, 1050

J. W. Jackson, 100, 100, 2500, 50, 300

W. Ronquist, 30, -, 300, 25, 50
Tim. Tinell, 170, 214, 3840, 100, 400
T. M. Lowry, 80, 28, 1100, 155, 425
W. H. Deals, 100, 200, 2500, 4, 70
Ralph H. Terrell, 60, 56, 1200, 12, 140
J. A. Hall, 50, 50, 1000, 5, 70
Thos. Chiles, 48, 2, 500, 5, 54
B. F. Baker, 45, 35, 640, 12, 50
W. J. Carpenter, 700, 440, 12000, 2000, 2000
N. E. Lowry, 250, 257, 3000, 250, 1000
Paulina Sims, 75, 115, 3080, 75, 400
J. W. Mills, 140, 58, 1980, 30, 360
E. L. Hall, 10, -, 100, 30, 200
C. C. Goodman, 60, 65, 1200, 35, 154
Alex. Goodman, 120, 61, 1500, 20, 150
A. M. Stanley, 50, 60, 600, 10, 200
R. H. Hall, 116, 5, 600, 50, 188
J. Stanley, 50, 10, 300, 20, 212
H. P. Stanley, 75, 5, 400, 20, 380
R. Grubbs, 68, 65, 665, 10, 154
Ann Maynard, 60, 120, 1440, 50, 474
R. B. Moody, 200, 150, 3500, 100, 350
J. M. Moody, 325, 108, 4320, 300, 934
J. C. W. Jackson, 70, 45, 1380, 10, 200
J. W. Harris, 75, 5, 400, 20, 125
B. P. Jones, 80, 80, 1600, 25, 530
E. Harris, 50, 10, 360, 20, 210
R. J. Stanley, 25, 12, 300, 20, 145
J. B. Buchanan, 80, 63, 1500, 50, 300
C. C. Butler, 75, 25, 600, 20, 275
D. Rice, 100, 35, 1300, 50, 318
J. Rice, 80, 7, 810, 75, 280
S. K. Harris, 100, 70, 1700, 100, 435
L. Higginson, 200, 100, 3000, 100, 600

H. S. Lowry, 100, 65, 2550, 10, 500
S. Pleasants, 150, 100, 3000, 90, 260
Ellen Turner, 45, 16, 610, 20, 143
J. A. Spindle, 200, 54, 3000, 500, 1022
S. O. Harris, 66, 33, 1000, 10, 150
G. C. Lumsden, 30, 33, 830, 10, 185
F. H. Thompson, 150, 137, 2875, 75, 300
C. S. Chisholm, 300, 200, 5000, 400, 1000
J. W. Moody, 75, 79, 1530, 110, 225
D. G. Jones, 60, 40, 1500, 100, 630
William Harris, 100, 150, 2240, 75, 150
James P. Holloway, 75, 25, 1200, 20, 300
John F. Stanly, 50, 100, 1200, 30, 210
William C. Stanly, 1, -, 50, 10, 75
Austin F. Ronquist, 40, -, 500, 15, 300
William W. Fulcher, 300, 800, 14000, 250, 1200
George Johnson, 500, 175, 10125, 450, 1400
Alex. R. Fontaine, 550, 226, 6000, 300, 1500
Lucius H. Minor, 700, 300, 20000, 500, 2280
Nuton West, 16, 1, 130, 25, 150
John Cooke, 500, 300, 20000, 600, 1800
C. L. F. George Noland, 400, 260, 12000, 350, 850
Landon C. Berkeley, 300, 300, 7200, 300, 925
Horate Stringfellow, 140, 31, 5000, 175, 800
N. W. Berkeley, 200, 260, 3680, 100, 581
R. F. Berkeley, 150, 95, 4500, 100, 425
Edmund Berkeley, 390, 250, 6400, 150, 420

P. H. Price, 800, 500, 16000, 600, 2000

Reuben Oliver, 75, 50, 1250, 30, 120

Wm. J. Kimbrough, 85, 132, 2500, 50, 225

Chas. H. Thacker, 50, 45, 800, 40, 220

William Duke, 50, 36, 850, 50, 200

Wm. Saunders, 150, 74, 2240, 10, 220

John C. Perkins, 60, 37, 1000, 175, 275

Wm. R. Price, 53, 51, 2500, 30, 400

Lewis T. Blunt, 100, 25, 883, 30, 275

Wm. J. Sacro, 50, 83, 924, 400, 250

Garland F. Stone, 35, 93, 966, 4, 75

J. R. Harris, 250, 194, 4160, 30, 830

Richd. Gwalthmey, 700, 400, 16000, 400, 1440

Joseph Norment, 800, 2400, 20000, 500, 1977

Saml. M. Baker, 300, 194, 5000, 200, 1060

Edmund Winston, 800, 800, 25000, 1000, 3586

David F. Waldrop, 275, 325, 6000, 200, 915

William Thompson, 250, 193, 8000, 75, 320

Francis G. Taylor, 75, 150, 2000, 50, 400

Albert Lane, 20, 20, 400, 5, 70

Thomas H. Perkins, 35, 30, 520, 35, 75

Reuben Southworth, 50, 80, 1000, 20, 83

Philip N. Mallory, 30, 186, 1725, 5, 100

William H. Vest, 150, 175, 1500, 100 700

James T. Butler, 15, 6, 161, 25, 190

J. W. Kimbrough, 150, 107, 2000, 75, 500

Milton C. Crenshaw, 150 70, 2000, 125, 620

John Ellett, 200, 217, 4170, 200, 664

Nathl. C. Crenshaw, 420, 280, 12000, 900, 1900

William H. Campbell, 600, 300, 15000, 500, 1600

William O. Day, 600, 585, 15000, 300, 1400

James Winston, 380, 220, 7200, 300, 1200

Thomas G. Bumpass, 150, 110, 2580, 175, 800

Charles L. Stanly, 25, 5, 300, 25, 100

Thomas J. Stanly, 80, 46, 1260, 50, 450

James Baughan, 100, 70, 1700, 20, 350

Thomas Cocke, 25, 75, 1200, 25, 175

John H. Wickham, 800, 700, 15000, 200, 2000

James W. Eddleton, 75, 23, 1000, 50, 165

Major J. H. Baughan, 30, 83, 1130, 25, 150

James W. Henry, 68, 35, 1000, 30, 180

Silas Shelburne, 228, 20, 2480, 100, 580

Thomas L. Page, 500, 600, 11000, 400, 1000

Oscar F. Chisholm, 100, 273, 6000, 200, 600

___ S. K. Waldrop, 500, 500, 12000, 300, 1000

E. T. Mann, 200, 33, 2500, 300, 600

Lander Baughan, 150, 175, 4000, 300, 700

Thomas Hardin, 600, 388, 19760, 588, 2200

Jno. Fleming, 900, 300, 18000, 700, 2180

A. Julian Welcher, 150, 50, 10000, 150, 700

Thomas F. Waldrop, 90, 60, 1500, 20, 250

Richd. Waldrop, 100, 113, 2130, 100, 300

John B. Jones, Jr., 120, 180, 3000, 25, 350

Wm. L. Jones, 50, 83, 1860, 100, 300

Betsy R. Carver, 140, 143, 1500, 10, 100

Jno. B. Jones, 600, 600, 15000, 350, 1124

Geo. Mason, 50, 33, 1000, 20, 300

Wm. H. Pleasants, 75, 21, 1000, 10, 270

Matthew A. Anderson, 225, 260, 7300, 300, 1450

J. R. Dandridge, 200, 260, 4600, 75, 600

Jno. D. Hardgrave, 30, 21, 400, 50, 280

Thomas Bourne, 166, 80, 2500, 100, 500

Wm. Glenn, 80, 33, 1130, 25, 125

P. E. Spindle, 200, 100, 3600, 150, 600

Thos. Jones, 250, 350, 3600, 150, 1325

Thos. Cockrane, 80, 100, 1800, 5, 85

B. F. Welcher, 75, 89, 1640, 25, 425

Jas. C. Wash, 140, 120, 3000, 250, 720

Chas. Nuckols, 100, 150, 2000, 1500, 200

J. C. Stone, 300, 100, 6000, 1000, 1500

P. Leadbetter, 400, 200, 7200, 50, 730

W. M. Massie, 180, 41, 3000, 300, 850

Chas. W. Dabney, 500, 100, 14000, 1000, 2200

W. Wingfield, 275, 177, 8700, 400, 1200

J. Patterson, 120, 80, 3000, 125, 250

P. N. A. Massie, 175, 230, 4860, 100, 520

J. Perkins, 50, 71, 825, 20, 65

Z. R Walton, 120, 125, 5000, 150, 450

W. P. Stone, 250, 150, 4000, 150, 700

R. Winston, 15, 15, 300, 50, 150

J. P. Tinnell, 350, 107, 12000, 300, 850

D. Winston, 75, 25, 1200, 25, 210

C. Childress, 65, 68, 1600, 50, 160

H. Apperson, 50, 43, 1000, 25, 200

E. Childress, 30, 46, 1000, 50, 275

E. A. L. Tiller, 85, 50, 1800, 50, 350

C. B. Jones, 45, 15, 400, 35, 125

N. Baughan, 100 75, 1700, 50, 400

R. Perkins, 100, 57, 1570, 30, 200

W. Waldrop, 230, 105, 3000, 40, 300

E. Morris, 600, 90, 18000, 500, 2500

Chas. Morris, 350, 600, 10000, 350, 1300

R. G. Meredith, 200, 356, 6700, 450, 1458

J. Leadbetter, 100, 106, 2500, 50, 475

F. Vaughan, 400, 223, 10000, 500, 1200

J. A. Smith, 40, 40, 1000, 100, 360

L. Strong, 100, 50, 1000, 10, 250

Jno. Jones, 65, 57, 1800, 100, 250

B. Harris, 50, 30, 1000, 30, 115

B. Vaughan, 308, 400, 10000, 500, 2400

Z. L. Jones, 200, 160, 4320, 50, 450

C. H. Vaughan, 150, 150, 2000, 100, 860

J. H. Blunt, 175, 89, 400, 150, 675

R. P. Mallory, 130, 33, 2000, 55, 400

R. B. Eubank, 225, 25, 2500, 25, 150

J. C. England, 80, 103, 2750, 25, 25

T. E. King, 25, 55, 800, 25, 250

S. H. Tinsley, 200, 100, 3000, 200, 700

J. F. Cross, 100, 75, 1735, 50, 450

J. F. Mallory, 50, 33, 830, 35, 300

J. T. Hughes, 75, 65, 1400, 75, 400

F. H. Hendrick, 40, 90, 1300, 56, 140

R. B. Hendrick, 270, 300, 5700, 100, 1070
W. L. Wood, 200, 236, 6000, 30, 540
S. Bumpass, 60, 55, 1150, 25, 500
J. H. Tiller, 75, 85, 1600, 5, 100
L. Allen, 35, 5, 400, 5, 25
T. Ellett, 150, 65, 1800, 50, 430
W. H. England, 60, 22, 820, 300, 360
H. H. Wood, 175, 225, 5000, 50, 525
W. Taylor, 180, 90, 3000, 1000, 400
L. P. Kimbrough, 130, 130, 2600, 50, 500
G. J. Brookes, 80, 80, 1600, 25, 250
E. Vaughan, 150, 150, 2400, 40, 300
M. Dillard, 80, 146, 2265, 25, 215
Alex. Barlow, 110, 325, 3000, 40, 350
E. Lowry, 80, 82, 850, 15, 175
W. A. Stone, 20, 60, 1000, 50, 210
W. Eddleton, 115, 115, 2500, 50, 250
L. Davis, 200, 100, 2000, 100, 500
W. Lumpkin, 300, 600, 7200, 350, 1400
R. Glazebrook, 100, 133, 2400, 30, 580
T. M. White, 100, 175, 3000, 50, 360
W. L. White, 225, 162, 5800, 150, 2000
P. Tinsley, 200, 200, 3000, 200, 600
S. Cross, 100, 37, 1500, 25, 175
S. B. Bullock, 500, 300, 9000, 200, 1000
M. Beasley, 65, 100, 1600, 25, 30
P. Woolfolk, 100, 280, 6000, 255, 775
J. R. Woodson, 100, 200, 3000, 25, 358
C. R. Montgomery, 50, 450, 2500, 100,825
J. T. Clough, 200, 500, 7000, 1000, 530
E. M. Tomkies, 200, 400, 10000, 100, 560
R. Lawrence, 30, 75, 525, 25, 125
T. C. Ellett, 200, 170, 2220, 30, 750

S. V. Cross, 225, 58, 2080, 100, 520
J. Donnella, 25, 75, 1000, 30, 435
C. B. Jones, 50, 54, 1000, 25, 400
C. P. Goodall, 580, 400, 10000, 300, 1370
C. Crenshaw, 100, 400, 6000, 500, 1100
R. B. Gilman, 300, 195, 3000, 300, 1120
E. White, 100, 75, 1750, 15, 225
W. Patmon, 400, 130, 5000, 500, 1020
J. Harris, 100, 150, 3000, 200, 600
B. F. Bryce, 75, 88, 1500, 20, 150
Alex. Wingfield, 160, 77, 2800, 100, 1175
T. F. White, 140, 60, 2000, 25, 150
G. Starke, 180, 63, 4300, 200, 460
G. Chewning, 140, 100, 3600, 50, 575
P. A. Snead, 150, 50, 2000, 100, 460
H. McDowell, 75, 85, 1600, 100, 470
G. Winston, 100, 87, 1800, 30, 360
H. J. Tinsley, 400, 540, 10000, 400, 1500
Z. Nash, 100, 30, 1300, 35, 300
H. Land, 200, 100, 3000, 100, 800
G. N. Parrish, 150, 150, 3000, 235, 500
J. Snead, 200, 200, 4000, 200, 1200
A. W. Nolting, 700, 1033, 17000, 400, 1500
Joseph Talley, 400, 448, 10000, 150, 800
W. C. Shelton, 530, 400, 10000, 450, 2000
R. P. Woodson, 260, 140, 3500, 50, 400
G. W. Doswell, 900, 847, 17470, 650, 2825
J. M. Nuckols, 280, 100, 3500, 200, 1100
L. R. Nuckols, 80, 97, 1800, 50, 450
T. J. Puryear, 175, 55, 2330, 100, 900

J. L. Rutnick, 280, 450, 7000, 200, 700

J. B. Saunders, 400, 100, 6000, 135, 200

W. J. Childress, 75, 75, 1500, 50, 300

J. E. Utley, 25, 79, 1000, 30, 125

F. P. Dabney, 235, 580, 7500, 50, 200

J. Nuckols, 200, 550, 6000, 25, 275

W. R. Ribble, 150, 150, 2500, 75, 400

B. B. Pleasants, 50, 38, 1000, 50, 400

T. Carter, 100, 92, 1500, 100, 460

P. F. Mosby, 180, 236, 3500, 100, 530

Geo. Powers, 200, 60, 2000, 55, 350

J. Bowles, 200, 100, 3000, 300, 1150

Jesse Bowles, 200, 125, 3250, 100, 1120

E. Perkins, 100, 84, 1500, 25, 220

Jos. Wade, 150, 150, 3000, 15, 120

Jno. Wade, 15, 5, 200, 45, 160

L. Atkinson, 60, 110, 1500, 50, 100

W. Hopkins, 152, 86, 2380, 100, 600

Roland Goodman, 350, 250, 12000, 500, 2000

P. Isbell, 125, 79, 2000, 100, 400

Wm. Nuckols, 225, 215, 4000, 100, 1160

W. Duggins, 155, 160, 3000, 50, 560

J. Woodson, 200, 154, 3540, 100, 700

Wm. Irbey, 200, 300, 5000, 100, 760

W. W. Nuckolds, 180, 60, 2000, 55, 500

E. Nuckols, 150, 180, 3000, 75, 535

J. Nuckhols, 335, 135, 5000, 100, 800

Bollman Bowles, 300, 325, 8000, 200, 1000

Andrew Nuckols, 100, 165, 1500, 150, 455

H. Leadbetter, 50, 40, 1350, 100, 400

J. Stanley, 60, 46, 600, 20, 160

R. Tiller, 15, 15, 200, 55, 300

Wm. Strong, 50, 40, 540, 20, 250

E. Bumpass, 50, 43, 600, 25, 155

Geo. Woodfin, 300, 155, 4000, 200, 1300

Wm. Nelson, 500, 300, 13000, 1000, 2300

Ben Vaughan, 104, 100, 2600, 75, 675

G. Stanley, 150, 140, 2100, 50, 500

H. C. Doswell, 550, 328, 8800, 1000, 3000

E. C. Taylor, 200, 172, 6000, 200, 900

Wmson. Talley, 600, 440, 14500, 250, 2150

W. S. Talley, 100, 70, 1360, 30, 400

B. B. Bumpass, 100, 65, 1000, 100, 345

Garland James, 120, 180, 3000, 100, 750

T. T. Taylor, 800, 687, 16031, 400, 1638

Thomas Lane, 70, 60, 1000, 10, -

Wyatt Wash, 37, 30, 600, 20, 150

F. M. Barker, 300, 150, 7000, 200, 700

Henrico County, Virginia
1860 Agricultural Census

The Agricultural Census for Virginia 1860 was microfilmed by the University of North Carolina Library under a grant from the National Science Foundation from original records at the Virginia Department of Archives and History in 1963.

There are forty-eight columns of information on each individual. Only the head of household is addressed. I have chosen to use only six columns of information because I feel that this information best illustrates the wealth of individuals. The columns are:

1. Name
2. Improved Acres of Land
3. Unimproved Acres of Land
4. Cash Value of Farm
5. Value of Farming Implements and Machinery
13. Value of Livestock

Robt. T. Pleasants, 130, 220, 5000, 150, 895
Joshua Wassener, 30, 10, 350, 40 300
Robt. M. Minson, 6, 3, 90, -, 10
Mitchell Fussell, 50, 20, 300, 50, 1700
Nathan Collins, -, 23, 130, 20, 120
Jno. C. Robertson, 100, 70, 2500, 50, 500
George Turner, -, -, -, -, 60
Thos. J. Blake, 2400, 3000, 260000, 15000, 17154
Robt. M. Taylor, 265, 164, 30000, 850, 2360
Mildred Gay, 63, 200, 2600, 50, 250
Thos. L. Lyne, 225, 225, 9000, 150, 634
Benj. Knight, 5, -, 100, 20, 200
Dorastus Pearce, 8, 42, 1000, 50, 300
George H. Cox, 20, 45, 1000, 20, 190
Moses Jonathan, 40, 30, 700, 10, 60
Albert W. Pearman, -, -, -, -, 15
Judith Scott, 6, 2, 100, -, 50
Thos. James 1, 11, 125, -, 600

Josiah Foster, 3, 1, 150, 50, 210
Jno. S. Barley, 30, 16, 1300, 50, 250
Henry Pearman, 35, 19, 1200, 30, 250
Thos. T. Galdsay, -, -, -, 20, 130
Berry B. Martin, -, -, -, 100, 450
Mary Burton, -, 185, 2000, 20, 200
Bernard O. Aikin, 100, 106, 500, 300, 745
Lee Scott, 5, -, 100, 10, 100
Alpheus Childress, 18, 20, 500, -, 100
Moses Throgmorton, -, -, -, 1, 100
Susan A. Chapin, 800, 500, 3500, 400, 2025
G. W. Mountcastle, -, -, -, -, 100
Albert M. Aikin, 800, 800, 70000, 5000, 4282
Albert M. Buffin, 150, 58, 2500, 100, 726
Thos. J. Yarbrough, 75, 105, 3000, 150, 800
Emily J. Grover, 85, 45, 4500, 100, 490
Robt. Johnson, 8, 15, 3000, 10, 20
William James, 20, 40, 3000, 20, 350

John H. Carr, 250, 400, -, 200, 940

Wm. Scott, 50, 24, 1000, 50, 600

Wm. H. Ammons, 150, 141, 12000, 200, 1162

Jessee F. Clarke, 30, 50, 3000, 40, 200

Wm. Stearns, 50, 16, 3000, 100, 270

Sallie A. Farrar, 36, 42, 1800, 30, 300

James R. Roper, 75, 82, 6000, 100, 300

Theophalus Tatum, 65, 24, 6000, 85, 400

Mary A. Gunn, 250, 160, -, 150, 1100

Ronald Mills, 200, 183, 11500, 600, 954

Martin Baker, 25, 31, 1500, 60, 355

Wm. M. Pearce, 60, 260, 5000, 125, 1310

Jessee F. Winfree, 100, 300, 6000, 150, 850

J. H. Goodman, 6, 16, 200, 15, 35

Wm. F. Gunn, 100, 42, 3000, 50, 200

Jno. Gathright, 200, 100, 7000, 100, 800

Jas. W. Binford, 200, 100, 8000, 300, 1500

Stephen B. Sweeney, 350, 150, 13000, 500, 1425

Benj. F. Dew, 500, 300, 24000, 200, 1300

L. H. Kemp, 250, 150, 10000, 155, 1325

E. J. H. Ladd, 225, 175, 8000, 300, 782

C. C. Gathright, 100, 114, 2500, 50, 381

Robt. T. Smith, 60, 15, 1500, 100, 480

G. W. Gatewood, 200, 125, 6500, 200, 750

John Wassener, 100, 162, 2500, 80, 350

J. J. Epps, 40, 40, 600, 25, 150

J. Heber Nelson, 150, 70, 1800, 200, 500

Wm. B. Woodfin, 125, 250, 4000, 150, 960

Jno. C. Gathright, 125, 177, 3000, 50, 620

Wm. Baker, 125, 150, 2700, 50, 425

Robt. F. Binford, 50, 50, 1000, 50, 500

Nathaniel A. Vest, 150, 125, 3600, 50, 250

Thos. Binford, 150, 150, 5000, 125, 900

S. P. Binford, 75, 45, 1300, 50, 525

Hampton Waldfaulk, 25, 57, 600, 20, 100

Nicholas Hobson, 100, 50, 1500, 20, 250

Robert Clarke, 17, 50, 600, 25, 125

Mary C. Hobson, 40, 60, 600, 15, 150

Cornelius Adkins, 30, 7, 200, 25, 90

Isaac Sikes, 25, 25, 300, 20, 150

Richard Sikes, 18, 2, 120, 20, 100

Thomas Wassener, 75, 30, 600, 10, 100

John Fisher, 150, 1000, 8500, 100, 1000

Thos. W. P. Goodman, 25, 86, 2000, 25, 150

Miles P. Wade, 8, 77, 800, 25, 200

Taply Irby, 130, 122, 1100, 10, 300

Henry Leonhauser, 22, 23, 400, 15, 75

Jno. W. Fussell, 200, 300 9000, 250, 1480

G. D. Pleasants, 70, 100, 5000, 75, 450

Edward Minson, 40, 60, 1600, 100, 350

Christopher Saunders, 7, 1, 500, 30, 75

Martin Laughlamer, 78, 30, 10000, 77, 1030

J. F. Bradley, 30, 27, 3000, 60, 300

A. Hernaman, 18, 20, 1000, 75, 150

S. Y. Landrum, 100, -, 10000, 50, 800

W. W. Ensonighty, 15, 15, 300, 10, 75

Nathan Ensonighty, 100, 100, 3000, 200, 619

Mary Picket & others, 560, 620, 60000, 1100, 2200

Robert Fallon, 80, 110, 6000, 300, 750

Mary Bosley, 20, 20, 400, 20, 120

Henry Cox, 1200, 1000, 80000, 3500, 4000

Liston Gatewood, 125, 140, 4500, 100, 700

Garland Narico, 200, 137, 6500, 100, 510

Thos. J. Melton, 50, 60, 250, 50, 550

Washington Bottoms, 30, 170, 3000, 100, 450

Wm. H. Throgmartin, 17, -, 250, -, 150

Thomas Burdon, 13, -, 130, 10, 75

J. B. Burton, 50, 40, 800, 25, 130

Mathew McComack, 20, 30, 400, 10, 28

M. Harvey, 20, 26, 450, 15, 125

Jno. O. Timberlake, 100, 170, 2500, 25, 70

Mildred A. Francis, 30, 20, 500, 25, 250

Wm. H. Jones, 50, 500, 5000, 10, 225

Wm. Jeffries, 30, 70, 1000, 40, 245

T. Hoppe, 30, 12, 800, 30, 301

A. S. Clarke, 70, 30, 2500, 50, 300

Carter Ball, 100, 70, 200, 15, 150

J. H. Makenzie, 150, 50, 6000, 500, 1060

Wm. H. Lowe, 10, -, 100, 15, 60

Joseph C. Nuckols, 8, 32, 400, 25, 60

Frances Lewis, 20, 21, 250, 25, 15

F. Parudo, 10, 11, 400, 1, 100

Richd. Vaughan, 10, 32, 800, 75, 200

Daniel Boone, 10, 30, 400, 40, 200

Francis Parudo, 40, 60, 1000, 10, 150

Winston Jordon, 30, 130, 2000, 200, 100

F. Fagundes, 12, 8, 200, 15, 75

J. R. Allen, 60, 140, 2400, 35, 325

Lanny Carter, 100, 85, 3000, 100, 550

Lanny C. Hobson, 80, 200, 4000, 150, 675

Joseph Purkee, 75, 175, 3500, 100, 200

John Curric, 160, 200, 7900, 500, 1200

J. J. Herbert, 12 ½, 8 ½, 11000, 25, 123

J. A. Goolsby, 30, 26, 1500, 30, 200

J. B. Kidd, 100, 106, 6000, 200, 250

William Guen, 40 32, 2700, 15, 230

Wm. L. Harrison, 185, 45, 18300, 200, 1025

R. W. Magruder, 14, -, 8000, 100, 800

J. W. Clarke, 6, -, 2000, 100, 500

Patrick Jordan, 40, 80, 2500, 150, 140

Samuel Boalts, 20, 70, 400, 10, 100

J. T. Childress, 70, 5, 8000, 300, 800

Franklin Stearns, 575, 100, 73000, 700, 2750

J. M. Taylor, 300, 450, 30000, 700, 2090

W. C. Knight, 900, 338, 50000, 500, 3000

Thos. Bailey, 3, 8, 110, 5, 20

Wm. B. Randolph, 500, 750, 75000, 1500, 3000

P. Horton Keach, 13, -, 12000, 10, 250

R. T. Brooks, 4, -, 4000, 60, 250

_. S. Atlee, 400, 1100, 24000, 1000, 1725

J. N. Davis, 140, 10, 10000, 200, 600

Richd. Cauthorne, 25, 35, 6000, 65, 350

J. H. Allen Jr., 2 ½, -, 1800, 25, 75

J. L. Eggy, 95, -, 15000, 100, 500
James Duke, 40, 45, 800, 50, 100
Bowles Eacho, 50, 50, 900, 25, 200
Wm. W. Tennent, 50, 310, 1200, 25, 100
J. H. Crittenden, 25, 95, 1000, 8, 190
Phillip Garthright, 125, 87, 2500, 50, 175
Jno. N. Doggett, 119, 100, 3000, 50, 500
E. F. Gathright, 75, 153, 2000, 50, 575
Jno. W. Gathright, 12, 31, 400, 30, 200
Thos. R. Howel, 50, 130, 120, 25, 150
Albert Fisher, 40, -, 200, 20, -
Ludwell Brackit, 100, 240, 4000, 60, 700
Robt. C. Baker, 30, 370, 2000, 12, 161
Susan Gathright, 75, 125, 1600, 20, 500
Littlebery Carter, 100, 100, 2000, 5, 200
Wm. Carter, 125, 85, 2000, 50, 300
John Carter, 175, 72, 3000, 60, 325
Moses Carter, 300, 500, 8000, 100, 964
D. B. Jordan, 120, 170, 3100, 150, 381
L. L. Carter, 288, 212, 10000, 300, 1345
Ed. N. Bradley, 189, 200, 6000, 34, 300
Jas. J. Camden, 70, 168, 3000, 30, 150
Wm. M. Carter, 100, 56, 1000, 50, 216
Martha Carter, 80, 90, 1000, 30, 200
Richd. Pollard, 30, 50, 800, 30, 200
W. H. Pryor, 200, 56, 3800, 150, 500
T. J. Carter, 250, 146, 10000, 400, 1600
L. S. Courtney, 50, 238, 7000, 150, 883

Ed. F. Gathright, 250, 350, 8000, 200, 1030
Philip Watkins, 6, 42, 600, 20, 200
Richd. Ritter, 70, 38, 3000, 175, 500
Robt. W. Irby, 50, 157, 1100, 25, 150
James E. Yarbrough, 40, 60, 700, 25, 175
Maria Allen, 100, 188, 2900, 50, 384
J. W. Sneede, 140, 90, 5000, 30, 525
Warren P. Southall, 60, 40, 10000, 50, 400
George Turner, 200, 400, 18000, 500, 3500
J. B. Bragg, 150, 140, 3600, 200, 1100
J. R. Ratcliffe, 73, 12, 4000, 30, 160
J. O. Austin, 60, 15, 15000, 300, 400
H. A. Watt, 80, 120, 18000, 500, 1400
B. W. Roper, 75, 228, 10000, 300, 650
Wm. Boulware, 750, 400, 35000, 600, 2210
Richd. Malone, 75, 25, 6000, 100, 930
John Hughes, 50, 102, 7000, 50, 320
John Poe, 200, 82, 14000, 115, 505
Mary Weber, 7, 2, 400, 150, 200
John P. Dickinson, 30, 15, 7000, 100, 1340
R. A. Major(Mayor), 25, -, 6000, 200, 1000
J. T. Huffman, 180, 53, 9000, 150, 655
Nathan S. Shipman, 34, 25, 1600, 90, 250
Wm. Palmer, 5, -, 300, -, 6
Thos. Whiteford, 40, 30, 2000, 100, 200
Peter W. Grubbs, 140, 130, 9500, 200, 925
Louisa Craddock, 75, 57, 3000, 50, 237
John Bridgewater, 200, 200, 4000, 200, 600

Mary Shurm, 25, 75, 2500, 50, 200
Thomas French, 50, 166, 3200, 100, 517
J. A. King, 10, 7, 170, 10, 157
Elvira Garrett, 15, 17, 800, 30, 160
Wm. W. Allen, 20, 160, 1800, 150, 600
John A. Eacho, 16, 12, 600, 50, 366
Geo. M. Savage, 300, 400, 18000, 400, 1160
Richd. Kidd, 75, 25, 1500, 30, 225
B. S. Crouch, 25, 101, 2000, 20, 300
Anna Crouch, 300, 330, 19000, 250, 1263
Truearnest Dudley, 100, 130, 4000, 200, 500
David Michie, 135, 165, 4000, 200, 600
Baylor Martin, 45, 100, 4000, 50, 300
Wm. Tignor, 21, 500, 1065, 40, 154
Eliza B. Allen, 45, 40, 850, 65, 239
Susan R. Allen, 100, 100, 3000, 100, 350
L. L. Fidler, 40, 44, 800, 20, 165
Peterfield Trent, 400, 400, 20000, 400, 1520
Simon Gouldin, 250, 300, 14000, 230, 840
Giles C. Courtney, 25, 130, 4000, 300, 505
Francis Adams, 108, 136, 4800, 150, 688
John G. Hyer, 40, 62, 1000, 30, 150
P. W. John Quarles, 40, 116, 4500, 120, 621
Oliver Gathright, 30, 70, 2000, 50, 150
Jas. M. Garnett, 400, 285, 17000, 400, 1010
John R. Garnett, 400, 200, 15000, 600, 1300
Mary Price, 275, 125, 10000, 150, 600
Harriet Jennings, 300, 290, 11800, 250, 1150

R. Y. Slater, 100, 55, 5400, 150, 250
Robinson Barker, 245, 200, 1100, 200, 650
Theodore Picot, 50, 55, 4000, 230, 1032
John E. Friend, 280, 30, 12000, 400, 1000
Chas. Y. Morris, 200, 100, 10500, 400, 705
Valentine Heckler, 8, -, 6000, 50, 500
B. D. Kinker, 9, -, 200, 50, 150
Peter Heckler, 10, -, 1800, 100, 200
Conrad Flitig, 10, -, 1000, 25, 150
Henry Voegler, 50, 15, 10000, 100, 575
J. J. Cosby, 16, 4, 3000, 50, 80
Josiah Dabbs, 120, 30, 15000, 200, 1200
Richd. Olphia, 20, 21, 2000, 25, 60
W. L. Timberlake, 10, -, 3000, 30, 110
Elizabeth Harward, 25, 20, 2500, 50, 200
Jas. W. Otey, 90, 70, 10000, 100, 560
J. C. Dashart, 5, -, 1000, 25, 300
Joseph Sharpe, 30, 52, 5000, 100, 350
S. Magruder Sims, 290, 70, 13000, 500, 1400
John D. Warren, 300, 119, 10000, 500, 1268
Jno. V. Hardwick, 2, -, 750, 15, -
John Harwood, 4, -, 5000, 20, 300
Bernard Ball, 20, -, 6000, 50, 300
Wm. Cullingenworth, 27, -, 1600, 100, 300
J. B. Bush, 75, 35, 5000, 30, 100
J. H. Acree, 33, -, 5600, 50, 500
D. Ball, 40, 80, 6000, 100, 500
J. F. Childress, 300, 200, 14000, 300, 850
J. W. Lewis, 20, 5, 1500, 70, 350
Jos. J. Pleasants, 35, -, 5500, 50, 500

Prudence R. Picot, 120, 10, 10000, 150, 1065

Joseph Reinhart, 9, -, 200, 10, 50

Sylvester Rubble, 8, 2, 2000 50, 150

James Snell, 85, 20, 10000, 100, 400

L. H. Haleway, 12, -, 2000, 20, 40

John Jacobs, 60, -, 12000, 200, 400

Wm. Copeland, 18, -, 3600, 50, 150

Lawrence Hubbard, 30, 30, 12000, 50, 120

Mary S. Schermahorn, 40, 50, 18000, 150, 350

James C. Schermahorn, 100, -, 20000, -, 60

James N. Shine, 30, 17, 7000, 75, 200

Ann H. Franklin, 14, 14, 1400, 20, 150

Leonidas Rosser, 60, 8, 10000, 200, 745

Richd. Philips, 20, 18, 300, 30, 200

James M. Carter, 120, 25, 14500, 300, 610

Charles Carmon, 55, 20, 7500, 120, 550

Josiah D. Smith, 200, 70, 16000, 60, 955

Salin Serger, 150, 17, 5000, 80, 1050

Wm. S. Buton, 10, 5, 1500, 30, 45

James Pearce, 60, 20, 10000, 50, 320

Lucy C. Tally, 80, -, 16000, 50, 300

H. Loften, 23, -, 3000, 150, 600

Samuel H. Taylor, 12, -, 3600, 30, 150

William Burton, 5, -, 5000, 50, 300

Lewis Keppler, 5, -, 200, 50, 150

Jno. M. Batts(Bates), 44, -, 8800, 75, 1000

John Belcher, 93, -, 23250, 200, 2150

Turner & Rowley, 25, -, 7500, 30, 300

David Couling, 10, -, 2100, 40, 450

J. M. Conrad, 5, -, 15000, 50, 1200

C. H. Clarke, 30, 127, 1000, 50, 450

John Bossieux, 100, 100, 3000, 50, 400

Peyton Johnston, 25, -, 10000, 1000, 1000

Joseph Sinton, 100, 233, 23000, 200, 800

Joseph L. Sinton, 40, 80, 4000, 50, 150

Thos. D. Corker, 13, 25, 2500, 100, 250

Martin S. Taylor, 120, 127, 14000, 500, 2000

John W. Beveridge, 60, 40, 3500, 50, 300

C. M. Terrill, 30, 70, 2500, 100, 300

Geo. Eubank, 20, 37, 2500, 75, 400

Thos. E. Tyler, 6, 8, 600, 5, 60

J. Lucius Davis, 250, 350, 25000, 500, 1800

M. E. Valentine, 16, 14, 700, 30, 130

J. G. Boudar, 120, 160, 11000, 400, 1100

Joseph Vondelehs, -, -, -, -, 40

John A. Tyler, 6, 21, 500, -, 70

Joseph Bernard, 110, 65, 6000, 100, 400

Wm. H. Saunders, 109, 62, 7000, 500, 600

Catharine M. Carter, 65, 35, 5000, 200, 800

Abner Hilliard, 220, 119, 18000, 500, 1400

Nathl. King, 175, 182, 11000, 200, 900

P. H. Waldrop, 40, 22, 3000, 100, 300

Price Lucas, 8, 8, 300, 10, 60

R. G. Walton, 18, -, 5000, 50, 300

Major. H. Newman, 4, 14, 500, 30, 200

Regina Tyler, 20, 10, 800, 20, 300

James More, 10, 28, 600, 10, 50

L. W. Glazebrook, 250, 150, 10000, 300, 1500

Wm. P. Darricott, 150, 50, 6000, 100, 500

M. S. Bowles, 50, 75, 2000, 40, 300
Billy King, 25, 26, 1200, 20, 190
Wm. Andrew, 80, 109, 2500, 20, 100
J. S. Walker, 50, 195, 4000, 200, 300
J. Seaton, 10, 18, 600, -, 100
S. C. Davis, 100, 50, 4000, 100, 600
Arabella Lownes, 90, 80, 5000, 100, 300
P. C. Kimbrough, 60, 40, 2500, 200, 400
Catharine Padget, 20, 40, 1000, 10, 250
R. R. Padget, 22, 20, 600, 50, 200
Benjamin Gentry, 8, 50, 700, -, 150
Spottswood Ford, 40, 140, 2000, 100, 600
John F. Willis, 8, 25, 300, -, -
Wm. Carroll, 10, 40, 500, -, 50
Wm. Blackburn, 15, 22, 500, 30, 150
Andrew S. Padget, 18, 54, 1400, 40, 300
Jacob Kick, 50, 22, 6000, 200, 900
M. W. Lawrence, 15, 5, 600, 10, 10
Robt. H. Brock, 40, 150, 2000, 50, 250
Benj. E. Anderson, 120, 60, 5000, 250, 500
A. H. Ford, 25, 75, 1500, 50, 400
Ryland Jennings, 12, 20, 300, 30, 100
Jessee B. Shepperson, 10, 25, 600, 10, 100
Frederick Strohman, 10, 2, 300, 20, 100
Wm. Lawrence, 60, 65, 3000, 100, 500
Wm. Shoemaker, 30, 57, 1000, 50, 300
M. W. Hutcheson, 250, 275, 20000, 250, 1200
Loftin N. Ellett, 80, 400, 16000, 100, 300
D. Cottrell, 30, 198, 2400, 30, 100
Levi Ford, 13, 12, 500, 20, 200
Lavenia Blair, 116, 16, 15000, 50, 300

Anthony Robinson, 100, 60, 60000, 200, 1000
Geo. H. King, 60, 10, 11000, 250, 1000
S. G. Waldrop, 150, 90, 8000, 100, 200
Robt. H. Henley, 150, 450, 6000, 200, 600
John E. Woodward, 30, 107, 2000, 80, 600
Art Morriss, 130, 34, 13000, 250, 700
Robt. N. Maxwell, 80, 100, 1100, 100, 200
Thos. E. Nuckols, 350, 350, 9000, 350, 1400
C. S. Worbriton, 5, 19, 500, 10, 60
C. C. Worbriton 10, 9, 1000, 50, 70
M. G. Seaton, 18, 32, 1500, 30, 130
John J. Brock, 18, 90, 1700, 80, 600
E. E. Orvis, 2, 42, 600, -, 100
Mosby Ford, 24, 126, 1800, 100, 500
Geo. Seaton, 18, 21, 900, 100, 350
A. M. Lawrence, 10, 70, 1000, 20, 70
Robert Ford, 100, 20, 3000, 100, 800
Agness Ford, 50, 30, 2400, 40, 150
Deetrick Bolton, 50, 50, 1000, 50, 150
Robt Orrick, 80, 50, 1800, 50, 150
Ryland Ford, 50, 40, 1500, 100, 500
G. G. Exall, 40, 86, 7500, 100, 650
Lucy Smith, 100, 300, 7000, 100, 600
Elijah Priddy, 120, 137, 6500, 50, 500
Jas. G. Francis, 80, 70, 1500, 75, 150
S. B. Throckmorton, 10, 5, 600, 20, 60
R. T. Cobbs, 25, 11, 1600, 70, 300
E. P. Chamberlayne, 395, 112, 12000, 300, 1700
D. S. Delaplane, 120, 160, 15000, 1000, 1500
J. B. Young, 160, 160, 30000, 300, 1500

N. L. Pelaske, 500, 300, 40000, 1500, 2400

John Stewart, 200, 130, 30000, 1000, 1500

Emily F. Hudson, 300, 75, 6000, 200, 1100

Anderson King, 300, 75, 6000, 300, 900

Jacob Fidler, 80, 30, 1800, 30, 50

M. B. Chamberlayne, 200, 157, 20000, 500, 2000

M. R. Gooch, 250, 20, 15000, 250, 1000

A. F. Gooch, 50, 12, 5000, 60, 400

A. S. Storrs, 120, 100, 10000, 200, 1200

R. T. Adams, 540, 260, 40000, 1000, 2600

S. C. Burton, 40, 66, 4500, 30, 400

J. D. Sheppard, 40, 130, 3400, 250, 800

G. B. Carter, 100, 200, 10000, 100, 600

Sophia P. Redd, 70, 100, 6000, 150, 900

Thos. H. Drew, 48, 7, 8000, 100, 700

A. Dill, 125, 75, 20000, 200, 700

B. F. Dickinson, -, -, -, -, 150

Wm. F. G. Garnett, 150, 20, 20000, 500, 2400

Chas. C. Hanes, 65, 28, 10000, 200, 500

Thos. M. Ladd, 95, 29, 6000, 500, 600

J. B. Crenshaw, 270, 100, 22000, 900, 2500

A. D. Johnston, 60, 33, 10000, 100, 300

Masena Beazley, 300, 100, 20000, 300, 1800

Henry P. Taylor, 40, 45, 6500, 150, 250

Thos. G. Goodin, 30, 30, 4000, 50, 300

Thos. B. Carter, 150, 126, 25000, 900, 900

L. W. Rose, 17, -, 3500, 50, 300

John Stansberry, 5,-, 1500, 20, 500

John A. Parker, 65, 14, 10000, 100, 400

Augustus Schutz, 12, -, 1000, 50, 200

Thomas Bruton, 5, -, 1000, 20, 200

Garland Harris, 175, 125, 30000, 800, 1100

Geo. W. Carter, -, -, -, -, 150

Josep Rennie, 34,-, 7500, 20, 400

Bernard Plagman, 3, 7, 1200, 30, 100

Jos. Rose, 9, 1, 1200, 30, 100

John Goddin, 66, 28, 12000, 100, 500

B. J. Walton, -, -, -, -, 300

James Lyons, 105, 55, -, 500, 800

Jessee Williams, 120, 5, 12000, 200, 700

C. J. Meriwether, 55, 50, 15000, 50, 500

T. W. Seward, -, -, -, -, 200

Jno. Darricott, 71,-, 15000, 500, 1800

James Gilmon, -, -, -, -, 600

R. A. Lancaster, 78, 13, 20000, 300, 800

Thomas Hooper, 2, -, 1000, 20, 50

Colin Jervis, 4, -, 2500, 150, 400

T. Hoffmann, 4, -, 400, 20, 200

Wm. A. Robinson, 95, -, 8000, 50, 800

Edward Griffin, 6, -, 4000, 20, 200

Henry Bruns, 10, -, 2500, 50, 300

Jas. Taylor, 10 9, 5000, 50, 200

Ann O. Cottrell, 30, 4, 2000, 20, 200

Isaac Williams, 15, -, 4000, 50, 100

G. G. Alberger, 16, -, 8000, 250, 500

S. Loveinstein, 15, -, 5000, 50, 400

S. Helstine, -, -, -, -, 150

L. Chamberlayne, 12, -, 5000, 150, 200

J. L. Woodson, 28, -, 4000, 30, 200

Peter Lawson, 55, 7, 10000, 50, 400

N. F. Bowe, 92, -, 40000, 1500, 1500

Fendall Griffin, 155, 100, 40000, 900, 1800

Peter H. Anderson, 117, -, 21000, 900, 1300

H. P. Poindexter, 28, -, 16000, 100, 800

Robert Courtney, 100, 100, 7000, 200, 600

R. G. Tunstall, 14, -, 10000, 50, 100

J. W. Gordon, 16, -, 3000, 100, 300

Richard P. Rose, 12, -, 4000, 100, 400

Alfred Shield, 40, 16, 5000, 50, 400

Wm. L. Cowadin, 40, 20, 10000, 40, 300

Wm. Smith, 3, -, 6000, 50, 2500

J. W. Yarbrough, 10, -, 7000, 50, 200

R. M. Thorne, 20, -, 6000, 50, 300

E. A. Hawks, 55, 35, 4000, 30, 100

Wm. B. Burriss, 20, 38, 1000, 40, 150

B. T. Baughan, 20, 15, 1500, 20, 50

S. E. Dove, 8, -, 4000, 30, 400

Thos. J. Dunnavant, 40, -, 4000, 30, 1100

Edward Newman, 8, -, 3000, 50, 100

Henry Krake, 6, -, 7500, 30, 800

Haxall & Brother, 200, -, 30000, 500, 800

L. H. Dance, 16, -, 20000, 200, 1000

Thornton Lipscomb, 20, 20, 2000, 20, 100

J. C. Hardy, 3, -, 2000, 20, 300

Wm. A. J. Smith, 11, -, 6000, 150, 300

Wm. Bennett, 4, -, 2000, 50, 200

Fleming Philips, 4, -, 3000, 20, 150

James Vea, 8, -, 5000, 10, 120

Morriss Lugnot, 24,-, 15000, 100, 200

Chas. Kelley, 4, -, 1000, 20, 150

Wm. H. Brown, -, -, -, 20, 100

Lucy Chamberlayne, 23, -, 5000, 80, 200

H. L. Gallaher, 100, 29, 16000, 100, 500

B. W. Haxall, 325, 88, 33000, 1000, 2000

James Guest, 12, -, 4000, 200, 200

Jos. Schepers, 48, 6, 8000, 100, 200

L. S. Squires, 22, -, 12000, 200, 150

A. J. Ford, 20, 4, 15000, 250, 500

Jas. Hill, 25, 5, 6000, 50, 350

J. N. Shields, 100, 40, 60000, 500, 1300

Wm. Jenkins, 12, -, 3000, 100, 200

J. C. Vass, 100, -, 6000, 500, 200

D. W. Carter, 18, -, 7000, 50, 200

J. A. Chevllie, 20, -, 8000, 60, 250

H. J. Christian, 16, -, 8000, 40, 120

E. J. Warren, 100, -, 8000, 250, 900

J. P. Ballard, 115, 15, 12000, 200, 200

M. Winkler, 45, 45, 7000, 40, 250

J. Baughan, 30, 35, 2000, 50, 400

Wm. P. Lawton, 130, 51, 10000, 300, 1000

B. Baughan, 25, 5, 900, 30, 100

Edward McConnell, 175, 15, 10000, 2000, 1100

B. W. Green Jr., 700, 100, 50000, 700, 3500

Wm. Baughan, 7, 14, 600, -, -

B. W. Green Sr., 800, 400, 50000, 1600, 4100

B. W. Green Sr., 250, 115, 10000, 500, 1500

E. G. Higgenbotham, 200, 100, 10000, 300, 2000

B. W. Green Sr., 600, 100, 50000, 1000, 3500

B. W. Green Sr., 500, 400, 40000, 600, 2200

Thos. G. Dicken, 80, 80, 9000, 50, 300

Thos. Gennett, 14, 32, 1000, 70, 200

R. D. Carter, 196, 37, 12000, 1000, 2600

S. G. Patterson, 200, 211, 16000, 150, 1500

J. F. Franklin, 50, 61, 2000, 50, 400

Thos. H. Crafton, 30, 64, 3000, 70, 230

Wm. H. Burton, 60, 52, 2500, 150, 500

R. W. Horner, 100, 25, 2500, 50, 130

L. W. Twickham, 600, 300, 35000, 1000, 4400

Thos. Duke, 150, 50, 2000, 50, 100

James Sheppard, 60, 75, 6000, 100, 650

John Ford, 20, 20, 500, 20, 50

Simeon Ford, 20, 20, 500, -, 80

Jessee Shepperson, 6, 7, 500, 50, 200

John Matthews, 30, 125, 3000, 30, 100

Miles Duval, 100, 300, 8000, 1000, 1000

Henry Austin, 75, 81, 5000, 50, 50

Crawford Alley, 12, 135, 3000, 100, 400

Flemming Ford, 20, 90, 1500, 100, 500

Josiah Blackburn, 35, 80, 2000, 100, 600

Thos. Gordon, 20, 41, 1200, 100, 500

N. N. Gordon, 30, 70, 1200, 100, 900

Chas. Cottrell, 150, 232, 6000, 300, 800

David Hubodward, 150, 134, 4000, 150, 750

Thos. Hudgins, 140, 35, 25000, 250, 950

Jno. F. Wren, 50, 85, 13000, 100, 325

Luzby H.Wade, 25, 125, 2000, 75, 440

S. N. Conway, 5, 2, 300, -, 100

Wm. Hutcheson, 60, 13, 10000, 50, 200

Reubin Cottrell, 35, 37, 2000, 50, 250

Reeves Tinsley, 400, 600, 15000, 100, 1200

Stepleton Coates, 150, 100, 5000, 100, 500

Henry K. Shaw, 15, 10, 500, 10, 100

John Wickham, 640, 160, 32000, 3000, 2400

Saml. S. Duval, 400, 350, 15000, 200, 160

R. Dabney, 75, 73, 3000, 30, 350

_. C. Brown, 250, 350, 6000, 300, 500

Elizabeth St.Clair, 15, 15, 600, 20, 70

Temple Redd, 100, 200, 2000, 100, 450

_. T. Jones, 20, 40, 1000, 20, 100

J. J. Werth, 100, 500, 6000, 200, 600

R. G. Crouch, 355, -, 6000, 100, 700

Wm. Smoot, 100, 387, 4000, 50, 600

Saml. Cottrell, 600, 410, 40000, 1000, 3000

R. H. Cottrell, 125, 75, 3000, 200, 500

_. S. Ellis, 16, 15, 700, 50, 80

M. S. Cottrell, 75, 100, 1000, 50, 120

Jessee Brown, 30, 45, 1000, 10, 150

Wm. E. Harriss, 200, 238, 7000, 100, 500

Seth Duval, 130, 200, 3000, 200, 400

Elisha Duval, 50, 20, 1000, 40, 200

Wm. H. Lacy, 25, 18, 600, 20, 100

Sarah C. Conway, 50, 8, 500, 10, 100

Joseph Ellis, 60, 44, 2000, 50, 200

B. F. Bowles, 40, 143, 2000, 100, 210

Smith E. Ellis, 25, 44, 1500, 150, 450

R. T. Ellis, 200, 250, 5000, 200, 700

Jane M. Nuckols, 175, 100, 3000, 120, 800

J. E. B. Jude, 50, 20, 1500, 40, 150

Henry Satterwhite, 175, 75, 3000, 200, 650

_. W. Alley, 20, 223, 3600, 100, 400

J. W. Leake, 30, 230, 4000, 150, 300

T. C. Leake, 100, 285, 5000, 200, 900

H. B. Tate, 12, 25, 250, 10, 140

Wm. J. Winston, 15, 45, 600, 10, 200

H. J. Jones, 30, 70, 1000, 30, 250

B. W. Thompson, 80, 94, 1000, 50, 350

Lucius Goyne, 100, 57, 1000, 50, 450

B. M. Leake, 75, 100, 2000, 50, 600

M. R. Leake, 200, 100, 3000, 200, 550

Wm. M. McGouder, 400, 600, 10000, 100, 1600

Wm. Ford, 8, 7, 2000, 50, 500

Julius Hase, 14, 14, 500, 75, 280

R. A. Henley, 40, 130, 2000, 130, 600

T. M. Montague, 40, 60, 2500, 50, 200

Ernst Postwig, 80, 120, 1600, 100, 750

John Barr, 1000, 2500, 60000, 150, 1200

Jesse Puryear, 75, 75, 1500, 80, 240

Jno. H. Holman, 75, 115, 2000, 50, 400

D. M. Harlow, 19, 10, 150, 10, 100

Larkin J. Harlow, 52, 52, 400, 30, 140

John A. Henley, 60, 140, 2000, 100, 550

Wm. A. Deetrick, 400, 275, 25000, 700, 2000

A. J. Blackburn, 120, 12, 4000, -, 150

Wm. A. Deetrick, 450, 230, 10000, 2000, 2000

Wm. C. Allen, 100, 5, 100000, 600, 1800

Danl. E. Gardner, 230, 90, 10000, 250, 1400

Saml. Rutherford, 20, -, 15000, 10, -

John A. Hutcheson, 180, 90, 15000, 500, 1300

John Clendening, 55, 46, 10000, 150, 250

Saml. P. Mitchell, 14, -, 7000, 200, 250

Alex. Kerr, 25, 20, 4000, 100, 700

Wm. H. Richardson, 100, 43, 15000, 200, 700

Geo. Nunally, 6, -, 500, 50, 100

L. D. Crenshaw, 44, -, 13000, 200, -

Jas. M. Darden, 28, -, 10000, 30, 120

J. B. Glazebrook, 30, 186, 4000, 200, 350

S. L. Myers, 41, 40, 5000, 100, 400

A. J. Terrell, 40, 25, 4000, 150, 800

Jno. N. Powell, 430, 200, 37000, 1000, 3000

P. T. Atkinson, 80, 45, 7000, 170, 600

Saml. Moran, 120, 45, 10000, 50, 200

Alfred Winston, 60, 106, 7000, 200, 1200

Fountain Blackburn, 8, 1, 500, 50, 160

John Staples, 7, 19, 1200, 25, 80

Jacob Snead, 15, 16, 1500, -, 100

Absalom Blackburn, 80, 268, 10000, 150, 600

Juan Pizzini, 12, -, 6000, -, 150

Henry County, Virginia
1860 Agricultural Census

The Agricultural Census for Virginia 1860 was microfilmed by the University of North Carolina Library under a grant from the National Science Foundation from original records at the Virginia Department of Archives and History in 1963.

There are forty-eight columns of information on each individual. Only the head of household is addressed. I have chosen to use only six columns of information because I feel that this information best illustrates the wealth of individuals. The columns are:

1. Name
2. Improved Acres of Land
3. Unimproved Acres of Land
4. Cash Value of Farm
5. Value of Farming Implements and Machinery
13. Value of Livestock

Geo. Richardson, 60, 528, 5000, 100, 500
James Lamkin, 80, 99, 2000, 50, 400
Joseph Gilley, 30, 40, 700, 40, 200
John Cayton, 20, 45, 650, 40, 150
John Hopper, 50, 75, 1500, 75, 300
John Cox, 20, 100, 1200, 50, 100
Wm. King, 45, 55, 1000, 20, 200
Drury Pulliam, 15, -, 150, 5, 75
Wm. Pulliam, 40, 95, 1500, 40, 300
Wm. Fretwell, 125, 375, 7500, 60, 200
Hugh Lewis, 25, 72, 800, 50, 150
Richd. McDaniel, 20, 49, 690, 5, 75
Edwd. Panky, 100, 130, 2500, 75, 625
M. M. Wingfield, 40, 60, 1500, 45, 320
Warren Norman, 235, 300, 6000, 50, 415
Theo. Armstrong, 125, 155, 2500, 75, 516
Reuben Mabes, 50, 60, 1000, 6, 195
Doctor Harris, 100, 345, 5000, 100, 700

Richd. G. Lamkin, 225, 300, 8000, 125, 1000
Rowland Majors, 20, 108, 1100, 10, 100
Courtney Norman, 30, 50, 800, 50, 100
Tho. Wilson, 6, -, 75, 5, 50
Charles Dolan, 28, 2, 300, 8, 50
Jno. Reamey, 200, 150, 2720, 150, 800
Wm. Bryant, 15, 5, 145, 2, 125
Arch. Mitchell, 48, 48, 880, 100, 150
Eliz. Mitchell, 14, 12, 1000, 30, 200
Wm. Shackleford, 100, 120, 2240, 150, 700
James Green, 12, -, 125, 10, 140
Isaiah Turner, 35, 155, 2500, 150, 250
C. H. Norman, 75, 87, 2000, 100, 450
Morgan Wilson, 150, 118, 2500, 30, 500
Dutton Norman, 35, 200, 2300, 50, 500
John Turner, 100, 100, 2400, 20, 565

Step. T. Turner, 60, 140, 2000, 12, 224

Munroe Turner, 10, -, 100, 3, 175

Jno. McDaniel, 10, 30, 600, 4, 150

Jos. Barker, 45, 95, 1200, 25, 125

Major McDaniel, 30, 80, 600, 10, 80

Abner Martin, 150, 180, 4000, 100, 260

Benj. Jones, 75, 300, 5000, 100, 300

Guill Barker, 15, 32, 350, 6, 150

Pleasant Nance, 22, 58, 400, 10, 100

Maria Barker, 25, 179, 1200, 60, 150

James Barker, 75, 56, 800 10, 365

James Wilson, 25, 225, 1500, 2, 70

John Tolbert, 20, 18, 200, 5, 175

Aaron Turner, 15, 10, 200, 8, 150

James Myers, 25, 15, 400, 12, 160

Wiley Murrell, 30, 40, 300, 6, 150

Polly Gilley, 20, 80, 800, 2, 50

Edward Gilley, 25, 15, 350, 6, 150

Wm. Wilson, 20, 4, 120, 29, 125

Wm. Gilley, 50, 6, 300, 5, 60

Edwd. Burgess, 30, 30, 480, 7, 65

Pryer Green, 120, 80, 2500, 100, 250

Madison Stone, 150, 75, 1640, 40, 300

Moses Turner, 25, 3, 300, 4, 55

Joseph Jones, 20, 25, 400, 5, 50

Step. L. Martin, 150, 250, 4000, 50, 750

Mary Davis, 225, 420, 5000, 400, 1000

Mary Scales, 600, 900, 20000, 100, 1500

Richens Brim, 30, 85, 900, 5, 30

Barna Cahall, 50, 52, 100, 100, 450

Nathan Murphy, 10, 90, 900, 60, 290

And. Newnum, 15, 65, 400, 35, 150

James King, 40, 40, 600, 15, 65

Preston Eans, 25, 75, 600, 25, 75

Azel Bateman, 50, 199, 2000, 100, 279

Jos. Robertson, 10, -, 100, 10, 75

Wm. P. Terry, 650, 974, 18000, 500, 2000

Jno. Shackleford, 40, 70, 1200, 5, 275

Henry Harris, 35, 75, 1650, 50, 130

Wm. H. Bouldin, 150, 262, 4000, 150, 680

Geo. Hairston, 600, 1100, 23000, 40, 475

Sarah Fontaine, 100, 116, 2000, 50, 340

Joseph Eggleton, 20, -, 200, 40, 425

Geo. Odle, 60, 40, 1000, 8, 350

G. W. Odle, 60, 40, 1000, 8, 150

Reynard Ray, 75, 25, 800, 75, 200

Mary Wilson, 200, 244, 5000, 40, 475

John D. Hanfield, 100, 128, 2000, 10, 200

David Hanfield, 50 77, 630, 25, 175

Polly Austin, 18, -, 50, 5, 70

James Hanfield, 45, 155, 1600, 28, 350

John Clopton, 5, -, 50, 5, 150

Wm. Pearson, 36, 100, 800, 12, 100

Saml. Griggs, 20, 200, 1600, 5, 135

Lucy Bray, 30, 20, 300, 5, 150

Ellis Wilson, 80, 29, 1000, 40, 175

Pleas Earls, 5, -, 50, 1, 175

Thos. Flippin, 10, 52, 312, 5, 260

Wm. Earls, 50, -, 400, 4, 100

E. Richardson, 100, 222, 4000, 100, 300

Edmd. Gravely, 200, 225, 3000, 100, 525

Saml. Woodall, 30, 100, 1000, 20, 175

Missouri Woodall, 40, 40, 800, 75, 425

Mike Lawrence, 150, 350, 4000, 50, 600

Darling Lawrence, 40, 110, 800, 20, 444

Geo. Motley, 100, 200, 2300, 30, 250

Elijah McGuire, 20, 64, 350, 10, 150

James Johnson, 20, 46, 350, 12, 60

Saml. Manning, 8, 84, 460, 10, 150

James Hankins, 15, 105, 700, 65, 125

Jefferson Austin, 66, 100, 300, 15, 190

Chris Hall, 15, 212, 1500, 30, 180

James Martin, 100, 50, 2000, 25, 125

Geo. Dickinson, 50, 84, 1200, 15, 600

Geo. Belcher, 130, 155, 2440, 200, 500

Valentine Martin, 80, 95, 1750, 25, 450

Paulina Johnson, 40, 40, 500, 15, 200

Wm. Martin, 65, 60, 800, 4, 225

Lucy Riddle, 30, 45, 600, 5, 50

Saml. Lawrence, 75, 75, 1225, 50, 250

Leftwich Gravely, 50, 130, 3000, 45, 350

John Hankins, 150, 25, 2000, 40, 375

Thos. Dickinson, 35, 35, 1200, 10, 100

Charles Martin, 130, 100, 3000, 12, 400

Jno. F. Pedigo, 160, 117, 4000, 30, 175

Wm. C. Hall, 25, -, 800, 15 25

Jno. Gravely, 150, 350, 8250, 175, 1000

Francis Minter, 112, 11, 2000, 50, 250

Jabez Gravely, 150, 344, 7460, 150, 1000

Granville Belcher, 25, 32, 1032, 25, 300

Stanley Morrison, 150, 200, 6000, 50, 525

Francis C. Gravely, 55, 130, 4000, 200, 525

Eliz. Thomas, 200, 150, 2800, 20, 425

Nathan Bell, 60, 80, 720, 21, 200

Ambrose Handley, 100, 100, 1200, 20, 375

Wm. Gravely, 150, 172, 2000, 125, 145

Tabitha Watson, 42, 100, 1000, 20, 250

Robt. Martin, 50, 61, 1200, 90, 400

Jno. Austin, 10, 54, 400, 4, 40

Wm. Cleft, 25, 100, 600, 50, 200

Thos. Donigan, 40, 100, 1000, 35, 125

Wm. Johnson, 86, 1000, 1200, 10, 200

B. H. Morrison, 200, 200, 10960, 65, 488

Wm. Minter, 180, 100, 2800, 225, 450

Eliz. Law, 257, 100, 2000, 100, 645

Wm. Hunt, 45, 200, 2000, 10, 165

Webster Hatcher, 12, 61, 730, 45, 55

Wm. Gilbert, 100, 467, 5000, 35, 475

Walker Terry, 5, 10, 140, 10, 25

Casy Curtis, 15, 25, 300, 5, 75

Marshall Law, 20, 197, 2000, 10, 60

John Burch, 30, 136, 700, 10, 200

Abner McCabe, 30, 200, 3000, 30, 300

Henry Carter, 20, 30, 450, 6, 165

Nancy Thomas, 300, 300, 6500, 50, 750

Peyton Gravely, 300, 350, 8000, 210, 1314

Granville Minter, 130, 85, 2000, 125, 450

_. Wingfield, 80, 58, 1380, 25, 275

Rebb Eggleton, 40, 20, 1000, 100, 425

Joseph Slayden, 700, 1200, 38000, 125, 3000

Peyton Gravely Jr., 30, 10, 400, 5, 130

Joseph Gravely, 40, 35, 800, 30, 200

John Gravely, 100, 100, 2000, 20, 450

Silas Minter, 67, 100, 2505, 15, 650

Francis Gravely, 27, 74, 1500, 8, 150

Ben. F. Dyer, 30, 20, 500, 50, 400

Geo. Davis, 45, 71, 1000, 5, 225
Joseph Dyer, 12, 2, 125, 5, 150
A. Richardson, 75, 75, 1500, 25, 315
Calvin Martin, 20, 96, 1200, 35, 400
Patrick Martin, 20, 30, 500 8, 125
Gerard Barch, 150, 150, 4500, 150, 525
Lewis Gravely, 200, 150, 3500, 100, 850
Willis Gravely, 300, 470, 15400, 100, 2000
A. Richardson, 50, 164, 1000, 12, 325
John Richardson, 375, 125, 5000, 50, 525
Geo. Terry, 60, 41, 500, 75, 150
F. Richardson, 65, 50, 1000, 300, 300
Agness Wyatt, 90, 33, 738, 10, 150
Silas Wyatt, 68, 100, 1000, 5, 75
Craven Wyatt, 28, 25, 500, 15, 150
D. Richardson, 110, 100, 2200, 25, 475
J. Richardson, 50, 65, 1000, 60, 425
Adam Stultz, 60, 73, 1360, 20, 300
Alex Hodges, 4, 20, 140, 2, -
James Minter, 50, -, 500, 10, 150
Lucy Stultz, 75, 75, 930, 15, 250
Vincent Wyatt, 100, 58, 800, 12, 225
Anderson Purby, 48, 50, 300, 45, 275
Jno. Eggleton, 60, 40, 1060, 25, 100
Step. Eggleton, 50, 300, 6000, 25, 200
Wash. Eggleton, 40, -, 600, 80, 200
Alexr. Eggleton, 50, -, 750, 15, 75
Geo. Eggleton, 20, -, 300, 12, 65
Geo. Gravely, 200, 100, 4500, 100,700
Silas Minter, 500, 200, 7000, 100, 1200
Granville Eggleton, 60, 65, 1400, 10, 150
Jefferson Harley, 40, 360, 800, 5, 75
John Gregery, 50, 50, 800, 25, 200

Anderson Stultz, 30, 75, 1000, 100, 275
Mich Eggleton, 300, 150, 2000, 20, 475
Leathy Pace, 5, 20, 200, 5, 35
Henry Stegall, 35, -, 200, 5, 80
John Adkins, 18, 140, 600, 5, 165
Zeph Stultz, 500, 500, 6000, 100, 1500
Joseph Stultz, 75, 58, 1330, 10, 140
Martha Eggleton, 35, 22, 570, 5, 125
Benj. Barrow, 200, 250, 7000, 100, 650
Wash. Flord, 300, 300, 6000, 300, 850
John Stovall, 50, 155, 2000, 75, 300
Bailey Pinkard, 65, 135, 2500, 25, 475
Lewis Pedigo, 25, 26, 205, 5, 112
Rhoda Lovell, 50, 100, 1500, 15, 70
Lucy Arnold, 50, 50, 800, 5, 40
Joseph Pedigo, 35, 100, 1000, 8, 150
Lucy Shumate, 33, 100, 921, 5, 75
Benj. Lester, 100, 50, 900, 10, 200
Wm. Lester, 100, 250, 2800, 50, 275
Wm. Pedigo, 30, 13, 301, 10, 80
Daniel Pace, 35, 18, 300, 15, 350
Nancy Self, 10, -, 100, 5, 110
Geo. Reynolds, 150, 129, 4100, 125, 250
Wm. Barrow, 40, 60, 1500, 140, 600
Thos. Lester, 50, 140, 2000, 10, 200
Betsy King, 50, 109, 954, 10, 125
James Bowles, 14, -, 85, 8, 115
John King, 100, 350, 3000, 30, 475
Martha Cabiness, 80, 60, 840, 25, 175
Lemuel Bowles, 150, 54, 2000, 20, 140
Joseph Stultz, 30, 100, 800, 2, 150
Sally Griggs, 120, 285, 2400, 50, 425
Ira Griggs, 40, 129, 1000, 10, 200
Marshall Houston, 2000, 4000, 90000, 500, 2115

Wm. B. Preston, 500, 500, 15000, 400, 1200
Hamlin Cole, 50, 100, 1000, 35, 325
Wesley Griggs, 40, 393, 3000, 25, 400
Charles Cole, 25, 10, 300, 7, 45
Geo. Gravely, 70, 110, 1500, 130, 275
Thos. Stultz, 75, 209, 2000, 40, 265
Geo. Dillard, 100, 140, 1920, 25, 450
Davis Stultz, 175, 75, 2500, 15, 900
Jno. Griggs, 35, 50, 1000, 25, 220
Alex. Seay, 60, 100, 1000, 10, 125
Nathan Harris, 40, 40, 500, 5, 75
Mathew Seay, 70, 71, 1000, 20, 85
James Seay, 45, 100, 800, 25, 125
Gid. Clark, 58, 50, 1000, 20, 200
Abram Wells, 20, 30, 400, 5, 75
Robert Leak, 30, 50, 480, 25, 250
John Wells, 75, 12, 800, 10, 225
William Wells, 75, 12, 800, 10, 225
Riley Thomasson, 350, 450, 16000, 30, 1000
Silas Self, 35, 70, 600, 28, 325
James Davidson, 30, 45, 500, 15, 15
Thomas Wells, 30, 88, 1180, 35, 200
Frank Wells, 40, 60, 800, 45, 165
James Gregery, 18, 3, 144, 12, 30
Eliza Salmon, 170, 200, 3700, 40, 625
Peter F. Griggs, 100, 279, 2800, 25, 365
Mary Jones, 100, 200, 4500, 200, 600
Charles Philpott, 100, 400, 3000, 25, 500
Hendren Philpott, 75, 75, 1000, 18, 275
Benj. Mann, 12, 20, 100, 5, 75
John Beal, 60, 45, 525, 5, 100
James M. Smith, 600, 700, 20000, 550, 1800
Ruth Redd, 450, 450, 22500, 150, 750
Henry Drury, 450, 550, 18000, 150, 900

C. Y. Thomas, 160, 260, 5460, 125, 375
O. R. Dillard, 150, 400, 6000, 400, 1400
Wm. J. Hamlett, 8, 25, 2500, 100, 150
Seaton A. Pearson, 50, 61, 1000, 75, 225
Wm. E. Randolph, 35, 475, 16500, 128, 1200
Drury Bresck, 81, 110, 3000, 150, 300
R. H. Hairston, 400, 1400, 36000, 200, 1500
Wesley Morris, 130, 270, 5000, 75, 500
Carter Barber, 60, 46, 800, 10, 100
Thos. H. King, 59, 100, 1200, 55, 350
Jos. Thomasson, 50, 130, 1500, 25, 165
E. B. Farmer, 200, 120, 5000, 200, 600
Perry Cahill, 300, 350, 8000, 150, 1000
John L. Jones, 40, 60, 1000, 70, 350
Mosebey Jones, 40, 60, 1000, 25, 100
John S. Jones, 55, 50, 1000, 50, 200
Ambrose Jones, 50, 60, 1000, 10, 30
Marshall Cahill, 200, 212, 3400, 20, 624
John W. Cahill, 75, 114, 2842, 135, 525
Stephen Nunn, 60, 42, 1000, 8, 130
Mary Nunn, 60, 10, 800, 10, 110
Riley Nunn, 30, 200, 2000, 15, 125
Joab Feazel, 30, 130, 1100, 30, 125
Alexr. Bowles, 100, 225, 2800, 10, 100
Jabal Bowles, 35, 74, 1500, 10, 100
Saml. Good, 75, 225, 300, 20, 300
Martha Feazel, 40, 40, 300, 5, 65
Tho. J. Edwards, 25, 62, 600, 5, 200
Albert G. Saul, 37, 200, 1500, 6, 200
Jesse Davis, 50, 150, 1650, 12, 200

Benj. Davis, 100, 300, 1000, 50, 375
Geo. W. Napier, 200, 340, 7700, 300, 2250
Wm. Philips, 70, 426, 4200, 20, 350
James Hunter, 20, 27, 400, 10, 50
John Hunter, 33, 39, 800, 10, 110
Saml. Shumate, 50, 30, 900, 20, 250
Benj. Davis, 100, 92, 1000, 30, 215
Margaret Lawrence, 50, 50, 1000, 18, 200
David Lacy, 30, 68, 1000, 35, 100
Wm. Wade, 12, -, 100, 5, 25
Floyd Lacy, 50, 162, 2000, 25, 200
Newson Pace, 70, 28, 2000, 200, 290
Joseph Pace, 100, 100, 4000, 40, 400
Dorothy Eggleton, 15, 15, 3000, 5, 125
James M. Nunn, 12, -, 50, 25, 250
Jesse Nunn, 90, 45, 1500, 30, 300
Phebe Dillion, 30, -, 300, 5, 150
John H. Bassett, 175, 225, 2000, 8, 225
John Hensley, 75, 150, 3000, 50, 160
Danl. Shumate, 50, 45, 1200, 40, 350
Armstead Eggleton, 35, 65, 1000, 15, 150
John M. Feazel, 30, 100, 1200, 16, 160
Pleasant Lacy, 15, 74, 700, 5, 225
Benj. Robinson, 25, 143, 500, 6, 150
John Philpott, 20, -, 200, 10, 175
Nancy Stone, 75, 75, 1200, 15, 100
Clayton Stone, 35, 100, 1000, 10, 150
Elisha Potter, 40, 20, 400, 5, 45
Balus Shumate, 12, 38, 400, 3, 30
Abram B. Ross, 150, 153, 2500, 80, 500
Edwd. Philpott, 100, 230, 2100, 25, 250
Thos. Whitticoe, 200, 200, 3200, 45, 225
Richd. Wells, 20, 239, 1000, 6, 125
Eliz Philpott, 60, 200, 1500, 45, 100
Constant Martin, 50, 70, 1200, 5, 350

Stephen Turner, 200, 400, 5000, 25, 325
Jeremiah Parris, -, -, -, 5, 45
Peter Smith, 50, 509, 2800, 100, 300
David Johnson, 33, 100 700, 8, 140
Peter Draper, 30, 195, 1200, 5, 100
W. Edwards, 50, 180, 1200, 15, 135
Josiah Doss, 30, 37, 600, 10, 100
Benj. Stultz, 50, 30, 800, 5, 150
Lucy Davis, 200, 200, 4000, 15, 225
Barna Hill, 45, 50, 470, 4, 25
Ralph Dyer, 10, 50, 500, 5, 25
Levi Hill, 30, 100, 600, 5, 40
James Patterson, 50, 450, 4000, 20, 325
David Davis, 30, -, 270, 4, 150
Anderson Wade, 450, 1050, 15000, 200, 1550
Thos. Spencer, 150, 244, 1500, 15, 175
Jese Barnett, 100, 136, 2000, 5, 235
Wm. Turner, 40, 93, 800, 80, 160
Sarah Turner, 50, 50, 1000, 10, 150
M. Turner, 40, 106, 1500, 10, 250
Clark P. Turner, 25, 100, 800, 50, 225
B. Hollandsworth, 50, 116, 1500, 15, 250
Allen Joyce, 40, 162, 500, 10, 100
Wm. Oldham, 35, 45, 100, 12, 450
Caleb Eans, 10, 28, 200, 6, 200
Joshua Faughan, 15, 15, 150, 5, 125
Barna Baleste, 50, 149, 1000, 5, 160
Mary Stone, 100, 150, 1250, 8, 100
B. Hollandsworth, 25, 460, 1500, 8, 200
Jane Stone, 100, 196, 1500, 8, 200
James Stone, 42, 100, 500, 6, 125
Josh Purdy, 40, 125, 400, 5, 135
Lee Baleste, 24, 100, 700, 8, 100
John H. Washbarn, 20, 117, 500, 40, 235
Wm. Dillion, 30, 50, 700, 8, 175
Baily Danul, 70, 513, 1100, 4, 300
Saml. Harris, 40, 135, 700, 15, 40
Lewis Martin, 10, -, 70, 5, 50

Sarah Ingram, 100, 200, 1500, 10, 300
Wm. Law, 15, -, 100, 5, 75
John Harris, 10, -, 50, -, 85
Thos. H. Watkins, 450, 552, 10000, 102, 1200
Eliza Hagood, 35, 35, 750, 5, 175
Mary Hagood, 60, 13, 750, 10, 200
R. Mitchell, 54, 100, 800, 10, 100
J. W. Carter, 66, 66, 800, 5, 250
John Purdy, 50, 50, 300, 5, 75
Abram Custer, 20, 10, 100, 3, 35
Stokely Martin, 50, 185, 1150, 10, 150
Mary Stone, 70, 70, 1000, 15, 300
Moses Stone, 15, 20, 250, 4, 30
John J. Philpott, 110, 495, 4000, 20, 450
G. Hollandsworth, 50, 60, 1000, 10, 100
D. Hollandsworth, 40, 200, 1100, 5, 150
Peyton Stone, 15, 135, 500, 5, 30
Wm. Hill, 75, 125, 1000, 18, 145
Edwd. Booker, 10, 10, 200, 15, 230
Robert Jarratt, 70, 70, 800, 10, 135
Jared Stone, 100, 260, 1000, 25, 175
Patrick Jarratt, 75, 113, 1000, 18, 350
Andrew Jarratt, 20, 20, 350, 5, 5
Jno. C. Mitchell, 50, 43, 500, 5, 330
Wm. B. Wells, 50, 100, 1000, 12, 200
Gregory Hagood, 37, 38, 700, 10, 250
Maria Craig, 150, 150, 2000, 45, 320
John A. Lee, 30, -, 300, 5, 165
James Craig, 50, 58, 1000, 6, 50
John Prunty, 160, 680, 4500, 40, 768
Charles W. Hardy, 100, 139, 1600, 40, 250
Jno. S. Koger, 25, 131, 1600, 6, 180
Joseph Koger, 150, 133, 1900, 28, 325
Carington Dillion, 50, 60, 800, 15, 200

Jesse Dillion, 20, 33, 200, 3, 65
Wm. Law, 42, 100, 1000, 12, 165
Andrew Arnold, 40, 60, 400, 5, 100
Jacob Koger, 35, 180, 1200, 25, 275
Robt. Via, 100, 400, 4000, 25, 600
John Koger, 100, 306, 4060, 25, 600
Edward Wells, 20, -, 150, 3, 45
John Scruggs, 75, -, 750, 40, 175
Barton Pyrtle, 75, 125, 2000, 40, 230
John P. Taylor, 80, 200, 2500, 25, 225
B. F. Mitchell, 100, 200, 4000, 30, 340
Henry Dillion, 30, 53, 800, 15, 100
Martin Dillion, 30, 53, 800, 50, 100
Nelly Banister, 25, -, 150, 3, 70
Robert P. Hitt, 30, 15, 350, 12, 110
Wm. Mitchell, 100, 300, 2000, 25, 400
John Mathews, 175, 125, 3000, 100, 725
Nancy Smith, 50, -, 250, 9, 130
Jos. M. Wells, 15, -, 125, 8, 100
James M. Dillon, 35, 50, 850, 12, 165
Peter B. Oakley, 50, 100, 1200, 10, 125
Wm. Clark, 75, -, 750, 10, 250
Thos. East, 60, 118, 2300, 60, 235
A. L. Jarratt, 200, 400, 4200, 75, 300
E. R. Mitchell, 35, 65, 1500, 25, 700
Wm. F. Mills, 150, 150, 2400, 30, 725
Marshall Finney, 100, 221, 2500, 5, 140
Susan Finney, 200, 250, 2000, 150, 550
David Philpott, 120, 260, 4000, 20,800
John McMelon, 30, 70, 800, 5, 30
Henry Coveins, 40, 13, 400, 6, 50
Catharine Baker, 75, 112, 1200, 80, 450
George Callaway, 290, 2635, 28000, 75, 975
Richd. Wells, 30, 60, 700, 28, 200

Wm. Wilkinson, 25, -, 250, 8, 150
Elias Chaney, 30, -, 300, 18, 110
Thos. Wilkinson, 45, -, 500, 10, 250
Hiram Watkins, 50, 90, 1400, 125, 425
Field Trent, 38, 85, 900, 16, 250
Wm. Turner, 65, 55, 700, 10, 190
Jas. Hollandsworth, 50, 50, 1000, 6, 175
Peter Hollandsworth, 43, 100, 1000, 7, 175
Thos. Stanley, 150, 173, 1230, 80, 375
Mat. Hollandsworth, 40, 107, 1000, 20, 120
Chris Mason, 70, 126, 1400, 20, 125
Zach Law, 30, 41, 800, 3, 130
Demarcus Franklin, 20, -, 200, 15, 110
John Jarratt, 40, 90, 1000, 8, 125
Wm. Morris, 200, 100, 1800, 60, 375
Henry Koger, 150, 228, 1900, 12, 325
Jas. M. Morris, 300, 400, 4250, 200, 850
James Seyars, 30, 75, 500, 5, 200
Geo. Parnnell, 400, 250, 8550, 300, 2700
Sarah Schoolfield, 100, 101, 1600, 60, 300
John Rangeley, 200, 400, 6000, 50, 865
Maria Waller, 130, 250, 3800, 60, 730
Martha Perkins, 250, 110, 6000, 100, 800
Thaddeus Salmon, 50, 36, 1200, 30, 135
I. J. Wray, 40, 60, 1200, 10, 175
D. W. Morris, 50, 233, 4275, 30, 550
Geo. Waller, 400, 550, 10000, 75, 620
Jno. R. Salmon, 80, 59, 1400, 60, 365
Jno. Cobler, 54, 100, 462, 10, 50
Elisha Dodson, 50, 200, 500, 10, 20

James Lawrence, 10, 30, 100, 3, -
John A. Hardin, 100, 98, 1000, 85, 365
Judith P. Hill, 100, 300, 4000, 110, 275
Alex. Bassett, 141, 100, 2500, 75, 500
Woodson Bassett, 80, 115, 1500, 100, 325
H. G. Mullins, 150, 1900, 1000, 850
G. B. Nuolds, 200, 267, 6400, 75, 850
Geo. Baker, 40, 80, 600, 50, 230
Mary Campbell, 125, 200, 1280, 45, 300
James H. Carter, 150, 183, 4000, 45, 450
Wm. W. Hill 100, 180, 3000, 125, 300
Starling Wells, 50, 290, 2700, 46, 150
Geo. Wells, 20, 5, 200, 8, 135
Robt. W. Wells, 20 117, 900, 5, 130
Jno. East, 200, 250, 3000, 50, 900
Jackson Smith, 40, 50, 600, 5, 150
Geo. Hanlen, 12, -, 100, 8, 65
Eliz. Abington, 150, 160 1550, 85, 530
James Clark, 30, 37, 335, 5, 40
Wm. Craig, 40, 40, 450, 10, 80
Wm. Wynn, 700, 2000, 21600, 100, 1300
Oliver Shelton, 450, 1550, 16000, 40, 850
Wm. A. Dandridge, 200, 500, 10000, 40, 750
James Coleman, 130, 158, 3500, 35, 400
Thos. J. Hughes, 250, 138, 6000, 100, 1000
H. C France, 800, 556, 14000, 100, 1235
Thos. Oakley, 80, 110, 1500, 5, 250
Wm. F. Morris, 250, 626, 7000, 100, 850
Robert Blackwell, 50, -, 500, 8, 80

Lorenzo Bousman, 40, 260, 2000, 50, 400

Joseph Minter, 50, 295, 1700, 25, 230

Jesse Purdue, 30, 180, 2000, 60, 550

Marvel Arington, 10, 23, 120, 3, 15

Wm. Payne, 15, 285, 2000, 4, 80

James Marshall, 10, 15, 150, 7, 75

Carter(Custis) Hardy, 20, 130, 1000, 5, 30

Ashford Dove, 20, 522, 3800, 10, 285

Wm. Ballard, 650, 235, 15000, 100, 1000

Henry C. Wootton, 300, 157, 8000, 140, 1000

Saml. Doyle, 100, 77, 1000, 70, 300

Wm. J. Salmon, 40, 50, 1700, 25, 82

Reed Ayres, 10, 600, 3500, 10, 1420

Robt. Mills, 6, 132, 500, 10, 150

Wm. Mills, 14, -, 100, 5, 75

Alex. Joyce, 75, 122, 2000, 75, 300

A. M. Shelton, 130, 269, 2000, 147, 1000

James Suttenfield, 70, 80, 900, 25, 150

James Taylor, 10, -, 80, 15, 128

Geo. W. Clanton, 300, 225, 6000, 200, 1000

Thos. J. Pratt, 40, 217, 2500, 6, 500

Green Watkins, 35, 98, 600, 8, 175

Joseph Smith, 30, 100, 500, 40, 330

Hardin Price, 50, 170, 2500, 75, 400

Jacob Hefflefinger, 50, 94, 1400, 50, 200

Coleman Davis, 60, 45, 1000, 75, 125

Abram Boaz, 30, 70, 1000, 4, 125

Thos. Nunn, 50, 172, 1300, 90, 300

Tho. Gilley, 50, 15, 400, 65, 150

Burwell Wilson, 40, 75, 350, 10, 100

Wilson Mahan, 10, -, 100, 6, 100

Jacob Hefflefinger, 40, -, 240, 10, 125

Jack Stratton, 40, 113, 900, 60, 275

Jno. D. Wade, 136, 200, 6700, 200, 800

Martin Joyce, 100, 735, 8000, 60, 350

Bennett Cox, 25, 100, 1200, 25, 150

Clifton Hayslip, 50, -, 300, 10, 75

Robert A. Joyce, 20, -, 200, 5, 130

Wm. Pulliam, 30, 170, 1800, 30, 75

David Covington, 25, 116, 1200, 35, 125

Wm. Cox, 20, 50, 350, 45, 175

Julia A. Taylor, 30, 70, 600, 10, 35

Wm. W. Wells, 10, 90 900, 5, -

Frances Gilley, 20, -, 200, 8, 75

John H. Burgess, 355, 400, 7600, 60, 1000

Jno. S. Reamsey, 250, 434, 16000, 100, 1000

Peter Hairston, 300, 930, 18600, 100, 1300

Cath. Cole, 200, 200, 6400, 90, 400

Sarah A. Redd, 250, 640, 14000, 150, 600

Wm. A. Sheffield, 600, 500, 12500, 150, 1450

Jno. W. Smith, 100, 110, 4000, 35, 585

Jno. C. Jones, 54, 150, 4000, 8, 225

Geo. Carter, 20, -, 200, 10, 75

Wm. S. Penn, 266, 400, 10000, 210, 1250

Saml. Hairston, 300, 887, 14600, 100, 1650

Nelson Cobb, 10, 15, 150, 5, 40

David M. Mathews, 15, 81, 1500, 100, 360

Eliza Mathews, 400, 236, 10000, 150, 1000

Geo. K. Jones, 175, 125, 4000, 45, 500

Robertson Anderson, 30, 160, 1200, 10, 150

David Anderson, 10, -, 60, 5, 50

Wm. Dodson, 25, 25, 500, 10, 80

Duke Price, 100, 197, 3000, 40, 650

Edwd. Seyars, 40, 70, 1100, 20, 200
Geo. W. Oakley, 30, 66, 500, 12, 100
Jared Patterson, 50, 82, 1300, 85, 250
Jno. P. Price, 300, 800, 10500, 100, 1700
Livingston Claiborne, 400, 636, 15500, 150, 585
Geo. H. Hairston, 700, 700, 27700, 200, 2300
Sarah Hairston, 200 1600, 6000, 100, 1200
Geo. T. Hairston, 280, 1142, 21300, 450, 1450
Benj. Jones, 100, 135, 2000, 60, 350
Wm. B. Trent, 125, 375, 9500, 150, 1450
Cath. Trent, 75, 225, 2500, 125, 250
Glover M. Trent, 200, 600, 10000, 50, 785
Kendall Vernon, 20, 10, 300, 12, 75
Thos. Edwards, 7, 244, 2000, 45, 375
Henry Vernon, 35, 104, 1500, 25, 200
Lewis G. King, 33, 200, 2700, 70, 300
David Nance, 9, 3, 75, 4, 30
Thos. Price, 20, 78, 800, 10, 150
Eliz. Roberts, 75, 158, 1400, 30, 150
Joseph Land, 30, 36, 800, 30, 90
John Garrot, 60, 85, 2000, 70, 278
Hardin Nance, 50, 50, 1500, 25, 350
James Odle, 62, 50, 1000, 15, 125
Milton Grant, 25, 92, 900, 35, 375
Munroe Wray, 35, 137, 1200, 6, 50
Burwell Flanigan, 40, 95, 1200, 30, 260
Wm. Abington, 100, 200, 6000, 100, 320
Jno, C. Mitchell, 200, 200, 4000, 100, 730
Wm. Linkous, 400, 1000, 14000, 250, 1290
John Wells, 12, 212, 1500, 25, 300
Wm. Grogan, 50, 360, 2000, 65, 260

Peter Shelton, 200, 700, 7000, 200, 680
James Coogan, 30, -, 500, 5, 100
Robert A. Read, 100, 50, 2000, 75, 535
John O. Redd, 350, 200, 8000, 100, 1250
John R. Dillard, 300, 300, 10000, 125, 850
David H. Spencer, 1500, 900, 47000, 150, 1600
Geo. Dillard, 450, 230, 12000, 80, 1230
Pinkney Spencer, 65, 125, 1330, 50, 515
Peter P. Penn, 400, 600, 10000, 200, 1425
Geo. Staples, 150, 277, 3500, 100, 900
Nathan Read, 125, 119, 3500, 50, 400
James Nance, 15, 62, 300, 6, 65
Silvester Webb, 14, 100, 798, 7, 100
Wm. M. Ayres, 50, 50, 800, 10, 375
James H. Lewellen, 40, 35, 600, 25, 130
Osburn Kindrick, 30, 100, 1000, 50 175
Leonard P. Ayres, 20, 20, 400, 10, 200
Thos. Koger, 75, 132, 2100, 75, 400
Edwd. Thomas, 50, 72, 1320, 60, 590
Danl. Harris, 15, 180, 800, 6, 40
James Athy, 500, 200, 2000, 300, 500
And. J. Leak, 230, 261, 5000, 80, 400
Philip Hill, 50, 100, 1000, 15, 100
Reuben Taylor, 100, 200, 2400, 50, 525
Danl. G. Taylor, 20, 130, 1000, 35, 220
Sarah Hay, 400, 1550, 17000, 300, 1100

Saml. Brown, 25, 55, 500, 5, 110
Geo. W. Taylor, 37, 100, 1000, 12, 250
James Purcell, 10, -, 80, 5, 100
Jno. H. Kendrick, 20, -, 200, 40, 125
Robert Cobler, 20, -, 200, 15, 100
Henry Lawrence, 10, -, 100, 5, 90
James Shelton, 30, -, 300, 75, 375
James Cayton, 30, 60, 450, 25, 75
John Grogan, 10, -, 50, 5, 70
John H. Jamerson, 200, 97, 4000, 130, 700
James Marshall, 35, 65, 800, 60, 325
Wm. S. Ivie, 30, 70, 525, 30, 175
James A. Pratt, 100, 163, 3000, 50, 500
Isham Fry, 40, 290, 2000, 40, 325
Tho. H. Rogers, 10, 50, 500, 10, 200
Francis Mills, 100, 200, 1500, 45, 500
Fred. H. Bouldin, 100, 474, 1700, 70, 560
Peter W. Watkins, 300, 1724, 19500, 140, 1360
Geo. Marshall, 20, 92, 600, 8, 210
Jno. Marshall, 60, 198, 2500, 35, 500
Jno. Hollandsworth, 2, 48, 300, 5, 50
James Stone, 20, 45, 550, 10, 80
Jos. Williams, 20, 30, 500, 5, 60
Peter H. Dillard, 200, 684, 8800, 150, 1056
Camellus King, 150, 210, 5800, 200, 560
John Davenport, 80, 100, 1500, 15, 350
Geo. L. Tush, 75, 155, 2000, 30, 550
Harrison Faris, 10, 90, 1000, 25, 175
Thos. Dunavant, 300, 500, 8000, 200, 1180
Eliz. Moore, 100, 688, 3850, 15, 100
Geo. W. Booker, 50, 147, 1000, 15, 550
Peter D. Watkins, 40, 50, 600, 10, 75
R. H. Watkins, 30, 220, 1000, 10, 300
David Mays, 50, 75, 750, 10, 350

David Smith, 20, 30, 350, 25, 70
And. J. Wilson, 100, 450, 4400, 20, 225
Geo. Dunavant, 20, 76, 400, 125, 300
Thos. Harbour, 150, 291, 8888, 20, 225
Eli Watkins, 100, 100, 800, 10, 250
Wm. Moore, 80, 64, 1150, 20, 100
Jno. W. Watkins, 12, 50, 450, 25, 150
Nancy Watkins, 54, 100, 770, 15, 150
Larkin Deshazo, 400, 400, 10000, 100, 1000
Pleas. Moore, 155, 355, 4300, 45, 880
Joseph Robinson, 12, 58, 350, 6, 90
Beverly Flanagan, 30, 90, 1800, 45, 450
Geo. Griggs, 100, 246, 4000, 40, 600
Wm. Gilley, 52, 100, 1200, 45, 500
Wilson Jones, 40, 177, 1700, 30, 250
James T. Odle, 40, 140, 1400, 90, 275
Richd. Odle, 25, 55, 650, 3, 125
Geo. E. Martin, 15, 93, 1200, 10, 175
John Hopper, 40, 52, 900, 50, 100
James B. Price, 75, 81, 1250, 35, 200
Nancy Cox, 50, 100, 1200, 10, 200
Danl. Jones, 150, 208, 2800, 60, 275
E. T. Starling, 400, 480, 15000, 150, 1550
James Hopper, 75, 193, 2000, 75, 400
James Nunn, 25, 25, 300, 100
Calvin Meeks, 15, 10, 60, 6, 80
Sarah Meeks, 20, 38, 200, 5, 35
James Meeks, 20, 30, 250, 5, 75
James B. Norman, 148, 148, 3000, 30, 600
Chas. Scruggs, 6, 10, 140, 25, 50
Joseph Brim, 30, 142, 1400, 50, 250
James W. Trent, 100, 266, 3600, 60, 600
Jesse King, 30, 100, 1000, 25, 400

Edward Miles, 35, 57, 550, 5, 30
John C. Coan, 50, 100, 2000, 55, 350
Wm. Martin, 400, 600, 20000, 150, 2500
B. F. Gravely, 300, 800, 12000, 200, 2100
Geo. Hairston, 1100, 1185, 25000, 500, 2500

Wm. Oakley, 20, -, 200, 10, 150
Granville Hundley, 20, 137, 1000, 10, 100
Tho. Higgs, 20, 166, 1200, 8, 75
Wm. Land, 50, 150, 1200, 75, 350
Jno. S. King, 10, 20, 300, 20, 100
Jno. H. Pedigo, 60, 190, 2000, 50, 500

Highland County, Virginia
1860 Agricultural Census

The Agricultural Census for Virginia 1860 was microfilmed by the University of North Carolina Library under a grant from the National Science Foundation from original records at the Virginia Department of Archives and History in 1963.

There are forty-eight columns of information on each individual. Only the head of household is addressed. I have chosen to use only six columns of information because I feel that this information best illustrates the wealth of individuals. The columns are:

1. Name
2. Improved Acres of Land
3. Unimproved Acres of Land
4. Cash Value of Farm
5. Value of Farming Implements and Machinery
13. Value of Livestock

Abell H. Armstrong, 280, 300, 7000, 25, 1620
Allen Armstrong, 200, 400, 8000, 12, 3784
James Beath, 34, 289, 1000, 15, 114
Thomas Beverage, 120, 165, 4000, 25, 400
James Jones, 170, 1000, 6000, 15, 1207
John Samples, 100, 63, 3000, 10, 359
Elijah Samples, 30, 376, 380, 10, 193
John Beverage, 60, 62, 2000, 10, 352
Isaac Seybert, 200, 1540, 6000, 50, 686
John Beverage Sr., 500, 230, 12000, 50, 1480
Solomon Waggoner, 500, 500, 21000, 200, 3094
Thomas J. Rimer, 50, 152, 2020, 60, 385
William Beverage, 160, 108, 4500, 20, 615
George Rimer, 27, 46, 365, 5, 75
Jacob Peck, 12, 8, 400, 3, 190

Abram Peck, 85, 85, 850, 5, 372
Joseph Patterman, 55, 45, 1000, 5, 199
Andrew Halterman, 40, 20, 720, -, 51
Jonathan Arbagest, 40, 58, 2700, 20, 201
Jacob Seybert, 200, 760, 9000, 60, 1060
Amos Gum, -, -, -, -, 129
Andrew Flersher, 300, 1000, 9000, -, 225
Solomon Flersher, 700, 700, 7000, 150, 1309
George Colaw, 70, 170, 2440, 15, 463
Ephraim Arbagest, 70, 363, 5683, 105, 506
George Mulinax, 60, 98, 1600, 129, 197
John Mulinax, 100, 544, 1030, 60, 455
Henry Waggoner, 89, 361, 1866, 70, 385
John Hansel, -, -, -, -, 350
John Vint, -, -, -, -, 127

Henry Wimer, 200, 199, 5227, 125, 800
John Chow, -, -, -, -, 162
Allen Colaw, 100, 325, 4700, 8, 810
Jonas Colaw, 150, 250, 4320, -, 120
Hezekiah Colaw, 100, 150, 2320, -, 760
Otho Gum, 90, 160, 4150, 110, 439
Jacob Newman, 60, 57, 3000, 80, 480
Cornelius Colaw, 175, 175, 7000, 90, 636
Levi Arbagast, 491, 982, 1200, 60, 278
George Colaw, 700, 3763, 10930, 100, 1187
John Hodge, 150, 463, 12000, 30, 1462
John Waybright, 35, 23, 600, 3, 195
Joseph L. Chew, 40, 257, 2000, 10, 297
Samuel Life, 70, 60, 2000, 50, 235
Peter Gum, 53, 100, 3000, 70, 251
Peter Gum, -, -, -, -, -
Solomon Nicholas, 49, 96, 711, 100, 443
George Mulinax, 150, 895, 4520, 90, 821
Lucinda Gum, 99, -, 2970, 5, 198
Henry Colaw, 32, 51, 1030, -, 432
Ephraim Colaw, 55, 55, 1100, -, 259
Benjamin Arbogest, 200, -, 4000, 5, 691
William Fox, 150, 30, 3000, 85, 400
Henry Arbagast, 160, 170, 3370, 25, 386
Cyrus Lantz, 200, 210, 3250, 20, 366
Jonas W. Chew, 100, 38, 3756, 100, 528
Benjamin Sovenker, 100, 295, 3590, 45, 657
William Hovener, 375, 32, 9410, 207, 2133
John Jack, 400, 362, 10000, 105, 881
David Swisher, 75, 125, 1600, 20, 668

Conrad Creamer, 50, 25, 750, 10, 183
Adam Miller, 100, 227, 6000, 80, 225
David Mawzy, 130, 184, 1608, 20, 767
George Hammer, 80, 52, 3500, 20, 373
Peter H. Kinkead, 259, 70, 10000, 24, 1780
William P. Kinkead, 168, 209, 6732, 18, 1095
John E. Gum, 150, 1650, 5640, 60, 857
Rachael Hull, 100, 100, 2000,-, 241
Frederick K. Hull, 100, 1477, 14535, 80, 3770
Jacob Hovener, 1400, 8824, 17343, 290, 3724
Kathanne Hovener, 200, 360, 2000, -, 30
David Snider, 295, 155, 6675, 130, 2159
George Beverage, 210, 165, 6725, 35, 653
William D. Hovener, 100, 300, 3570, 25, 1007
Adam Folks, 80, 384, 3000, -, 698
John Folks, 66, 60, 1262, 60, 372
John Snider, 800, 2161, 18064, 155, 1579
Henry Hovener, 150 200, 3687, 60, 1358
Samuel Rexroad, 25, 25, 600, 6, 103
Jacob Shoneburg, 100, 299, 2500, 10, 600
Samson Waggoner, 60, 311, 667, 5, 199
William Lunsford, -, -, -, 15, 152
William Simmons, -, -, -, 100, 412
Mark Simmons, 75, 177, 2562, -, 422
Henry White, 160, 260, 2241, 80, 345
John White, -, -, -, -, 65
Jacob White, 200, 534, 4544, 10, 338

Benjamin Rexroad, 300, 200, 10000, 10, 497

Henry Gragg, 32, 94, 1000, -, 375

Solomon Rexroad, 100, 1100, 4800, -, 174

Adam Gragg, -, -, -, 110, 324

Jacob Hull, 230, 9355, 1329, 200, 2342

John Gum, 50, -, 600, 40, 414

Isaac Gum, 5, -, 600, -, 287

Daniel W. Wilfong, 100, 350, 2250, 8, 446

Elias Wilfong, 75, 365, 2000, 45, 485

Charles Collins, 20, 734, 2000, 5, 128

Henry Wolf, -, -, -, 10, 104

Adam Waggoner, 14, 326, 300, 10, 24

John C. Bright, 12, 29, 150, 5, 102

John W. Davis, 50, 450, 1500, 10, 168

Jacob Hovener, 12, 213, 225, 25, 42

James E. Gum, 30, 303, 1665, 5, 167

Elizabeth Gum, 20, 30, 600, -, 40

Zachariah Tomlinson, 100, 53, 2295, 75, 154

James Abagast, 30, 73, 700, 100, 241

Charles M. Gum, 30, 83, 1000, 6, 213

John Gum, 35, 53, 2500, 10, 174

John Hines, 100, 150, 3000, 200, 576

Matthew Gwin, 200, 139, 3500, 30, 342

David R. Rickman, 800, 2500, 18170, 200, 3580

Alexander Gilmore, 873, 3026, 17942, 80, 2363

John Dover, 260, 1285, 9810, 100, 960

Jacob Brisco, 40, 34, 1400, 5, 221

George H. Bord(Bird), 400, 300, 8300, 50, 1139

John C. Bord(Bird), 100, 990, 2120, 25, 460

John F. Shemuts, 73, 310, 1720, 8, 211

Abraham Wades, 75, 93, 1746, 15, 337

Peter Ginn, 75, 125, 2000, -, 213

Stuart C. Sloven, 100, 65, 2970, 20, 846

John S. Pullin, 26, 229, 2800, 60, 390

Henry Folks, -, -, -, 5, 228

David Clark, -, -, -, 20, 509

James Terry, 200, 391, 4790, 70, 1222

Samuel Hines, -, -, -, 80, 275

Bennet Hines, -, -, -, 60, 346

John McGlaughlin, 170, 1092, 5889, 30, 601

William Hix, 75, 332, 1628, 65, 221

James A. Gardner, 13, 133, 146, 20, 46

Ann Doyl, 60, 282, 2000, -, 119

William Thorurson, 70, 190, 2000, 65, 366

Samuel Gwin, 125, 275, 3200, 60, 458

John Gwin, 125, 275, 3200,-, -

James Gwin, 115, 245, 3500, 10, 380

James McGloughlin, -, -, -, 5. 220

Danuel McNulty, 100, 264, 2069, 20, 805

John Lightner, 346, 480, 10860, 150, 2517

James Gay, 200, 3250, 9737, 130, 848

Edgar Campbell, 300, 300, 13787, 110, 909

James Woods, 104, -, 3121, 10, 406

John Hull, 514, -, 6945, 30, 785

David Gwin, 200, 100, 9000, 10, 270

Michael Trainer, 150, 238, 3039, 140, 776

David Groves, 13, 133, 146, -, 231

Benj. B. Campbell, 419, 1800, 8690, 60, 1374

William W. Benson, 100, 690, 3600 30, 343

Susan Benson, -, -, -, -, 312
James Carry, -, -, -, 5, 213
James Terry, -, -, -, 60, 434
William C. Bird, 68, 1, 1856, 5, 376
Charles Wade, 150, 794, 3500, 100, 745
Saml. B. Campbell, 110, 100, 3510, 10, 536
Wesley Buzzard, 150, 343, 5500, 12, 165
David Gwin(of Jno), 200, 362, 5000, 120, 835
Sarah Gibson, 300, 1144, 4121, 60, 1036
William D. Gibson, -, -, -, -, 141
Joseph G. Chesnut, 100, 352, 1760, 62, 425
Elisha Wright, 100, 208, 3076, 60, 609
Peter Bird, 22, -, 440, 5, 193
A. H. Campbell, 100, 85, 1753, 65, 776
Alexander Campbell, -, -, -, -, 120
Thomas Campbell, 750, 5394, 200, 3765
Austin W. Campbell, -, 325, 200, 80, 260
William Swadly, 15, 200, 1188, -, 110
John M. Rexroad, 100, 95, 1935, -, 242
John Bird, 329, 165, 1000, 90, 1520
James Bird, -, 150, 100, -, 90
George Folks, 90, 475, 2735, 100, 232
Valentine Folks, -, -, -, -, 111
John Campbell, 217, -, 5781, -, 699
Matthew Hull, 50, 100, 3000, 10, 613
Cornelius Gwin, 75, 75, 1800, 10, 452
Jesse Slaven, -, 334, 1000, -, 88
Isaac Gwin, 125, 508, 6942, 170, 1091
Adam L. Gwin, 300, 414, 5000, 70, 686

Peter Gwin, 70, -, 1750, -, -
Abram Gum, 200, 400, 6300, 15, 765
Susan Wade, -, -, -, -, 60
James Wade, 432, 100, 3397, 50, 1626
William G. Chesnut, 100, 454, 1900, 10, 201
Volintine Bird, 120, -, 2894, 10, 258
Aaron Bird, -, -, -, -, 172
William Bird, 170, 1645, 3135, 40, 268
David H. Bird, 90, 990, 3518, 68, 521
Benjamin Ervin, 70, 214, 2000, 95, 468
James May, 10, 10, 100, 10, 92
Abram May, 10, 10, 100, 10, 121
Anson Wade, 200, 689, 7000, 100, 1717
Adam Lightner, 400, 1372, 10108, 110, 1384
James H. Rides, -, -, 100, -, 275
Levi Mathaney, 60, 53, 2500, 75, 627
Robert Mathany, 150, 223, 3125, 60, 703
Eli Doyle, 47, 47, 1500, 10, 145
William Lightner, 300, 1138, 6802, 50, 1095
Edward Ervin, 100, 77, 1800, 12, 92
John Woods, 200, 23, 3000, 70, 603
Polly Ervin, 20, -, 8000, -, 148
William M. Campbell, 236, 236, 5908, 40, 1550
Michael Wise, 300, 800, 7000, 120, 2389
William Wilson, 800, 900, 20000, 130, 1890
Elizabeth Sheridan, 200, 300, 5000, 50, 475
Lanty W. Hickman, -, -, -, -, 580
David G. Kincaid, 200, 270, 7000, 40, 587
St.Clair Turner, -, -, -, 7, 119

Jacob C. Doyl, 31, 266, 1000, 10, 152

Susan E. Stephenson, 200, 336, 4000, 40, 588

David Stephenson, 150, 186, 4000, 50, 796

Thomas T. Brown, 250, 50, 3000, 30, 417

Robert Carpenter, 40, 208, 3000, 10, 103

Mary Turner, -, -, -, 5, 161

Godleau Hingardner, 40, 153, 1000, -, 70

James P. Bead, -, -, -, 5, 175

Morgan Carpenter, 40, 153, 3000, 5, 77

Marthe Carpenter, 40, 58, 600, 5, 198

William Townsend, 70, 100, 4000, 5, 224

David Carpenter, 40, 100, 1800, 5, 178

David McNulty, 60, 83, 1800, 10, 375

Jared M. Folks, 60, 162, 1200, 10, 285

William D. Kelly, 50, 60, 1100, 5, 285

John Sharp, 100, 500, 10000, 100, 341

William J. Griffin, -, -, -, 10, 141

Lewis Davis, 150, 100, 3000, 2,287

Paschal D. Williams, 200, 434, 2824, 40, 971

Harry Hicklin, 80, 82, 2000, 30, 423

Peachy H. Stuart, 100, 96, 1568, 70, 479

John Cobs, -, -, -, -, 202

Samuel Pullin, 200, 1535, 9000, 160, 1505

Henry Seybert, 3990, 1330, 22755, 275, 3493

Andrew J. Rexroad, -, -, -, -, 176

William Marshal, 40, -, 800, 20, 185

John Vandevender, 60, 73, 1200, 5, 295

George Vandevender, 60, 73, 4000, 150, 486

Job Puffenbarger, 100, 148, 4000, 85, 1630

Philip Wimer, 50, 89, 560, 5, 479

Daniel Lantz, 70, 132, 1000, -, 64

Michael Rexroad, 17, 106, 369, 5, 29

George White, 150, 250, 3000, 70, 431

Solomon White, 100, 170, 2500, 10, 483

Harman White, 100, 101, 3000, 5, 423

Peter White, 150, 200, 4000, 10, 1285

Benjamin Flusher, 200, 2000, 4000, -, 254

Henry J. Flusher, 100, 96, 5000, 100, 715

Jacob Sybert Jr., 51, 52, 1804, 35, 248

Mary Varner, 60, 149, 3000, 85, 185

George Flusher, 205, -, 6000, 110, 573

Daniel Varner, 90, 17, 1804, 65, 361

David Varner, 100, 69, 2000, 10, 437

James Trimble Sr., 200, 694, 4500, 150, 1402

Henry McCoy, 170, 95, 2645, 200, 443

Henry J. McCoy, -, -, -, -, 90

Henry J. Blogg, 70, 82, 1520, 5, 244

Naomi Wilson, 70, 30, 1000, 4, 319

John Blogg, 150, 214, 2313, 130, 793

Samuel Blogg,-, -, -, 80, 254

Townsend Price, -, -, -, 50, 214

Isaac N. Botkin, 50, 137, 654, 5, 106

Thomas J. Meadows, 30, 30, 300, 5, 132

John Rolston, 30, 30, 300, 5, 136

George Botkin, -, -, -, 5, 160

Jonathan Sivon, 250, 896, 9500, 90, 1744

Josiah Rolston, 30, 30, 500, -, 232

Thomas Douglas, 40, 238, 1000, 7, 136

John Dover Jr., 9, 21, 750, -, -

Edward E. Crony Jr., -, -, -, 8, 442

George M. Hooks, -, -, -, -, 189

Henry Propat(Prossat), 100, 301, 1027, 60, 468

John Malcomb, 50, 185, 2000, 5, 157

James A. Hodge, 25, 13, 228, -, 216

Robert Settenglove, 1000, 2289, 30000, 100, 3431

Margaret A. Stuart, 100, 357, 3267, 40, 508

Benoni Wilson, -, -, -, 10, 247

Jane Malcomb, -, -, -, 90, 371

Furgerson Malcomb, 75, 290, 2679, 85, 832

Joseph E. Malcomb, 150, 134, 2679, 60, 852

Henry Ruleman, -, -, -, 100, 343

Jacob Sivon, 150, 320, 3000, 100, 487

Joseph Sivon, 702, 1406, 18000, 250, 2104

William Curry, 50, 50, 1500, 5, 32

Samuel Wilson(of W. J.), 89, 118, 6000, 60, 654

William Hiner(of H), -, -, -, -, 410

Charles W. Wilson, -, -, -, -, 313

Samuel C. Eagle, 300, 75, 7500, -, 360

Jared Armstrong, 150, 150, 6000, 60, 668

Josiah Armstrong, -, -, -, -, 225

Joseph Hiner, 250, 308, 3808, 20, 78

Josiah Hiner, 500, 444, 10000, 160, 1385

George Wine, 75, 75, 1500, 10, 288

Samson Jordan, 70, 130, 2000, -, 288

Jacob Crummet, 70, 430, 2500, 60, 280

George Whistleman, 40, 220, 1000, 5, 198

Joseph Rexroad, 60, 300, 1109, 60, 428

Sarah Sims, 36, 56, 800,-, 198

David & Peter Michael, 100, 70, 1232, 40, 253

James Botkin, -, -, -, 5, 115

Christian Simons, 62, 100, 1387, 10, 750

John Botkin, 70, 70, 2000, 160, 501

Samuel E. Armstrong, 130, 104, 2000, 60, 309

Jared Armstrong, -, -, -, -, 146

Benarnn Armstrong, -, -, -, -, 33

Conrad Siples, -, -, -, 12, 258

George Armstrong, 1560, 205, 2500, 100, 449

Oliver Armstrong, -, -, -, -, 147

Samuel Botkin, 96, -, 1000, 5, 224

Joseph Botkin, 100, 150, 1000, 5, 429

Robert S. Hooks, 300, 747, 3525, 120, 792

Benjamin Hooks, -, -, -, 5, 1198

John M. Hooks, 131, 131, 1500, 10, 462

John Leach, 300, 277, 4000, 36, 765

John T. Armstrong, 281, 281, 1000, 130, 1463

Henry J. Flusher, 200, 315, 3500, 100, 475

Joseph Siples, -, -, -, -, 182

William F. Curry, 30, 194, 1500, -, 183

James Davis, 300, 8546, 4060, 20, 636

Harvy S. Davis, 720, 40, 3132, 80, 798

John H. Pullins, 200, 1500, 7000, 60, 1013

Henry Pullins, 100, 89, 4500, 80, 867

James Moyer, 500, 2680, 19500, 300, 3509

Rufus Oaks, 100, 120, 3000, 80, 652

Robert Pullins, 10, 230, 240, 10, 379

James Pullins, 12, 88, 600, -, 146

Jacob Stuart, 288, 1333, 10000, 80, 2094

George Vance, 640, 695, 12000, 130, 2928

James Hicklin, 150, 586, 2000, 70, 648

Ervin Dover, 550, 600, 10000, 75, 2329

William McClung, 1340, 6776, 28000, 160, 10419

Edward Stuart, 200, 20, 5000, 100, 1119

George Carlisle, 250, 250, 5000, 90, 752

Christian Haruff, -, -, -, 5, 325

Thomas Graham, 250, 589, 8000, 30, 1291

George Rivercomb, 60, 452, 8000, 350, 1601

John Rivercomb, -, -, -, -, 850

John Bradshaw, 130, 1264, 4500, 60, 810

James Wright, 30, 424, 400, -, 426

Rebecca Hamilton, 200, 200, 8000, 100, 1159

Franklin Bradshaw, 143, 287, 5000, 50, 850

Elizabeth Lockridge, 75, 40, 3000, -, 402

Peter Savage, 150, 119, 2500, 12, 367

John T. Byrd, 250, 2237, 5000, 385, 2349

Andrew H. Byrd, 340, 448, 13887, -, -

Charles Stuart, -, -, -, 35, 177

Argil Stephenson, -, -, -, 35, 218

William Stuart, 120, 390, 4677, 60, 1100

Jared M. Stuart, 25, 31, 560, 100, 381

John E. Marshal, -, -, -, -, 311

Amanda Kincaid, 90, 200, 1846, -, 112

Samuel M. Marshal, -, -, -, -, 238

David N. Kinkead, 140, 304, 1344, 40, 393

William Kinkead, 90, 200, 1845, 40, 227

James Hussman, -, 205, 40, 50, 133

Peter Hussman, 700, 1558, 8014, 135, 1787

John C. Gwin, 200, 667, 2380, 30, 525

Moses Givin, 200, 461, 3310, 90, 502

John P. Bishop, 15, 65, 100, -, 65

William K. Gwin, 70, 537, 1058, 70, 386

George H. Benson, 80, 370, 3000, 100, 194

Hamilton Benson, -, -, -, 15, 105

Sarah L. Kincaid, -, -, -, -, 120

John Gwin, -, -, -, -, 110

George Rowf, -, -, -, 7, 300

William Lockridge, 170, 1747, 2327, 50, 624

Robert Lockridge, 200, 837, 1500, 70, 940

Andrew Kinkead, -, -, -, -, 116

Robert A. Stuart, 110, 189, 3000, 50, 442

James Brown, 50, 67, 1000, 10, 258

William Helsel, 60, 133, 1800, 5, 300

St. Clair Stuart, 500, 1600, 10000, 200, 1516

Margaret Stuart, 86, 655, 3000, 95, 421

Samuel Johns, 40, 23, 630, 60, 182

Joseph Church, -, -, -, -, 70

William Johns, 56, 108, 540, -, 159

W. W. Johns, 40, 122, 675, -, 239

William Vint, 80, 268, 1200, 60, 267

Jesse Chew, 80, 172, 1500, 60, 275

John G. Wilson, -, 74, 25, 60, 134

Samuel Wilson, 120, 620, 1800, -, 240

William Wilson, 15, 185, 400, -, -

David Givin, -, -, -, 4, 39

John Devonicks, 130, 98, 1900, 20, 542

Allen H. Devonicks, 15, 245, 130, 5, 240

John Hodge, 20, 244, 201, 260, 211

Sarah Morton & Co., 16, 28, 250, 5, 94

Alenn Wilson, 30, 66, 300, -, 15

Robert McCray, -, -, -, -, 51

Stephen J. Reynolds, 2, 258, 1200, -, 989

John Burk, 50, 360, 1900, 40, 195

William Church, 20, 159, 800, 25, 118

Jared D. Ervin, 100, 250, 3000, 20, 656

James Smith, 160, 402, 1167, 5, 145

John Wilson, 50, 116, 1400, 100, 348

Amos Deal, 68, 400, 1400, 20, 256

William Ervin, 100, 145, 1800, 25, 205

John Leach, 150, 650, 3800, 20, 357

Decatur H. Jones, 81, 162, 1500, 10, 295

Henry W. Wilson, 30, 50, 1000, 5, 215

William Grogg, 20, -, 50, -, 77

Henry Jones, 640, -, 5000, 75, 674

Andrew J. Jones Jr., 500, 2000, 8000, 100, 2316

William Ervin Jr., 40, 40, 300, -, 70

George W. Hull, 800, 2100, 43000, 660, 5129

George Eagle, 270, 270, 5000, 30, 506

Robert Botkin, 200, 200, 2000, 5, 189

Joseph Hiner, 20, 60, 800, -, 115

John Trimble, 300, 900, 7200, 130, 672

William Bowers, -, -, -, 5, 150

William Trimble, 70, 137, 3560, 10, 306

Adam Stephenson, 560, 1492, 20000, 200, 5178

William W. Fleming, 513, 1817, 9500, 110, 1333

Harvy Trimble, 230, 177, 6642, 170, 1175

John Lukins, 100, 377, 1700, -, -

Andrew Seybert, 150, 2450, 8000, 100, 600

Houston F. Givin, -, -, -, 5, 398

Leonard Rexroad, 125, 20, 1800, 10, 190

Archibald Rexroad,-, -, -, 5, 113

Adam Snider, 75, 249, 1800, 5, 197

Joseph A. Beath, 150, 144, 1751, 125, 498

Andrew Rexroad, -, -, -, 50, 218

Frances Pullins, 1000, 2683, 11000, 200, 1280

Samuel Pullins, -, -, -, -, 117

F. H. Hull, 2000, 4500, 32000, 500, 2500

Charles Stuart, 600, 1000, 23000, 55, 2531

Joseph Bishop, -, -, -, -, 119

Joseph Layne, 150, 107, 2500, 70, 614

James Trimble, 189, 847, 7627, 100, 540

Adam Snider, 270, 530, 10000, 20, 2330

Isle of Wight County, Virginia
1860 Agricultural Census

The Agricultural Census for Virginia 1860 was microfilmed by the University of North Carolina Library under a grant from the National Science Foundation from original records at the Virginia Department of Archives and History in 1963.

There are forty-eight columns of information on each individual. Only the head of household is addressed. I have chosen to use only six columns of information because I feel that this information best illustrates the wealth of individuals. The columns are:

1. Name
2. Improved Acres of Land
3. Unimproved Acres of Land
4. Cash Value of Farm
5. Value of Farming Implements and Machinery
13. Value of Livestock

John D. Chalmers, 150, 150, 4500, 50, 700

Wm. C. Edwards, 250, 300, 8000, 150, 1155

Francis M. Boykin, 250, 1300, 15000, 300, 400

Thomas B. Bacesue, 120, 190, 8000, 300, 950

Wm. H. Casey, 200, 256, 10000, 350, 600

Wm. H. Stephenson, 50, 106, 2000, 50, 255

Geo. W. Britt, 53, -, 2000, 100, 100

Jas. B. Southall, 250, 170, 7500, 350, 1350

R. H. Whitfield, 150, 200, 7000, 200, 80

Octavius Goodrich, 135, 200, 6000, 200, 1215

W. H. Jordan, 225, 375, 70000, 200, 800

Wm. H. Jordan, 50, 350, 8000, 400, 1300

E. G. Stringfield, 200, 260, 6000, 100, 500

James Thomas, 60, 40, 10000, 100, 750

Levy Ashby, 16, 34, 800, 25, 100

Wm. B. Milley, 50, 50, 2000, 50, 300

John Tynes, 75, 75, 1500, 75, 250

John Edwards, 125, 400, 3500, 20, 700

Armstrong Turner, 75, 125, 1200, 25, 300

Wm. Whitley, 50, 70, 700, 40, 300

N. F. Pascaell, 200, 300, 5000, 100, 1000

R. A. Todd, 90, 160, 5000, 200, 400

T. H. Southall, 175, 295, 5500, 50, 560

Watson D. Jordan, 500, 900, 2500, 1000, 1455

John W. Cawson, 250, 343, 7000, 500, 1000

L. M. Spratley, 200, 300, 7000, 250, 8000

Chs. F. Wrenas, 500, 400, 25000, 800, 15000

Wm. T. Doyle, 80, 55, 2500, 200, 700

Robert F. Deck, 80, 40, 800, 10, 100

Mary Shelley, -, -, -, -, 50

Robt. R. Wilson, 350, 850, 12000, 300, 1484

Joseph T. Fraizer, 100, 52, 1500, 50, 500

Benj. Word, 170, 180, 3000, 120, 1000

G. W. Carroll, 600, 1400, 20000, 1000, 2000

Robt. W. Gibbs, 300, 600, 15000, 150, 1000

Ann McCallister, 100, 600, 5000, 25, 250

F. M. Boothe, 100, 230, 5000, 50, 200

Benj. G. Washington, 75, 60, 1500, 125, 600

Caroline Pedin, 90, 90, 6000, 100, 400

J. J. Wail, 16, 32, 2000, 75, 75

John D. Wail, 60, 85, 2500, 30, 400

Jno. T. Cowan, 150, 100, 5000, 30, 225

Lewis Knox, 100, 300, 6000, 45, 700

Nelson D. Hall, 75, 150, 5000, 50, 900

Jas. S. Buckley, 60, 100, 1600, 25, 300

Giles Daniel, 50, 150, 2000 50, 300

Joseph Hallsay, 100, 300, 6000, 25, 500

Joseph Shepherd, 40, 70, 1500, 25, 200

Geo. W. Milby, 50, 50, 1200, 25, 250

Mial McClenney, 60, 140, 2000, 25, 200

J. N. Cutchen, 70, 60, 1300, 75, 350

Edmond Pitt, 75, 200, 3000, 25, 350

Palley(Polley) Burkley, 50, 50, 1000, 10, 500

George. F. Channel, 125, 150, 5000, 50, 1000

Essez Halliway, 40, 11, 500, 20, 40

Geo. T. Nelms, 130, 75, 5000, 100, 1000

Roland Bell, 40, 60, 1200, 25, 300

Edwin Massisson, 70, 117, 1500, 100, 350

C. E. Reynolds, 25, 65, 800, 25, 100

Jas. F. Scott, 50, 80, 800, 20, 125

Wm. G. Green, 35, 45, 800, 25, 100

S. M. Cutchens, 40, 119, 1500, 10, 60

Josiah Thomas, 200, 348, 8000, 100, 1042

Willis J. Buckley, 150, 285, 5000, 100, 885

S. W. Southall, 150, 50, 4200, 100, 800

A. A. Whitehead, 180, 80, 5000, 75, 528

J. B. Hodesden, 100, 230, 5000, 50, 700

Wm. P. Channel, 100, 89, 1500, 40, 350

Mills H. Parr, 20, 30, 500, 50, 200

James T. Edwards, 150, 200, 3000, 100,800

Elias Hill, 20, 20, 400, 50, 52

Richd. H. Randolph, 200, 246, 5000, 100, 600

Emiline Wright, 133, 200, 6000, 75, 300

Jim Pastrick, 12, 5, 175, 30, 75

Wiley G. Bagnall, 50, 50, 2000, 50, 450

Henry Turner, 80, 127, 1000, 25, 195

Jno. W. Edwards, 90, 60, 1000, 35, 190

Jno. Godwin, 100, 200, 1200, 100, 300

Joseph Parkerson, 100, 115, 900, 50, 225

N. P. Wills, 250, 310, 7000, 150, 720

Zach. Turner, 90, 220, 1600, 100, 300

J. H. Hampton, 20, 5, 300, 20, 50

Wm. H. Parker, 100, 100, 4000, 150, 420

Chs. Chapman, 240, 360, 7000, 250, 1130

Geo. W. Andrews, 150, 50, 5000, 50, 675

S. J. Cofield, 100, 100, 2000, 100, 350

N. J. Williams, 150, 175, 3000, 25, 483

Jas. T. Shepherd, 40, 60, 1000, 15, 175

E. H. Dardan, 90, 100, 2350, 20, 100

Jno. W. Chapman, 130, 130, 7000, 150, 860

M. R. Minton, 70, 105, 1400, 150, 495

E. H. Ridick, 300, 300, 15000, 300, 2345

Jas. A. Chapman, 100, 250, 5000, 50, 750

J. A. Parker, 200, 460, 7000, 250, 1105

C. D. Jordan, 200, 200, 8000, 100, 750

W. P. Jordan, 200, 224, 8240, 75, 888

Sarah Newman, -, -, -, 40, 283

M. A. Taylor, -, -, -, 50, 75

J. C. Norsworthy, 150, 358, 6000, 200, 976

Martha Drian, 50, 62, 2000, 50, 220

Norman Hines, 62, 118, 2000, 50, 200

T. N. Goodson, 30, 70, 1200, 30, 300

Wm. Hines, 150, 170, 5000, 75, 400

Henry Bagnall, 100, 50, 1200, 25, 400

Jno. T. Burkley, 50, 50, 2000, 50, 825

Geo. Minyard, 25, 45, 700, 20, 125

Jos. Goodwin, 275, 416, 8000, 150, 750

W. H. Day, 340, 370, 15000, 300, 1783

Danl. Joyner, 10, -, -, 10, 15

Konce Pruden, 100, 131, 740, 25, 125

Saml. J. Hargrave, 50, 104, 860, 25, 213

R. T. Deck, 100, 140, 2000, 100, 290

Geo. J. Purdie, 200, 200, 15000, 25, 325

Augustus Burkley, 120, 120, 3000, 100, 860

Octavia Chapman, 17, 17, 450, 30, 20

Edwin Bunkley(Burkley), 75, 125, 2500, 50, 420

Joshua Thomas, 6, -, 80, -, 25

E. L. Eley, 75, 175, 1000, 30, 232

Geo. Blacksom, 55, 100, 1500, 25, 320

Moses Tynes, 40, 160, -, 10, 50

Jno. W. Roberts, 25, 55, 500, 50, 200

Jno. R. Holloward(Hallomard), 150, 146, 2500, 30, 340

Jos. A. Stallings, 100, 90, 1500, 300, 459

Nathl. P. Johnson, 100, 800, 3000, 100, 330

Jos. Bradshaw, 75, 75, 2000, 75, 530

Jas. P. Jordan, 125, 150, 4000, 200, 800

Mills Godwin, 75, 200, 2500, 50, 256

Jno. Holland, 50, 50, 800, 10, 130

David Atkins, 50, 125, 1000, 10, 150

Jno. T. Nelms, 35, 35, 600, 30, 175

Emanl. Turner, 8, 100, 700, 12, 100

B. W. Spivy, 12, 120, 250, 15, 125

Thos. H. Saunders, 75, 200, -, 25, 135

Timothy Edwards, 100, 100, 800, 30, 250

Jno. R. Watkins, 50, 100, 800, 50, 200

T. F. P. P. Cowper, 140, 630, 8000, 150, 1160

A. A. Jordan, -, -, -, -, 400

Meret Bracy, 75, 225, 2200, 6, -

Geo. Halliway, 7, 43, 600, 20, 85

Joe. Davis, 6, 9, 150, 15, 175

Jno. E. Adams, 200, 200, 4000, 350, 1172

Jno. J. Murphey, 125, 200, 2000, 100, 172

Wm. Norsworthy, 50, 72, 1000, 40, 200

Henry Tynes, 35, 90, 2000, 30, 250

Lewis Judkins, 20, 23, 500, 5, 65

Joseph Tynes, 30, 30, 300, 10, 50

Manning Halliway, 70, 73, 1500, 30, 275

Everett Wilkerson, 25, 30, 700, 20, 100

Jno. Johnson, 10, 5, 200, 10, 20

Elizabeth Stringfield, 133, 200, 3000, 40, 364

James Outland, -, -, -, -, 35

H. T. Brack, 60, 75, 2000, 100, 300

H. P. Howell, 25, 35, 500, 15, 75

E. P. Epps, 40, 95, 1500, 20, 250

Sarah Davis, 40, 60, 400, 20, 80

F. T. Vail, 3000, 540, 8000, 1000, 1175

M. A. Garisson, 75, 175, 700, 25, 135

Livinia Battin, 70, 139, 700, -, 80

Sally Jones, 670, 41, 700, 20, 175

Wm. C. Gray, 40, 60, 500, 10, 40

Adolphus Parr, 20, 3, 400, 12, 98

Anslum Crocker, 30, 20, -, 9, 40

Jim Green, 15, -, -, 15, 90

Wm. W. Chapman, 50, 30, 500, 30, 200

A. J. Villines, 200, 200, 4000, 100, 600

Jacob Bailey, 30, -, -, 10, 12

Wm. Gray, 75, 75, -, 50, 210

Burnet Brock, 50, 80, 700, 25, 195

Watson White, 200, 350, 4000, 200, 860

Augustus White, -, -, -, -, 138

Mary White, -, -, -, 30, 145

J. H. Gwathney, 50, 110, 900, 12, 80

L. W. White, 40, 110, -, 25, 85

Jos. Turner, 50, 20, 350, 3, 50

Patsey Little, 100, 100, 1000, -, 30

Josiah Bell, 75, 267, 1100, 50, 230

Robt. Bailey, 100, 100, -, 10, 184

Nancy Barlow, 85, 100, 900, 20, 75

Wilson Hallerman, 270, 434, 12650, 125, 430

James Gray, 200, 100, 3000, 100, 650

E. H. Gray, 150, 200, 2500, 150, 500

Shad. Barlow, 60, 220, 600, 20, 150

Joseph Miltingloso, 100, 75, 1000, 25, 200

Nancy Clark, 40, 40, 300, 15, 100

Jno. H. Morris, 50, 75, 400, 10, 70

Mallory Jones, 20, 30, -, 10, 50

Edwin Little, 60, 140, 500, 25, 350

A. B. Gwalthney, 40, 60, 1000, 15, 140

A. G. Moody, 250, 200, 5000, 350, 831

F. H. Atkins, 75, 40, 600, 20, 200

Geo. J. Barlow, 90, 100, 800, 25, 180

Josiah Edwards, 100, 175, 200, 25, 150

Thos. M. Pinkarm, 100, 75, 600, 25, 100

Joseph Stephenson, 80, 30, 800, 50, 255

Willis Wilson, 300, 900, 20000, 300, 1873

E. G. Harrasson, 30, 70, -, 30, 174

Harman Haine, 50, 100, 1200, 10, 150

Pleasant Gale, 15, 85, 500, 10, 30

Moses Atkins, 100, 140, 1000, 50, 155

Jno. F. Battin, 40, 130, -, 20, 75

John Scott, 60, 30, 500, 10, 130

Jno. C. Handcock, 20, 20, 250, 10, 56

E. T. Holland, 25, 35, -, 10, 60

Jno. R. Holland, 60, 140, -, 25, 125

Saml. P. Jordan, 125, 75, 7000, 150, 600

Michael Murphey, 100, 336, 1500, 20, 308

Robt. Blackwell, J. W. Cook owner, 700, 1800, 27000, 350, 3000
Eliza. Hill, 30, -, -, 10, 44
Thos. Cofield, 130, 228, 3000, 150, 553
Richd. H. Latimer, 60, 140, 1500, 100, 350
E. M. Hallerman, 100, 40, 2000, 25, 245
Simon Atkins, 125, 265, 3000, 150, 470
Wm. Gray of Hardy, 25, 38, 500, 20, 321
Jas. J. Gray, 8, -, - ,5, 38
Geo. Crocker, 40, -, -, 15, 35
Dawsey Daws, 100, 50, 1500, 40, 464
Thomas Gwathney, 30, 29, 600, 15, 102
Caroline Delk, 30, 135, 600, 25, 154
Liz___ Chapman, 400, 400, 7000, 100,700
Jno. E. Thomas, 1000, 400, 15000, 500, 2300
Shaderack James, 40, -, -, 5, 75
Jos. T. Gale, 10, -, 200, 20, 155
Burvin Jones, 225, 425, 8000, 200, 600
Thomas Griffin, 40, 100, 500, 25, 250
Jas. H. Pruden, 30, 70, 500, 30, 200
Eley Parkerson, 30, 24, -, 15, 50
Jno. W. Gray, 200, 230, - 50, 325
Joseph Turner, 50, 100, -, 25, 50
Mary Garner, 60, 140, 1000, 25,208
David Garner, 30, 270, -, 5, 70
Jonathan Godwin, 40, 135, 1000, 35, 175
Jones Griffin, 50, -, -, 10, 100
Thomas E. Cutchen, 25, 28, 500, 8, 25
Charles W. Pitt, 20, 40, 400, 8, 148
Sarah Wrench, 40, 120, 500, 20, 142
Thos. H. Pitt, 40, 90, 800, 30, 240
Edwin G. Pitt, 30, 112, 300, 15, 220
Nathan Newby, 15, -, -, 5, 10

Mary F. Gale, 43, 60, 600, 20, 145
Henry Pruden, 150, 183, 1665, 50, 358
William M. Johnson, 25, 60, 3500, 30, 500
Thos. Carroll, 35, 95, 600, 15, 123
Jno. M. Shivers, 700, 1500, 15000, 500, 2531
R. C. Dugins, 40, 210, -, 10, 45
Thos. H. Pitman, 100, 47, 500, 50, 290
Wm. Powell, 20, 25, 100, 20, 50
F. E. N. Wills, 150, 162, 1500, 50, 410
Levy Rix, 50, 150, 1000, 30, 330
E. C. Pruden, 120, 300, 3500, 250, 523
Dr. T. Page, 2, 1800, 16000, -, 1000
Jonathan Adams, 75, 75, -, 50, 126
Benj. Turner, 40, 60, -, 10, 40
Martha Scott, 25, 150, -, 10, 30
Micajah Butler, 75, 150, 600, 12, 20
Joseph Spencer, 45, 75, -, 15, 20
Joseph Pruden, 75, 125, 1000, 25,152
Thomas Ross, 36, 30, 250, 20, 125
Edmond Archer, 40, 265, 250, 15, 125
John Whitehead, 75, 136, 3000, 25, 380
Jim Ash, 35, 65, 275, 15, 125
Richd. H. Baker, 60, 50, 900, 35, 128
R. D. Marshall, 150, 250, -, 60, 425
M. N. Jones, 25, 23, -, 25, 100
Isham Davis, 10, -, 200, 10, 160
Lawanda Jameca, 100, 200, 1000, 10, 200
Wm. H. J. Wilson, 135, 265, 5000, 200, 1170
Jno. B. Underwood, 55, 185, 1500, 60, 300
Mills E. Marshall, 200, 210, 6000, 100, 520
John Parkerson, 30, 70, 300, 20, 87
Miles Edwards, 300, 400, 1800, 100, 580

Abra. Villines, 100, 50, 2000, 100, 290

M. H. Womble, 80, 100, 640, 30, 85

James W. Womble, 70, 30, 1500, 40, 400

Nancy J. Stagg, 50, 50, 500, 100, 110

Wm. Chapman, 60, 91, 1500, 57, 350

Jas. R. Edwards, 80, 63, 1200, 20, 180

Peter Turner, 100, 100, -, 25, 70

Catharine Edwards, 30, 50, 300, 4, 40

Henry Whitley, 40, 60, -, 15, 80

Allen P. Parkerson, 50, 25, 400, 10, 90

Wm. B. White, 75, 125, 1500, 30, 300

Jno. W. Word, 110, 340, 3000, 100, 675

M. E. Jones, 60, 10, 800, 10, 100

C. E. Delk, 100, 245, 2000, 75, 335

Robt. T. Jones, 200, 358, 3000, 100, 570

Wm. A. Stott, 200, 140, 3500, 75, 600

Martha Turner, 40, 200, 500, 20, 167

A. V. Goodson, 40, 40, 500, 25, 105

A. R. Ellis, 40, 60, 500, 5, 46

Lucretia Ellis, 40, 60, 500, 5, 46

Jacob Jordan, 25, 27, 400, 20, 87

Isaac Halliway, 50, 116, 500, 60, 100

William A. Moody, 25, 115, 300, 15, 45

Richd. Reynolds, 70, 30, 1500, 50, 280

Bennit E. Lane, 90, 75, 500, 50, 300

E___ Bracy, 75, 225, 1500, 100, 432

B. F. Roberts, 100, 250, 600, 10, 160

Jesse A. Whitley, 75, 125, 2000, 100, 402

Saml. Betts, 80, 125, 2500, 75, 174

Thomas Jolliff, 4, 21, 150, 15, 50

Winerfred Babb, 110, 100, 3000, 40, 385

Julious O. Thomas, 240, 1032, 10000, 300, 1000

Wm. M. Cawper, 30, 66, 300, 10, 18

Elizabeth Banks, 30, 20, 400, 15, 25

Henry D. Banks, 8, 37, 125, 6, 65

Daniel Tynes, 16, 5, 100, 5, 70

B. J. Gray, 100, 73, 1600, 50, 595

Benj. H. Gwaltney, 70, 70, 1500, 25, 300

Algernon Carroll, 30, 30, 1200, 25, 450

Jas. H. L. Cofer, 200, 220, 3500, 350, 525

James Crocker, 25, 25, 200, 20, 103

Burwell Green, 25, 38, 150, 15, 50

A. W. Barlow, 20, 27, 150, 6, 75

P. F. Crocker, 40, 35, 300, 25, 30

Benj. Bailey, 25, 25, 250, 10, 75

Joshua Bailey, 40, 71, 700, 40 204

Geo. W. Wiley, 20, 55, 300, 25, 220

John Jones, 80, 80, 1500, 50, 312

P. D. Barlow, 100, 50, 500, 15, 50

Geo. F. Hall, 400, 750, 11600, 550, 1700

Archebald Roberts, 100, 280, 3000, 50, 500

Madison Edwards, 60, 82, 600, 50, 405

Robert Saunders, 65, 135, 750, 50, 300

Geo. F. Whitley, 58, 202, 800, 25, 200

M. R. Urquhart, 1200, 2600, 25000, 150, 3650

Nancy Cofer, 65, 65, 650, 25, 242

Matthew Turner, 100, 30, 1500, 25, 254

Jere Delk, 400, 11000, 10000, 100, 580

Burwell Brock, 60, 55, 1200, 50, 250

Eltingson Betts, 60, 40, 500, 25, 290

Noah Stephenson, 100, 160, 1200, 50, 250

Sarah D. Gwaltney, 112, 100, 1300, 75, 436

Joel Cook, 150, 150, 2500, 25, 380

Robt. Binford, 125, 125, 3500, 25, 520

Willis Pretlaw, 50, 30, 500, 25, 65

Junious Barlow, 77, 10, 700, 25, 246

Eliza Sykes, 50, 200, 400, -, 68

John Bland, 40, 35, -, 25, 40

E. P. Womble, 200, 100, 3000, 120, 620

Elizabeth Edwards, 150, 150, 2000, 250, 560

Wiley Balton(Batton), 40, 50, 1000, 100, 362

Clemmons Batton, 250, 250, 3000, 100, 600

Thos. Mountford, 80, 200, -, 20, 140

Wm. P. Pasaell, 100, 160, 1000, 100, 280

Robt. Resson (Besson), 60, 40, 800, 40, 150

L. B. Edwards, 100, 50, -, 25, 375

Sarah T. Edwards, 100, 44, 1500, -, -

Wm. Hall (of Thos), 200, 300, 3500, 225, 640

Richd. H. Pulley, 70, 180, 800, 100, 265

Jno. Tomblin, 80, 200, 1000, 25, 130

Thos. Underwood, 30, 90, 225, 10, 50

Chs. H. Dardan, 200, 400, 3000, 400, 813

Nicholas Edwards, 100, 50, 400, 15, 45

Wm. H. Dardan, 300, 200, 7000, 150, 1000

Jas. H. J. Johnson, 100, -, -, 50, 265

Robert J. White, 150, 500, 2500, 60, 456

Wm. N. Gray, 50, 50, 400, 12, 170

Mary Hall, 75, 135, 1500, 25, 278

Susanah Allnnard, 75, 75, 1000, 50, 265

Riddick Butler, 4, 106, 350, 10, 65

Wm. Grace, 20, 40, 600, 25, 200

Langston Pearce, 30, -, -, 15, 142

Jas. S. Turner, 200, 220, 3500, 40, 610

Elizabeth Gaskins, 75, 26, 400, 10, 175

Mills Barrett, 150, 156, 2500, 50, 436

Jno. W. Byrd, 16, 9, 125, 15, 90

Britton Edwards, 50, 30, 350, 15, 135

Jumcus O. Cofer, 125, 59, 2700, 150, 290

C. E. Jordan, 25, 160, -, 10, 65

Thos. M. Saunders, 50 700, 900, 35, 450

E. A. Crumpler, 60, 40, 400, 60, 250

Jno. B. Boykin, 8, 42, 200, 10, 75

A. J. Busby, 20, 80, -, -, 5

Wm. W. Jaynes, 300, 345, 5000, 150, 750

N. P. Young, 240, 331, 8000, 150, 1275

F. C. Latimer, 100, 200, 2000, 25, 173

Nancy Whitley, 40, 20, 600, 50, 100

Willis Spacy, 80, 269, 2000, 100, 519

Josiah Britt, 50, 27, 750, 50, 187

Thomas Hall, 300, 1000, 8000, 200, 900

Benj. H. Griffin, 50, 220, 800, 15, 115

Mary Hall, 100, 112, 1200, 20, 167

Dempsey Mauestford, 15, 29, -, 7, 25

Stephen A. Eley, 100, 60, -, 25, 250

Richd. Edwards, 150, 400, 5000, 150, 600

Elizabeth Barrell, 80, 20, 500, 20, 110

Mills H. Carr, 35, 65, 600, 40, 170

Jincy Hall, 75, 100, 700, 20, 100

William Pearce, 35, 15, 75, 12, 50

Thomas Pearce, 60, 75, 400, 10, 130

Levy Hart, 30, 70, 600, 10, 56

Handcock Barrett, 40, 140, -, 10, 40

John Barrett, 50, 100, 800, 50, 240

David Roberts, 200, 200, 3000, 200, 800

Wm. P. Wright, -, -, -, 25, 164

E. L. Ballard, 180, 130, 1500, 200, 600

Wm. E. Barnes, 150, 350, 2500, 50, 725

Nathan Carr, 25, 25, 250, 20, 100

Washington Bridges, 60, 70, 600, 25, 75

Martha Carson, -, -, 3000, 40, 450

Elijah Butler, 40, 20, 250, 20, 154

Elija Holland, 30, 10, 200, 15, 95

Elisha Bradshaw, 235, 25, 360, 20, 217

Exum Carr, 60, 40, 400, 50, 270

Allen Hodgepeth, 200, 200, 1200, 25, 220

David Butler, 20, 4, 50, 15, 50

James Cones, 15, 42, 400, 5, 50

Merit E. Dardan, 150, 200, 1200, 20, 600

Jesse D. Councill, 175, 175, 2000, 40, 540

John Parker, 50, 50, 300, 25, 78

Geo. W. Councill, 100, 230, 2000, 75, 400

John Carr, 100, 300, 1000, 20, 210

Wm. G. Bradshaw, 75, 100, 1000, 40, 300

John H. English, 10, -, -, 5, 60

Daniel Bradshaw, 40, 100, 300, 30, 150

Robert Butler, 50, 85, 1800, 5, 255

Nathan Turner, 40, 30, 350, 15, 153

Archiball Fowler, 75, 25, 660, 20, 280

Wm. Carr of Jno., 30, 31, 250, 30, 120

Jno. Dardan, 40, 63, 500, 25, 135

Wm. H. Lawrence, 100, 220, 2000, 100, 465

Edgar Rawls, 60, 11, 2500, 20, 250

James M. Councill, 160, 170, 4000, 50, 130

Edward Rawls, 120, 200, 5000, 40, 52

Wm. Bule, 40, 20, 200, 50, 250

Eliza Cross, 100, 113, 1500, 25, 195

Lovincia Saunders, 40 22 ½, 400, 15, 260

Wm. Wyatt, 50, 50, 800, 20, 165

Riddick Daughtrey, 50, 50, 500, 30, 161

Abraham Fowler, 20, 110, 500, 25, 140

Wm. H. Stephenson, 50, 223, 700, 50, 300

John Vaughan, 75, 50, 1000, 50, 250

Wm. N. Outland, 100, 300, 1000, 50, -

Jno. L. Hines, 40, 60, 500, 25, 100

Jethero H. Butler, 20, 51, 300, 15, 50

Benj. D. Councill, 75, 50, 400, 10, 75

James Councill, 50, 50, -, 10, 300

Wm. Jennings, 41, 600, 300, 25, 100

Thos. R. Judkins, 200, 407, 6000, 100, 800

Sol. Webb, 40, 45, 500, 25, 250

D. H. Holland, 75, 325, 3000, 40, 400

Joshua Daughtry, 60, 49, 1500, 50, 300

Wm. H. Cutchins, 75, 75, 1800, 40, 300

J. G. Cutchins, 50, 50, 1000, 50, 385

Mills Eley, 75, 150, 600, 20, 300

J. J. Cutchins, 40, 10, 500, 50, 350

Josiah Cutchins, 60, 40, 1000, 50, 300

Timothy Duck, 150, 350, 3000, 50, 500

John Whitfield, 15, 150, 900, 10, 60

James M. Norfleet, 15, 135, 1000, 20, 100

R. J. Johnson, 100, 133, -, 20, 225

N. B. Gardner, 100, 108, 2000, 50, 300

Johnson Langford, 50, 50, 500, 10, 75

Elisha Langford, 50, 94, 700, 25, 50

Thomas Saunders, 60, 20, 900, 25, 200

James P. Saunders, 30, 20, 400, 75, 50

Henry Drake, 50, 150, -, 40, 200

R. C. Willeford, 15, 11, 100, 25, 50

Hardy Chapman, 600, 400, 12000, 150, 1700

James Johnson, 75, 50, 800, 30, 260

Arthur Johnson, 75, 50, 600, 25, 135

Jas. Davis, 80, 80, - 10, 60

Jno. B. Butler, 30, 20, -, 5, 70

Frances Niblet, 75, 130, 3000, 20, 350

Geo. W. Crumpler, 50 90, 700, 20, 181

Jno. A. Carr, 40, 70, 250, 25, 220

Dempsey Bowden, 6, 6, 200, -, 15

Tempy Bowden, 100, 130, 1000, 100, 250

Hartwell F. Powell, 150, 250, 3000, 100, 610

Elias Britt, 40, 17, 400, 30, 205

Patsey Eley, 150, 250, 1500, 50, 450

Mills Carr, 30, -, -, 20, 86

Nathan Mountford, 100, 160, 2000, 100, 470

Wm. E. Mountford, 50, 100, -, 10, 100

John C. Thomas, 150, 150, 3500, 200, 770

David Edwards, 200, 250, 5000, 300, 2175

Edwin Gray, 30, 120, 500, 10, 50

Shaderock Bracy, 150, 300, 2000, 10, 200

Robert R. Pope, 200, 400, 4000, 200, 632

R. F. Chapman, 100, 600, 2500, 50, 614

Jno. R. Andrews, 45, 50,-, 25, 150

M. T. Atkinson, 30, 100, 200, 10, 115

Madison Red, 60, 30, 250, 20, 100

Jno. Braswell, 30, 75, -, 15, 50

Jas. A. Atkinson, 250, 420, 10000, 500, 1388

Geo. Wilkerson, 20, -, -, 20 90

A. Atkinson, 800, 4200, 30500, 600, 1830

H. H. Womble, 60, 40, -, 30, 30

Geo. R. Atkinson, 80, 97, 3700, 125, 300

Chs. Halliway, 20, 10, 300, 10, 35

Catharine Womble, 80, 27, 1500, 30, 150

N. J. Hargrave, 75, 125, 3000, 12, 20

J. B. Whitehead, 150, 200, 9000, 250, 1140

Richd. C. Edwards, 60, 80, 300, 100, 450

C. B. Simpson, 150, 150, 3000, 75, 250

Elizabeth Pope, 20, 50, -, -, 32

Benj. Brock, 30, -, -, 15, 150

James Parr, 80, 100, 600, 20, 185

Joel Brock, 120, 137, 1500, 25, 445

Joseph Nuoley, 15, 10, 150, 15, 57

James Matthews, 30, 70, -, 12, 206

Willis Godwin, 50, 60, 300, 20, 125

Lewis Johnson, 60, 75, 1500, 25, 230

W. J. Allenson, 90, 200, 1000, 20, 190

C. H. C. Baily, 100, 160, 600, 50, 375

Ishmell Johnson, 75, 90, 2000, 50, 380

L. B. Roberts, 200, 200, 2500, 50, 610

Ishmell Bland, 20, 10 200, 5, 125

Jas. G. Saunders, 40, 112, 600, 10, 150

Elias Saunders, 40, 58, 500, 20, 291

Thos. J. Marshall, 100, 116, 1300, 50, 300

Wm. W. Roberts, 20, 33, 300, 20, 95

Elizabeth Butler, 50, -, 600, 20, 90

Jno. M. Boykin, 40, 160, 800, 10, 230

Gale Johnson, 40, 260, 2000, 25, 225

Mills Crumpler, 30, 30, 600, 15, 209

Mallary Vaughn, 200, 300, 1500, 25, 300

Allen Johnson, 30, -, 100, 5, 50

Exum Stephens, 8, 62, 300, 15, 65
Peter Ashburn, 40, 220, 1200, 30, 223
Anthony Parsons, 200, 550, 3000, 100, 800
Jno. S. Daughtery, 166, 200, 1500, 50, 440
Norriss Johnson, 80, 343, 1000, 20, 430
Jno. Johnson, 50, 20, 350, 25, 165
Jos. Johnson, 75, 175, 750 25, 105
Stephen Holland, 25, 101, 400, 15, 138
Meredith Butler, 30, 67, 500, 15, 92
Jos. Carr, 15, -, -, 10, 40
Jere Holland, 40, 40, 300, 10, 100
Augustus Outland, 100, 181, 1500, 50, 560
Martha Hodgepeth, 32, 40, 300, 15, 125
Polly Holland, 144, 200, -, 25, 225
Tolbert Harriss, 75, 75, 1500, 15, 240
Hezekiah Butler, 40, 100, 400, 20, 100
S. B. Johnson, 100, 140, 1000, 25, 250
Martha Freeman, 150, 150, 1200, 25, 273
Elum Rhodes, 100, 190, 1000, 30, 275
Abram Carr, 25, 30, 200, 7, 75
Medilina Avis, 100, 47, 800, 5, 100
Jos. M. Holland, 35, 40, 225, 15, 55
Elizabeth Barr(Baer), 40, 5, 200, 6, 75
James Carr, 40, 30, 250, 15, 60
Jacob Turner, 30, 58, 300, 15, 60
Jas. Dardan, 35, 40, 250, 15, 100
Elisha Carr, 5, 43, 300, 15, 50
Saml. Turner, 15, 55, 470, 10, 92
Jordan Turner, 12, 13, 50, 2, 50
Norfleet Pearce, 90, 400, -, 100, 470
Irvin W. Duck, 700, 1400, 20000, 400, 1556
Jordan Johnson, 30, 170, -, -, 50

Jincy Stephens, 60, 50, 400, 25, 140
Sol. Carr, 40, 100, 1000, 40, 283
Sally Jinkins, 40, 100, 600, 10, 160
Lucy A. Whitehead, 75, 75, 1400, 20, 275
Elisha Gay, 100, 116, 1000, 30, 269
Jacob T. Barton, 75, 30, 1200, 40, 250
Odem Bradshaw, 50, 150, 300, 15, 130
Jos. Mountford, 40, 60, -, 15, 65
Nancy Johnson, 30, 25, 250, 15, 65
Josiah P. Gay, 300, 800, 6750, 125, 992
Joseph Spicey, 30 70, 300, 15, 56
Jno. Duck, 40, 50, 400, 30, 205
Benj. Spicey, 20, 15, 150, 15, 66
Wm. Duck, 40, 35, -, 5, 19
J. W. Holland, 140, 143, 4000, 20, 280
Benj. Eley, 100, 116, 2500, 50, 480
Spikes Beal, 40, 152, 800, 30, 245
Geo. W. Holland, 15, 50, 300, 5, 25
Elizabeth Holland, 40, 20, 400, 12, 100
Joseph Bradshaw, 8, 36, 300, 15, 65
Hillary West, 35, 50, 1200, 40, 215
M. H. Daughtey, 200, 418, 4000, 50, 700
Thos. Spicey, 20, 10, 75, 20, 50
Jas. R. Purvis, 150, 900, 4000, 400, 400
Martha Johnson, 20, 10, 100, 10, 18
Sallie Lester, 75, 75, 700, 25, 300
Alford R. Butler, 97, 292, 2500, 50, 384
Ervin Beeler, 50, 10, 600, 50, 250
Jno. M. Holland, 200, 298, 5000, 150, 920
Charles Westy, 25, 10, 150, 20, 125
Timothy Hays, 150, 850, 3000, 25, 525
Alfred Gardner, 35, -, -, 3, 8
R. W. Taylor, 45, 68, 4000, 50, 365
Wm. Bracy, 40, 20, 200, 25, 75
Benj. Whitfield, 67, 68, 300, 40, 75

Wiley Councill, 50, 100, 1500, 5, 450

Abel Rogerson, 75, 145, 500, 40, 224

John Rose, 40, 60, 500, 20, 160

Frank Rose, 20, 57, 500, 20, 100

Henry Mountford, 30, 45, 350, 20, 100

Jos. S. Holland, 600, 663, 5400, 150, 440

Ludlow Lawrence, 130, -, -, 100, 1385

Miles Lester, 30, 40, -, 20, 75

Uriah Vaughn, 15, 110, 500, 10, 150

Elihu Langford, 50, 140, 1000, 10, 209

Elisha Bradshaw, 25, 18, 100, 10, 25

Bervely R. Vaughn, 75, 100, 1500, 60, 335

Henry Pruden, 75, -, -, 25, 60

Wm. Lester, 50, 40, 400, 10, 75

J. H. Duck, 400, 1200, 5000, 400, 1260

Wm. T. Barrott, 50, 100, 350, 15, 50

Jacob Carr, 4, 2, 50, 15, 50

Abram Duck, 30, 15, 300, 15, 165

Wm. H. Vaughn, 100, 300, 2000, 60, 418

Parker D. Howell, 80, 300, -, 15, 100

Jas. C. Key, 50, 50, -, 15, 55

Margaret Mountford, 40 110, 300, 25, 118

Jesse B. Johnson, 100, 200, 1500, 20, 231

D. D. Watkins, 150, 200, 2000, 150, 530

Saml. English, 8, -, -, 5, 40

Execum Hines, 10, 10, 125, -, 13

Zachariah Spicey, 25, 30, 400, 10, 75

Jaro. Crumpler, 10, 20, 200, 13, 60

Marcella Spicey, 12, 10, 100, 5, 30

Meredith Crumpler, 20, 30, 250, 15, 115

Jacob Spicey, 30, 30, 200, 7, 100

Jethero Rhodes, 100, 200, 1100, 40, 350

David Dardan, 40, 160, 500, 15, 75

A. M. Ballard, 300, 459, 5000, 100, 839

R. H. Ballard, 200, 400, 5000, 100, 612

Geo. W Parker, 75, 425, 8000, 150, 850

Jno. R. Todd, 175, 625, 15000, 100, 1650

D. H. Battin, 300, 400, 10000, 500, 1400

Pleasant Baley, 35, 58, 1000, 15, 100

M. N. Matthews, 30, 170, -, 15, 250

John Shepherd, 40, 32, -, 25, 300

Cyrus Tynes, 30, 120, 1000, 25, 180

Eliza Kimbell, 50, -, -, 10, 75

Lazarus Whitfield, 75, 50, 500, 25, 150

Alexd. D. Callcote, 150, 140, 3500, 40, 510

Jas. T. Cocks, 12, 57, 700, 25, 150

Henry Barvau(Barvan), 50, 20, 700, 20, 110

Emily Gale, 60, 40, 1500, 50, 450

A. G. Spratley, 200, 200, 8000, 100, 1325

Josiah Cutchins, 40, 73, -, 25, 125

Partrick Pearce, 100, 55, 700, 30, 246

Jas. M. Saunders, 25, 120, -, 30, 85

Josiah Eley, 50, 133, 2000, 100, 600

Cary Eley, 20, 30, 350, 12, 65

Wm. S. Roberts, 40, 60, 800, 15, 200

Priscilla Roberts, 100, 112, 1500, 25, 125

Etheldridge Demston, 100, 100, 2000, 50, 400

Wm. H. Gay, 100, 150, 1000, 50, 186

Jno. P. Johnson, 40, 70, 1000, 40, 256

Stephen G. Dardard(Dardan), 30, 53, 3000, 100, 225

David Gay, 20, -, -, 10, 40

Nerverson Saunders, 70, 130, 500, 25, 13

Wm. H. Powell, 200, 303, 3000, 80, 365

Meredith Watkins, 200, 300, 7500, 100, 640

Deron W. Kitchen, 75, 275, 5000, 200, 480

Wm. H. Johnson, 10, -, -, 10, 75

Wm. K. Johnson, 60, 340, 1500, 30, 275

Mills W. Roberts, 50, 130, 2000, 25, 187

S. J. Roberts, 20, 40, 600, 20, 30

Mary__ Roberts, 30, 15, 500, 20, 275

B. E. Dardan, 100, 200, 1500, 25, 445

Jno. T. Phillips, 175, 200, 3000, 75, 500

Jno. Underwood, 35, 340, 3600, 12, 90

Martha S. Chapman, 15, 15, 500, 10, 75

Putman Davis, 150, 150, 2600, 75, 500

Sion Gay, 120, 220, 4000, 1056, 384

John Urquhart, 50, 60, 200, 50, 130

James Eley, 150, 100, 4000, 100, 400

Melicia Dardan, 75, 75, 2500, 30, 375

Thomas Pinner, 50, 40, 1200, 40, 225

Jas. R. J. Corbett, 70, 100, 1500, 40, 230

Jno. E. Denson, 300, 300, 4000, 50, 300

Wm. Doyle, 30, 33, 350, 25, 250

Jas. J. Tynes, 77, 100, 2000, 30, 200

Edwin Mingo, 5,-, -, 15, 75

Richd. Fentriss, 75, 444, -, 10, 65

Geo. W. Morriss, 70, 100, 1000, 20, 130

Jacob N. Edwards, 10, -, -, -, 25

Mariah Lane, 60, 15, 300, 16, 50

Henry W. Murphey, 150, 75, 1000, 50, 300

Henry Pope, 20, 20, 350, 15, 60

Willis Butler, 30, 56, 250, 20, 50

John Babb, 50, 200, 800, 30, 150

Jno. P. E. Eley, 10, 35, 200, 20, 60

Stephen Wright, 40, 30, 500, 50, 150

Sallie Scott, 50, 57, 250, 20, 100

Eley Pearce, 50, 52, 1200, 20, 150

Jno. B. Saunders, 60, 100, 800, 20 150

Thomas Jayner(Joyner), 130, 130, 200, 25, 215

Wm. Hall of Jno., 200, 600, 2000, 50, 440

John Marshall, 25, 235, 1700, 20, 250

Jno. H. Beatton, 60, 158, 1000, 25, 150

Wm. J. Womble, 30, 30, 1000, 25, 100

Olivia Stephenson, 10, 60, -, 10, 75

James J. Moody, 50, 222 2700, 15, 400

Thomas M. Dashield, 500, 1000, 12000, 200, 1126

Geo. W. Stephenson, 209, 339, 2500, 150, 275

Edwin Bradshaw, 35, 155, 600, 20, 80

E. H. Vallentine, 125, 265, -, -, 200

S. J. Wilson, 60, 300, 1500, 200, 900

John Johnson, 13, -, 800, 30, 225

Margaret Goodson, 90, 100, 2000, 25, 375

James L. Crider, 150, -, -, 30, 150

R. M. Bucktent, 16, 1, 3500, 100, 290

R. T. Clower, 4, -, 400, 25, 545

W. W. Rust, 23, -, 3000, 4, 725

R. M. Garrett, 40, -, 12000, 500, 800

R. H. Armistead, 10, -, 2000, 100, 300

R. P. Taylor manager For E. L. Oseylima, 10, -, 2000, 200, 500

Geo. Gaines, 4, -, 400, 10, 150

Dr. E. Casuns, 6, -, 600, 200, 350

James City County, Virginia
1860 Agricultural Census

The Agricultural Census for Virginia 1860 was microfilmed by the University of North Carolina Library under a grant from the National Science Foundation from original records at the Virginia Department of Archives and History in 1963.

There are forty-eight columns of information on each individual. Only the head of household is addressed. I have chosen to use only six columns of information because I feel that this information best illustrates the wealth of individuals. The columns are:

1. Name
2. Improved Acres of Land
3. Unimproved Acres of Land
4. Cash Value of Farm
5. Value of Farming Implements and Machinery
13. Value of Livestock

Cyrus A. Branch, 100, 75, 4000, 125, 675

Littleton T. Waller, 185, 675, 8000, 500, 950

C___ Vaiden, 200, 300, 3000, 10, 260

Robert T. Binns, 90, 41, 2500, 80, 380

Wm. G. Johnson, 150, 266, 5000, 100, 1000

Robert. P. Taylor, 125, 105, 4000, 165, 456

William Bush, 150, 1850, 10000, 300, 950

Virginia A. Gatewood, 6, -, 500, 15, 60

Saml. T. Slater, 100, 65, 6000, 400, 510

Felix Pierce, 70, 60, 2500, 75, 350

Beverly Slater, 60, 233, 2000, 130, 285

Thomas J. Taylor, 150, 700, 5000, 75, 436

Dandridge W. Marston, 200, 333, 8000, 300, 750

George Hankins, 550, 550, 20000, 500, 1060

Eliza H. Piggott, 300, 300, 6000, 250, 450

George E. Geddy, 150, 325, 6000, 260, 750

A. J. E. Jennings, 150, 550, 4000, 150, 850

Warner Enos, 75, 280, 3500, 200, 650

Ann S. Yates, 60, 38, 1500, 50, 400

Warner T. Enos, 60, 40, 1000, 25, 185

Cary Wilkinson, 250, 263, 4000, 100, 520

Drucilla Cowles, 400, 1000, 1500, 250, 1000

Daniel Janes, 200, 250, 5000, 100, 750

Horace Edwards, 200, 300, 5500, 40, 550

Pleasant Hicks, 400, 100, 1000, 30, 125

John Manning, 125, 350, 2500, 35, 250

George W. Stewart, 60, 175, 1200, 75 275

William H. Barnes, 100, 320, 200, 65, 275

Frances E. Roswell, 100, 550, 5000, 175, 565

Overton G. Slater, 150, 132, 2500, 85, 350

John Timberlake, 100, 295, 3000, 260, 750

Littleberry Spraggins, 75, 125, 1200, 50, 450

Geo. W. E. James, 130, 300, 2300, 20, 300

George P. Hazlewood, 100, 340, 2000, 140, 325

William O. Hockaday, 240, 750, 5000, 300, 1775

Jno. A. Hockaday agt., 100, 140, 1275, 65, 430

George Nelson, 40, 10 1/3, 250, 20, 195

Christopher Past, 75, 490, 5000, 200, 450

Eliza T. Richardson, 250, 400, 10000, 250, 800

John E. Richardson, 50, 60, 1000, 20, 80

Richard H. Whitaker, 120, 257, 5000, 115, 758

Elizabeth Richardson, 150, 950, 20000, 25, 320

Simon Pitts, tenant, 30, -, -, 20, 140

Stanhope Richardson, 95, 370, 6000, 10, 235

Pink T. Garrett, 125, 200, 2500, 10, 235

William A. Mconly, 120, 150, 3500, 140, 620

William B. Vaiden, 75, 175, 2500, 132, 400

William J. Lindsey, 18, 10, 600, 10, 130

Martha A. Richardson, 60, 220, 1800, 20, 190

William Deal, 65, 255, 5000, 85, 665

Albert W. Hankins, 400, 600, 10000, 140, 1010

Geo. E. Richardson, 80, 120, 2500, 50, 600

Alex. H. Hankins, 220, 320, 7500, 235, 940

Allen Richardson, 250, 150, 5000, 80, 460

Aaron Chapman, 100, 340, 2500, 25, 360

Wm. H. Rogers, 35, 65, 800, 25, 310

Thomas W. Taylor, 40, 70, 600, 10, 72

Madison J. Mustin, 45, 120, 1500, 50, 375

Edward R. Coke, 220, 180, 5000, 210, 810

Richard E. Taylor, 200, 125, 5000, 165, 855

Francis W. Hammond agt., 80, 120, 2000, 75, 400

Richardson Henley, 250, 210, 4500, 160, 620

John T. Marston, 300, 200, 8000, 230, 1170

James T. Farthing, 120, 100, 2500, 100, 550

Danl. P. Hawkins, 200, 350, 4000, 120, 675

Pink. A. Taylor, 100, 158, 2500, 75, 340

Nathl. Piggott, 400, 2370, 18500, 230, 1450

Charles M. Hubbard, 180, 100, 6500, 120, 950

Benj. T. Piggott, 100, 217, 4000, 208, 630

Moses Moore, 50, 75, 1100, 20, 300

Elizabeth B. Warren, 150, 400, 5000, 110, 624

Jas. R. Warren, 300, 1200, 15500, 235, 650

Marshall M. Martin, 120, 75, 3000, 100, 350

William Simpson, 20, 10, 60, 15, 45

Robert Greenhow, 30, 20, 600, 10, 170

Robert Wallis agt., 20, 45, 240, 40, 240

Jonathan Canaday, 40, 20, 300, 20, 240

Upton M. Spencer, 100, 650, 2625, 55, 100

E. P. W. Apperson Agt., 250, 75, 3000, 40, 300

William S. Minor, 50, 156, 2000, 160, 570

William L. Spencer, 250, 1050, 9100, 200, 955

Lafayette A. White agt., 300, 415, 8000, 90 975

John W. Canaday, 15, 70, 350, 15, 240

John Nettles, 30, 190, 1000, 16, 340

Garrett Knight, 150, 367, 4500, 140, 640

George W. Miner, 75, 265, 4000, 50, 375

Richardson Braune, 55, 225, 650, 41, 150

Allen Davis, 100, 250, 2000, 80, 525

Thos. D. Harris, 140, 260, 2500, 55,692

Robert Wasburton, 130, 309, 4390, 90, 520

Burwell Harrell, 95, 522, 4000, 375, 1090

Littleberry G. Waddell, 48, 400, 3000, 105, 970

John W. Jones, 200, 1000, 9700, 465, 1200

John S. Morris, 100, 219, 3000, 135, 490

William H. Kerby agt., 270 200, 9000, 400, 1250

James E. Small agt., 40, 800, 20000, 550 2295

Henly L. Taylor, 90, 118, 2000, 80, 380

Elizabeth D. Taylor, 30, 45, 1800, 40, 110

James E. Maynard agt., 320, 280, 10000, 300, 1200

James Sanfer agt., 220, 380, 10000, 405, 1525

W. F. Browning agt., 150, 700, 6000, 185, 850

George W. Garrett agt., 550, 650, 12000, 320, 1000

Virginia Spencer, 75, 62, 1000, 45, 225

Geo. D. Williams agt., 30, 90, 500, 20, 60

Benj. Taylor, 35, 56, 800, 30, 240

James A. Banks, 45, 370, 1700, 30, 286

Thomas P. Marston, 120, 210, 3200, 45, 858

Geo. W. Hicks, 55, 70, 1700, 22, 135

Wm. P. Enloe agt. 400, 2600, 30000, 660, 2540

Wm. M. Buck agt., 180, 120, 10000, 366, 1550

Juliana Dorsey, 250, 150, 14000, 245, 1120

Wm. Martin, 400, 600, 12000, 135, 1050

Joshua Morris, 75, 144, 2500, 90, 460

Wm. R. Emery agt., 650, 1850, 50000, 380, 2800

J. C. Gibson agt., 430, 1170, 40000, 1600, 4650

Thos. V. Buck agt., 600, 400, 40000, 685, 5151

John Wortham agt., 370, 500, 16000, 800, 2380

Goodrich Durbry 50, 50, 7000, 100, 730

Martha Wilson, 50, 51, 1500, 160, 55

Geo. C. Richardson, 100, 100, 1200, 10, 285

David S. Cowles, 228, 863, 10000, 595, 2600

Jas. M. Clarke, 105, 75, 2400, 60, 455

Parke Jones, 700, 900, 23000, 550, 4155

Saml. A. Perry agt., 200, 300, 7000, 95, 800

John N. Pettitt, 250, 150, 5000, 75, 584

John H. Casey, 15, 25, 300, 20, 216

P. W. Hudgins, 250, 550, 12000, 650, 1160

Benj. S. Ewell, 200, 330, 7000, 500, 420

John J. Clowes, 35, -, 3500, 90, 295

William Lynus, 200, 98, 10000, 575, 1065

William D. Powers, 475, 774, 30000, 960, 4860

Churchill J. Adams, 1000, 2100, 55000, 900, 5200

James D. Scarborough, 215, 145, 5000, 300, 1260

William B. Wynne, 20, 106, 7000, -, 470

Lewis Ellison, 300, 600, 18000, 1120, 3470

Nat. Taylor Jr., 260, 525, 10000, 350, 1810

George Blow, 620, 1380, 36000, 1070, 400

William B. Hubbard, 175, 125, 6000, 275, 890

Richard C. Wynne, 240, 614, 13000, 550, 2260

John D. Mumford, 200, 122, 15000, 500,950

John T. Martin, 75, 75, 3500, 115, 390

The Agricultural Census for Virginia 1860 was microfilmed by the University of North Carolina Library under a grant from the National Science Foundation from original records at the Virginia Department of Archives and History in 1963.

There are forty-eight columns of information on each individual. Only the head of household is addressed. I have chosen to use only six columns of information because I feel that this information best illustrates the wealth of individuals. The columns are:

1. Name
2. Improved Acres of Land
3. Unimproved Acres of Land
4. Cash Value of Farm
5. Value of Farming Implements and Machinery
13. Value of Livestock

Benjamin Hales, -, -, -, -, 85

Charles H. Ashton, 400, 170, 7000, 150, 700

William E. McClahahan, 350, 120, 7000, 200, 1085

David Coakley, 200, 100, 4000, 100, 425

Leonard Tricker, 250, 50, 6000, 250, 915

William N. Jett, 450, 56, 13500, 200, 1363

Joel Stokes, 150, 46, 2000, 50, 312

Charles Staples, 42, 8, 750, 25, 100

James L. Quisenberry, 700, 100, 13000, 200, 1116

William G. S. Fitzhugh, 75, 25, 2000, 50, 188

Gustavus B. Alexander, 500, 600, 16500, 200, 1076

C. Lewis Jones, tenant, 2, -, -, 40, 141

Nancy P. Perd, tenant, 200, 142, 2395, 75, 275

William McDaniel, 200, 217, 6000, 125, 668

Sanford Morgan, tenant, 100, 50, 1500, 35, 60

John Gibbons, 137, 100, 2500, 60, 200

Jane Roach, 300, 60, 4000, 150, 370

Richard Holbut, 100, 32, 1800, 40, 250

Charles A. Berry, 200, 125, 2500, 150, 721

Geo. Scrivener, tenant, 100, 89, 1518, 75, 355

Lewis A. Ashton, 300, 250, 8000, 200, 1300

Thomas I. J. Grymes, 1000, 352, 27000, 300, 1500

George E. Grymes, 425, 250, 10500, 1000, 1765

Elizabeth M. Hoor, 425, 251, 10500, -, 275

Dangerfield Lewis, 1000, 500, 15000, 600, 2700

Margaret S. Lomax, 1200, 600, 50000, 600, 3310

Richard H. Stuart, 1200, 1100, 60000, 800, 5633

John L. Crismond, 500, 1000, 30000, 200, 2225

Giles Boonbury, 100, 250, 2500, 50, 250

Albut G. Owens, 18, 2, 200, 50, 50

Henry C. Pinkins(Jinkins), 125, 57, 2000, 60, 100

William E. Stuart, 140, 140, 3500, 75, 670

John Bryan, tenant, -, -, -, 5, 65

Nancy Brice, 50, 50, 2000, 25, 260

John S. Washington, 50, 46, 1500, 250, 500

Henry Rollett, 50, 67, 1500, 50, 225

George D. Ashton, 200, 270, 5000, 250, 500

James Strother, 40, 114, 1500, 20, 165

George W. Turner, 200, 175, 5625, 100, 700

Charles Pomeroy, 150, 37, 2500, 50, 360

William Jones, 250, 67, 4500, 150, 230

Thomas Landy, 95, 5, 1500, 25, 180

Charles Prior, tenant, 100, 97, 2000, 35, 125

Henry Fitzhugh, 350, 250, 10000, 300, 1400

Benjamin B. Arnold, 175, 50, 3000, 50, 650

John A. Billingsley, 75, 50, 2000, 30, 520

George K. Stuart, 150, 150, 5000, 100, 500

Richard Potts, 200, 200, 4000, 150, 700

Thomson Potts, 300, 35, 5000, 200, 700

Thomas W. D. Massey, 140, 30, 2000, 75, 320

John Taylor, 400, 100, 10000, 1000, 1000

Oswald Ferril, 35, 45, 700, 30, 150

Thomas Jett tenant, -, -, -, 5, 75

Sarah A. Boner, 40, 20, 2000, 100, 250

Ruth Borchell tenant, -, -, -, 5, 200

George W. Lewis, 300, 200, 10000, 300, 1454

John H. Rawlings, 50, 20, 1000, 15, 50

Addison J. Davis tenant, 50, 36, 500, 15, 60

Enoch P. Rose tenant, -, -, -, 5, 163

Harrison G. Howland, 400, 400, 20000, 600, 1400

John Dickens, 1000, 500, 40000, 1000, 3550

Wiley R. Mason, 500, 330, 15000, 600, 1000

John Arnold, 700, 300, 12500, 500, 1815

William Owens, 250, 100, 8750, 100, 700

R. Henry Hudson, 50, 112, 1625, 60, 140

W. Henry Griffin tenant, 40, 29, 672, 30, 80

Thomas L. Hunter, 200, 20, 6000, 250, 617

Madison Clift tenant, 50, 56, 854, 10, 119

Lexington Clift, 30, 70, 500, 10, 77

Drury B. Fitzhugh, 340, 100, 8000, 350, 1200

L. W. Griffin, 670, 300, 20000, 600, 1312

Elizabeth Fitzhugh, 100, 33, 2660, -, 80

Robert T. Fitzhugh, 59, 25, 1680, 200, 500

William H. Scrivner tenant, 40, 60, 800, 50, 200

John L. Tricker, 30, 24, 540, 25, 100

William Sharkling tenant, -, -, -, 25, 160

Robert A. Welch, 70, 85, 1000, 40, 235

David Redley tenant, 60, 60, 1000, -, -

Emily Hudson, 25, 5, 500, 10, 90

E. Poinsett Taylor, 350, 150, 10000, 300, 1475

Joseph S. Greenlaw, 350, 150, 10000, 200, 876

Joseph Lee, 100, 11, 2500, 100, 400

William A. Arnold, 140, 20, 3000, 250, 400

Hunter B. Brracke, 120, 45, 2500, 200, 665

Catherine Rollet tenant, 20, 25, 400, -, -

George Dunlop, 22, -, 500, 100, 260

Joseph F. Billingsley tenant, 153, 100, 3000, 15, 110

Joseph A. Billingsley, 200, 100, 5000, 350, 835

Reuben Rollett tenant, 100, 30, 650, 15, 135

Richard Ellis tenant, 25, 25, 250, 10, 50

George M. Miffleton, 75, 31, 1000, 25, 250

James Phillips, 200, 100, 5000, 150, 672

William Taylor tenant, 100, 30, 650, 25, 180

Ashton Stuart, 250, 50, 6000, 50, 624

William L. Bryan tenant, 100, 40, 1520, 25, 162

Hezekiah Potts, 200, 127, 6500, 100, 1200

George B. McKenney, 40, 29, 2000, 75, 367

Phillip N. Jett tenant, 30, 22, 500, 5, 20

Sarah L. D. Jett, 75, 35, 1500, -, 140

John L. Jett tenant, 100, 36, 1088, 5, 25

Edward V. Jones, 75, 25, 850, 30, 186

Sally Atwell tenant, 25, 20, 270, 5, 50

Hiram Carver, 65, 8, 1000, 50, 300

John Clift, 125, 75, 3000, 50, 225

Thomas Sorne tenant, 60, 43, 1000, 6, 50

Mark Arnold, 225, 75, 6000, 250, 1350

Hugh M. Tenant, 425, 150, 8000, 600, 1500

Abner B. Price, 400, 250, 2000, 800, 4000

William J. Jenkins tenant, 100, 54, 1500, 20, 175

Francis A. Lunsford, 70, 68, 1000, 15, 100

Peter Evans tenant, -, -, -, 20, 160

Henry Bailey, 50, 56, 1000, 25, 270

William L. Bryan tenant, 56, 200, 2848, 25, 400

Edmond Hoomes, 75, 125, 2000, 10, 75

Thomas Hoomes tenant, 25, 75, 1000, 10, 25

Thomas E. Payne, 150, 40, 4000, 50, 620

Richard H. Turner, 600, 320, 37000, 300, 2000

John Baker, 75, 15, 2000, 80, 400

Robert Scott, 20, 3, 300, 5, 75

Charles Inscoe tenant, 60, 43, 1030, 25, 65

William Pollard, 94, 500, 750, 500, 165

George C. Marshall, 40, 10, 500, 50, 450

Lawrence B. Edwards, 57, 67, 1400, 60, 415

John A. Edwards, 60, 51, 1200, 40, 125

Thomas Pollard, 100, 35, 750, 75, 225

Charles Mason, 1200, 600, 36500, 2000, 4700

Thomas B. B. Baber, 1180 694, 32000, 800, 3000

Stephen Owens, 20, 5, 500, 5, 20

Robert Stoore, 60, 30, 500, 5, 75

William B. Coakley, 70, 16, 2500, 250, 600

James Mahorney tenant, 125, 125, 3000, -, -

Robert Coakley, 125, 76, 3500, 200, 860

Lavinia Bramican tenant, 200, 50, 3000, -, 30

William Fuel tenant, 70, 134, 3000, 10, 43

Albert B. Suttle, 100, 25, 1500, 50, 375

Reuben Jenkins, 25, 28, 530, 15, 235

James Jett, 80, 24, 936, 50, 75

William D. Thompson, 60, 42, 1200, 50, 210

Stephen Clarke, 100, 150, 3500, 200, 450

John P. Robb, 600, 199, 25000, 400, 1873

Robert Wallace, 600, 1200, 20000, 500, 3300

S. J. S. Brown, 600, 400, 12000, 500, 1740

Joseph Crismon tenant, 100, 100, 2000, 15, 80

Edwin D. Brown, 150, 150, 4000, 125, 1000

William H. Bullard, 200, 170, 3700, 50, 575

James M. Edwards, 400, 50, 5000, 150, 900

Edward Spilman, 218, 200, 5000, -, 260

Overson Clarke, 100, 100, 3000, 50, 715

John D. Owens, 75, 25, 2000, 50, 308

John Coakley, 35, 27, 500, -, 30

Virginia Washington, 800, 200, 17000, 250, 1800

William D. Hove(Hore), 1500, 500, 4000, 10, 250

R. Ernest Hoor, 250, 310, 10500, 25, 475

Augustine McClanahan, 375, 150, 6500, 300, 1440

Patterson Marks, tenant, -, -, -, 25, 62

James McClanahan, 130, 40, 2000, 150, 528

Richard V. Tiffe_, 200, 25, 5000, 100, 350

David Jones, 40 60, 500, 15, 100

Samuel Staples, 30, 30, 300, 20, 90

William Staples tenant, -, -, -, 10, 80

John Roach, 100, 95, 1500, 50, 216

James Staples tenant, 75, 25, 500, 25, -

Rush Marshall, 300, 300, 10000, 310, 1000

William Marshall, 40, 56, 800, -, -

Henry S. Jones, 50, 59, 1500, 40, 175

Ferdinando F. Fairfax, 250, 486, 10000, 250, 1416

John Rose, 75, 132, 1400, 20, 182

Stanfield Jones, 125, 125, 3000, 75, 200

____ T. Dishman, 175, 51, 3616, 250, 1033

Gulvey Farms, 200, 62, 3884, -, -

Pine Hill Farm, 225, 75, 4500, -, -

Sarah McKinney, 22, 22, 500, 10, 100

Alexander Frank tenant, 100, 200, 1500, 40, 230

Mabon W. Potts, 150, 150, 1500, 50, 435

Jams C. Treacle, 20, 140, 2000, 60, 175

William Jenkins, 100, 70 1500, 10, 75

Jane Hutt, 24, 24, 600, 5, 100

Vinten A. Dickerson tenant, -, -, -, 50, 85

John H. Ridman, 60, 48, 1500, 50, 225

William B. Allensworth tenant, -, -, -, 25, 75

George C. McKinney, 375, 125, 6000, 250, 1150

Isaac Wilkenson, 200, 211, 4000, 100, 788

Henry A. Suttle, 60, 53, 1200, 100, 255

Joshua Frank, 40, 18, 600, 15, 180

Nicholas Farmer, 20, 236, 2000, 100, 159

Joseph J. Reamy tenant, -, -, -, 25, 80

James Overton tenant, -, -, -, 25, 55

Baldwin Owens, 80, 62, 1500, 50, 210

Frederick F. Ninde, 150, 90, 4000, 90 240

James Baker, 50, 44, 800, 10, 33

Samuel J. Berry, 50, 31, 400, 5, 53

Winnifred Berry, 20, -, 200, -, 80

Elizabeth Cox, 30, 6, 300, 15, 125

Cad. W. Cuismmond, 100, 36 1500, 15, 260

Nicholas Quisenberry, 500, 125, 20000, 500, 1660

Caroline Shih, 300, 200, 5000, -, -

William Mullen, 340, 92, 1780, 100, 2150

John Washington Plentiful Farms, 400, 33, 5199, -, -

Abram D. Hooe(Hoor), 800, 583, 20000, 800, 4624

Henry Crannage, 600, 177, 15000, 250, -

William H. Crannoss, 77, 130, 2500, 50, 275

Martha C. Stuart, 300, 100, 5600, 300, 2360

John H. Stuart, 200, 116, 3798, -, -

John T. Washington, 480, 100, 10000, 500, 1475

John F. Dickerson, 450, 300, 23000, 500, 2000

William H. Dickerson, 450, 40, 17000, 310, 700

Henry B. Lewis, 550, 50, 12000, 300, 925

William T. Smith, 1140, 200, 50000, 3000, 4500

James Porter, 130, 40, 3000, 100, 400

Enoch Edwards, 600, 600, 12000, 300, 1532

Nancy J. Payne, 120, 140, 3000, 50, 350

Lorrison Dishman tenant, 140, 140, 2680, 50, 50

Edward O. Greenlaw, 500, 150, 16000, 200, 435

Thomas P. Greenlaw, 200, 82, 5000, 100, 276

William Green, 75, 115, 1500, 10, 90

James Spilman, 10, 90, 500, 15, 60

Nancy Potts, 60, 10, 700, 25, 21

Joseph Spilman, 20, 50, 700, 10, 75

Frederick McGinniss, 120, 25, 2000, 250, 400

Robert Hall, 200, 158, 3500, 100, 480

William P. Lunsford, 25, 25, 400, 10, 125

Lucy Miller tenant, -, -, -, 15, 100

James Grigsby, 125, 55, 2000, 35, 300

Duke Pitts, 40, 50, 500, 30, 225

John Rollins, 15, 35, 1000, 20, 150

John T. King, 275, 125, 4000, 250, 800

Jane E. Lunsford, 100, 50, 1500, 25, 135

Hazlewood Thompson, 250, 250, 2500, 150, 560

Reuben N. Wilkerson, 250, 175, 5000, 300, 1000

James Wright tenant, 60, 75, 750, 15, 70

Virginia A. Taylor, 896, 420, 40000, 500, 2300

William P. Quisenberry, 150, 92, 4000, 50, 310

George Turner, 735, 125, 45000, 2500, 2700

Augustine Fitzhugh, 600, 500, 20000, 250, 1400

Carolinus Turner, 630, 100, 30000, 800, 1125

Baldwin Lee, 970, 250, 40000, 700, 1125

Stephen Marders, 25, 25, 300, 14, 125

Hiram Johnson tenant, 15, 15, 500, 25, 116

George W. Marshall, 75, 125, 2000, 150, 360

Robert Dishman tenant, 150, 150, 3000, 100, 165

Linsey Green, 100, 50, 1500, 25, 195

Bandridge Green, 20, 5, 100, 15, 60

Henry Winkfield, 30, 20, 250, 15, 50

Sarah Rowley, 200, 191, 3142, 10, 130

John Rogers, 75, 25, 750, 40, 200

Austin Rollins, 100, 20, 1500, 50, 250

John Edwards, 75, 25, 1000, 40, 250

Samuel Atwell, 100, 45, 1015, 25, 325

James Bruce, 100, 200, 6000, 75, 500

John Lucas tenant, 50, 150, 4000, 75, 135

John E. Owens, 425, 175, 12000, 500, 1400

Sidney Burchell tenant, -, -, -, -, 25

Mahala Owens tenant, -, -, -, -, 85

Sarah Jones tenant, -, -, -, -, 105

Martha A. Lee tenant, -, -, -, 20, 80

Daniel Miffleton tenant, -, -, -, 25, 138

Anthony Miffleton tenant, -, -, -, 25, 60

William Sorrel tenant, -, -, -, 25, 375

John Jeter, 225, 275, 8300, 500, 1275

Elizabeth Minor, 400, 230, 6300, 125, 860

Addison Haneford, 500 474, 25000, 3000, 2370

Julia A. Haniford, 350, 250, 8000, 50, 950

Reuben Suttle tenant, 50, 50, 1000, 15, 135

Charles Miffleton, 125, 125, 3000, 75, 775

Charles Dodd tenant, -, -, -, -, 60

Reuben Martin tenant, -, -, -, 25, 75

Nathaniel Elkins, 125, 75, 1600, 100, 375

William Colton tenant, 200, 200, 4000, 100, 400

Ellen Clift, 24, 20, 268, 15, 65

Alexander Walker, 50, 39, 537, 30, 315

Robert Bullard tenant, 175, 125, 1500, 25, 300

William Barker tenant, -, -, -, 10, 75

Daniel Thorpe tenant, 50, 250, 2000, 5, 75

John H. Rollins, 85, 45, 780, 75, 300

Gustavus B. Newton tenant, 50, -, 500, 40, 125

James H. Henderson, 25, 75, 500, 25, 75

John W. Allen tenant, -, -, -, 10, 145

Lucy P. Bowen, 350, 150, 5000, 150, 650

Landon McCarty, 15, 35, 300, 15, 55

Sidney Jones, 40, 10, 500, 50, 65

Edward Jones, 10, 6, 160, 25, 85

Burket McCarty, 20, 30, 2000, 10, 35

Madison Rollins, 100, 48, 1400, 65, 210

Thomas Acres, 50, 50, 500, 25, 100

Henry Henderson, 20, -, 200, 10, 195

James Acres, 50, 50, 500, 10, 100

Lewis Cross, 500, 200, 3500, 200, 450

James Finney tenant, 125, -, 625, 15, 50

James Bullard tenant, 125, -, 625, 25, 150

Neri Canada tenant, 125, -, 624, 5, 100

Arthur Henderson tenant, 125, -, 625, 5, 120

Thompson Henderson tenant, 125, -, 625, 25, 160

Robert Henderson tenant, 125, -, 624, 25, 205

Gustavus Henderson, 10, 23, 165, -, -

Edward Satterwhite tenant, -, -, - -, 100

Richard Randall, 273, 100, 3438, 100, 675

Thomas Lee, 100, 38, 2500, 150, 273

Ashley C. Tricken tenant, -, -, -, -, 130

John Self tenant, 52, 150, 2020, 15, 190

William L. Pratt, 120, 200, 3200, 100, 600

Robert Jones tenant, 8, 2, 150, 15, 100

Azel Truslow tenant, 10, -, 75, -, -

James L. Henderson, 50, 50, 500, 15, 60

Robert Randall, 75, 50, 900, 30, 215

John Henderson tenant, 30, -, 300, 10, 100

Catharine Henderson, 30, 5, 200, 15, 50

Richard A. Jones tenant, 100, 100, 1000, 10, 175

William Henderson, 40, 21, 500, 25, 200

Alexander Bowin, 70, 28, 1200, 75, 210

William G. R. Carter, 175, 80, 2000, 75, 360

Elizabeth West & 4 others, 125, 175, 4000, 50, 300

Reuben D. Bullard, 180, 70, 2500, 100, 465

Thaddeus Jenkins, 90 153, 1500, -, 75

Beverley Jenkins, 80, 71, 1500, 5, 282

William B. Jenkins, 200, 45, 2500, 150, 600

Bazil Jones, 100, 28, 1000, 25, 210

Bazil Jones manager, 550, 404, 28620, 200, 1300

Ann Meredith tenant, 100, 181, 2529, 25, 217

Frank L. Dade manager, 550, 157, 20000, 250, 985

King and Queen County, Virginia
1860 Agricultural Census

The Agricultural Census for Virginia 1860 was microfilmed by the University of North Carolina Library under a grant from the National Science Foundation from original records at the Virginia Department of Archives and History in 1963.

There are forty-eight columns of information on each individual. Only the head of household is addressed. I have chosen to use only six columns of information because I feel that this information best illustrates the wealth of individuals. The columns are:

1. Name
2. Improved Acres of Land
3. Unimproved Acres of Land
4. Cash Value of Farm
5. Value of Farming Implements and Machinery
13. Value of Livestock

John Pollard, 500, 200, 16240, 350, 1600
Lucy Pines, 10, 3, 150, 10, 125
Lewis Taylor, 40, 25, 500, 15, 115
William Brown, 75, 58, 1200, 25, 161
Maria A. Hart, 300, 205, 11000, 200, 946
Henry Didlake, 75, 26, 1200, 12, 144
Matilda Guthrie, 200, 101, 3900, 50, 372
Claudius C. Guthrie, 90, 175, 2000, 15, 240
William T. Guthrie, 75, 65, 1100, 25, 160
Philip Gibson, 75, 160, 1500, 100, 375
Thomas H. Jeffries, 200,180, 4500, 127, 600
William D. Walker, 6, -, 120, 5, 24
Philip G. Burnett, 37, 9, 1000, 10, 170
Philoman Bird, 300, 225, 6000, 250, 1199
Fendal Gregory, 500, 331, 16600, 100, 630

Alfred Smith, 35, 15, 1000, 20, 236
Henry Cox, 150, 263, 3344, 37, 455
Kary Kemp, 80, 103, 2745, 15, 201
Volney Walker, 300, 260, 8000, 300, 1389
Edward C. Fox, 250, 145, 4000, 200, 697
John Temple, 300, 300, 6000, 50, 846
James H. Fogg, 40, -, 400, -, -
William Dew, 450, 130, 12000, 275, 956
Robert T. Gwalthney, 150, 200, 7000, 100, 627
Edward S. Acree, 50, 60, 800, 15, 345
Thacker Muire, 100, 340, 6600, 150, 439
Albert Hill, 600, 400, 12000, 200, 946
John C. Crump, 200, 114, 7000, 100, 600
Albert G. Sale, 300, 250, 12000, 250, 1327
Archibald Pointer, 100, 277, 12000, 100, 600

William A. Taliaferro, 600, 300, 5000, 400, 1312

Benjamin F. Dew, 1200, 513, 24800, 400, 1500

Susan F. Dew, 500, 534, 20000, 175, 1090

Alexander C. Martin, 100, 40, 1500, 25, 255

Priscilla Segar, 150, 285, 3000, 50, 445

Ann J. Motley, 1000, 500, 15000, 300, 1242

Franklin B. Hall, 50, 13, 2000, 25, 175

George W. Cooke, 40, 17, 570, 30, 170

Ransome Harris, 150, 350, 5000, 100, 140

Phil Bluefoot, 20, 40, 400, 50, 50

Lorenzo Robinson, 175, 75, 700, 50, 152

John Kaufman, 35, 40, 600, 25, 140

John S. South, 110, 96, 4000, 200, 812

Bernard H. Walker, 275, 105, 10000, 200, 926

Benj. W. McLellard, 130, 60, 4000, 100, 421

George M. Pendleton, 350, 152, 7500, 500, 1072

William C. Anderson, 80, 30, 2000, 40, 315

Lorenzo D. Brown, 259, 259, 6200, 40, 592

Richard Longest, 100, 67, 2500, 40, 320

Edward Clayton, 128, 100, 2500, 50, 220

William B. Bird, 200, 40, 5000, 150, 469

William G. Wright, 320, 100, 8000, 150, 842

Catharine Wright, 80, 20, 1400, 25, 400

Thos. R. Gresham, 400, 210, 15000, 260, 1431

William D. Gresham, 600, 600, 30000, 500, 1317

John C. Walton, 100, 137, 3000, 25, 229

Nancy Turner, 150, 66, 3000, 40, 230

Robert Lumpkin, 100, 117, 3000, 50, 355

William S. Cooke, 100, 75, 4000, 50, 250

Robert H. Land, 72, 20, 2700, 75, 320

Julia Todd, 350, 200, 11000, 250, 1412

Cornelius H. Carlton, 400, 200, 10000, 375, 934

Walter R. Carlton, 300, 240, 10000, 100, 754

Mortimer Smith, 450, 300, 15500, 500, 1200

Joseph T. Henley, 350, 165, 12000, 350, 1260

Rebecca Pendleton, 200, 100, 4000, 100, 475

Temple Walker, 300, 150, 8000, 100, 475

Joanna S. Walker, 500, 425, 12000, 250, 1131

Catharine A. Pendleton, 35, 45, 1600, 25, 180

Lavinia Shelton, 36, 13, 1000, 15, 135

Temple Clarke, 50, 20, 1200, 50, 382

James Kemp, 37, 15, 275, 10, 75

Joseph Basket(Burket), 45, 20, 650, 25, 142

Joseph Watkins, 80, 27, 2000, 50, 198

Fountain W. Cooke, 200, 75, 4000, 75, 357

Ambrose Acree, 60, 30, 1600, 25, 150

Joseph Ryland, 600, 400, 15000, 300, 2093

James W. Haynes, 30, 22, 1200, 10, 95

Thornton Pemberton, 60, 42, 1000, 21, 139

John N. Ryland, 900, 300, 20000, 1000, 2716

Malinda Allen, 50, 60, 750, 10, 90

Robert Noel, 200, 100, 3000, 50, 439

Catharine B. Reynolds, 30, 34, 600, 5, 30

William Lumpkin, 165, 165, 3000, 15, 218

Richard Lumpkin, 100, 67, 1600, 20, 175

Reuben M. Garnett, 600, 500, 15000, 300, 1068

John Cooke, 30, 10, 500, 5, 56

Richard H. Pollard, 500, 290 12000, 200, 1222

Joseph L. Pollard, 400, 265, 7000, 50, 599

John Gatewood, 175, 87, 2187, 20, 175

Sarah Howerton, 150, 50, 2718, -, -

Sarah Howerton, 58, 30, 1000, 100, 700

Washington Mahon, 100, 44, 1152, 10, 156

Calvin Daws est., 1200, 500, 20000, 500, 1039

Elija Schools, 165, 96, 3500, 25, 484

Lawrence Muse, 200, 80, 3000, 75, 620

John Lumpkin, 700, 241, 12000, 250, 1170

Griffin Fauntlery, 450, 250, 12000, 300, 1197

Ann C. R. Fauntleroy, 350, 200, 10000, 200, 882

James A. Saunders, 40, 20, 400, 25, 150

Benjamin Fleet, 900, 400, 30000, 400, 2767

Columbia Cooke, 200, 75, 2750, 25, 175

Robert Watkins, 175, 50, 1800, 20, 306

Richard B. Lyne, 400, 400, 10000, 300, 1263

Bernard Eubank, 175, 68, 1944, 50, 438

Mary R. Minor, 300, 100, 5000, 200, 598

Lucy A. Saunders, 300, 200, 10000, 300, 1002

John H. Minor, 125, 50, 1600, 25, 245

Isabella Noel, 30, 30, 480, 10, 400

William Taylor, 45, 38, 400, 15, 45

William H. Gatewood, 100, 104, 3500, 30, 375

Claiborn Gatewood, 90, 40, 1000, 15, 170

S_hreshley Stokes, 154, 154, 3080, 120, 490

Samuel P. Wilson, 200, 100, 2400, 30, 376

Thomas W. L. Fauntleroy, 600, 544, 18304, 400, 1290

William W. Cox, 20, 85, 500, 15, 230

John Walker, 450, 193, 7500, -, -

John Walker, 400, 123, 9500, 300, 1815

Philip Brooke, 160, 60, 5000, 80, 430

Henry S. Nunn, 75, 25, 1000, 50, 230

Thomas C. McLellard, 250, 137, 10000, 250, 742

Alexander C. Carlton, 175, 62, 2500, 40, 279

John W. Shackleford, 200, 185, 7000, 100,710

J. H.C. Jones, 200, 100, 5000, 150, 807

Polly Cooke, 45, 15, 600, 25, 123

James Howell, 75, 50, 660, 25, 200

Robert G. Verlander, 70, 38, 860, 27, 50

Mary Kemp, 20, 15, 210, 20, 85

John S. Crow, 40, 40, 800, 20, 50

James C. Council, 175, 162, 6000, 150, 440

Edwin Watkins, 100, 105, 1640, 30, 267

Joanna D. Acree, 23, 9, 250, 10, 106

Elizabeth B. Carlton, 60, 40, 490, 8, 100

William H. Burkley, 120, 90, 1680, 30, 175

Mary Skelton, 175, 175, 2800, 25, 203

Augusta Edwards, 100, 100, 2500, 40, 106

Priscilla B. Smith, 600, 400, 10000, 200, 1624

Richard Watkins, 160, 81, 2000, 50, 355

Joseph H. Skelton, 250, 85, 5000, 150, 1000

Edwin Alexander, 33, 30, 500, 20, 75

Elizabeth Hutchinson, 125, 25, 2000, 40, 369

Martha Cauthorn, 120, 50, 3000, 30, 379

Mary E. Gatewood, 115, 50, 1200, 50, 285

Carman H. Gatewood, 60, 60, 720, 20, 328

Washington Broach, 50, 50, 700, 20, 100

Sally Minor, 60, 48, 756, 15, 195

John Watkins, 60, 52, 1000, 20, 107

Luroy Hutchinson, 130, 70, 2000, 30, 343

Muscoe W. Watkins, 80, 60, 1500, 30, 357

Robert Minter, 130, 70, 2000, 30, 243

Isaac Williams, 40, 7, 250, 8, 108

Richard Longest Jr., 60, 40, 500, 10, 75

Charles Prince, 36, 65, 400, 12, 50

Isaac Williams, 40, 60, 600, 15, 45

Philip Tate, 75, 25, 550, 15, 75

Baylor Cook, 25, 75, 300, 5, 50

Claiborn Longest, 50, 50, 800, 15, 60

James Longest, 65, 35, 550, 10, 60

George H. Trice, 80, 20, 450, 8, 60

Albert Minter, 70, 30, 500, 5, 40

Elizabeth J. Harper, 35, 65, 500, 9, 90

Peter Tombs, 300, 160, 6000, 300, 1005

William L. Rowe, 70, 50, 1400, 10, 204

Oliver White, 300, 130, 9000, 250, 689

Thomas E. Carlton, 100, 63, 1600, 50, 410

James E. Eubank, 100, 160, 3000, 40, 294

James Gibson, 16, 32, 600, 10, 200

Lydia F. Carlton, 150, 100, 2500, 50, 295

Susan Carlton, 50, 45, 960, 15, 15

William B. Carlton, 75, 57, 1000, 40, 160

John Marshal, 25, 15, 210, 30, 90

William Martin, 34, 75, 7000, 150, 1345

Mary Motley, 130, 130, 2600, 40, 356

William E. Hundley, 175, 55, 4000, 100, 633

Richard Gilmore, 41, 40, 400, 15, 63

John R. Furgurson, 40, 33, 1000, 15, 70

Heritage H. Cauthorn, 100, 145, 3050, 100, 361

Francis Prince, 50, 15, 400, 15, 75

Mary A. Hawes, 400, -, 10000, -, -

Mary A. Hawes, 400, 300, 12000, 500, 1690

Richard Gresham, 60, 70, 700, 25, 200

Catharine Smith, 80, 80, 1300, 60, 500

Nancy Brown, 100, 45, 1400, 5, 150

Elija P. Nunn, 400, 200, 5000, 20, 358

Tazwell W. Jones, 75, 40, 1500, 100, 210

Richard Carlton, 150, 135, 2000, 25, 213

William Bourne, 100, 77, 2000, 40, 173

John Radford, 35, 39, 500, 15, 70

John H. Smither, 80, 14, 900, 20, 202

Thomas B. Hart, 250, 200, 7000, 200, 767

Levi Carlton, 200, 160, 4000, 100, 367

Richard Carlton, 25, 25, 500, 20, 112

Garrett Carlton, 150, 100, 2500, -, -

Garrett Carlton, 30, 30, 300, 75, 440

Thomas Wyatt, 200, 65, 1500, 30, 308

James M. Davis, 90, 90, 1080, 20, 123

Lamberth Hundley, 90, 29, 2500, 30, 429

Beverly D. Roy, 500, 291, 12000, 200, 755

Alexander Dudly, 350, 50, 12000, -, -

Alexander Dudly, 150, 100, 3000, 500, 1125

Jacob W. Turner, 170, 30, 1000, 25, 140

James R. Garrett, 120, 40, 2500, 30, 352

Mary E. Spencer, 100, 50, 2000, 25, 275

Thomas K. Savage, 645, 70, 12000, 250, 606

Lewis Jeffries, 170, 170, 5250, 100, 555

Richard M. Smith, 160, 150, 5000, 125, 715

John R. Haynes, 150, 120, 2970, 65, 741

Major Brookes, 20, 19, 1300, 25, 101

Thomas Latane, 500, 300, 9600, 500, 1400

Ralf Mitchell, 20, 10, 294, 7, 60

James Nunn, 100, 105, 3075, 20, 319

William Coleman, 100, 74, 2620, 191

Robert Brookes, 30, 30, 750, 10, 75

Richard Prince, 24, 26, 700, 15, 74

John Jessie, 35, 4, 600, 15, 122

George R. Finch, 150, 309, 4000, 300, 404

Benjamin T. Taylor, 300, 189, 4800, 200, 790

Sylvanus Gresham, 150, 100, 2500, 25, 345

William Hoskins, 300, 166, 4600, 75, 584

Thomas W. Garrett, 325, 90, 7000, 160, 604

Robert Bland, 400, 130, 7950, -, -

Robert Bland, 400, 80, 7000, -, -

Robert Bland, 500, 350, 8500, -, -

Robert Bland, 500, 280, 10360, -, -

Robert Bland, 700, 470, 14040, -, -

Robert Bland, 1000, 400, 11700, -, -

Robert Bland, 200, 67, 1050, 400, 2425

Zackariah Harris, 30, 30, 600, 12, 40

William F. Bland, 700, 250, 25000, 600, 1904

William J. Smith, 400, 700, 10000, 100, 911

Lewis O. Smith, 100, 25, 3450, 20, 275

Norman Yarrington, 50, 43, 930, 25, 98

James W. Courtney, 90, 58, 3370, 200, 762

John W. Bulman, 105, 245, 4000, 100, 365

Thomas J. Dunn, 200, 97, 1788, 50, 288

Benjamin Lumpkin, 200, 95, 3000, 25, 190

Robert Collins, 200, 100, 1800, 50, 541

George Collins, 125, 175, 1800, 30, 240

John P. Folliard, 50, 50, 600, 25, 150
Lawson Revere, 90, 50, 2000, 4, 389
Edward Gibson, 45, 45, 270, 10, 71
Allen R. Hilliard, 60, 55, 570, 10, 76
Samuel Tunstal, 300, 300, 6000, -, -
Samuel Tunstal, 250, 300, 6000, -, -
Samuel Tunstal, 60, 250, 1230, 600, 965
Luroy C. Bulman, 150, 67, 3000, 25, 412
John D. Muire, 30, 23, 500, 25, 241
Daniel Thurston, 50, 68, 590, 10, 60
James Estiss, 60, 55, 2000, 25, 96
Henry Walton, 40, 93, 550, 30, 135
Stage Davis, 100, 63, 1100, 175, 342
Benjamin W. Walton, 20, 30, 200, 20, 52
Ofelier R. Dillard, 80, 73, 600, 25, 151
Philip Bluefoot, 20, 20, 300, 15, 50
Christopher W. Carlton, 20, 7, 200, 10, 60
Thomas C. Garrett, 25, 15, 200, 15, 141
George Cardwell, 25, 75, 500, 30, 94
John L. Lawson, 40, 30, 700, 15, 54
John Richerson, 120, 80, 1000, 10, 165
James L. Wyatt, 25, 18, 400, 5, 195
Eliza Oglesby, 10, 2, 100, 10, 90
William Birch, 50, 50, 650, 12, 161
Ira B. Cauthorn, 40, 53, 450, 25, 213
George C. Nunn, 700, 560, 13000, 500, 1300
Catherine Wyatt, 40, 40, 500, 5, 86
Elizabeth W. Garrett, 300, 250, 5000, 50, 516
John Motley, 325, 355, 10000, 290, 1200
James M. Birch, 20, 4, 150, 10, 68
Fernando Garrett, 50, 90, 480, 30, 231
Quintin Lumpkin, 60, 116, 700, 15, 145
William A. Colly, 125, 98, 1100, 27, 139

Robert B. Briton, 75, 196, 2000, 30, 288
John Y. Burton (Benton), 75, 196, 1100, 20, 277
John H. Ben, 30, 20, 500, 20, 175
Thomas Carr (Cow), 45, 425, 18800, -, -
Thomas Carr (Cow), 75, 270, 1200, 150, 889
William W. Shenen, 300, 200, 5000, -, -
William W. Shenen, 350, 60, 5000, 200, 596
John Spencer, estate, 258, 250, 4000, 150, 650
Danforth Butrick, 40, 40, 2000, -, -
Absolum Simeon, 50, 50, 1150, 10, 203
Robert V. Hart, 200, 88, 8000, 100, 400
David A. Farinholt, 125, 125, 4000, 100, 500
James G. Mills, 300, 200, 6000, 100, 905
James W. Mitchel, 60, 25, 1000, 25, 180
Jefferson S. Purcell, 110, 60, 3000, 50, 671
Robert R. Garrett, 150, 150, 1800, 50, 257
Thomas R. Gresham, 350, 250, 6220, 100, 800
John A. Watkins, 700, 400, 12000, 300, 1710
Edgar B. Montague, 100, 30, 3000, 40, 370
Evelina Brooker, 50, 200, 1000, -, -
Evelina Brooker, 143, 143, 1978, -, -
Evelina Brooker, 500, 500, 14000, 400, 1323
James W. Garner, 54, 55, 700, 25, 84
Wm. R. Didlake, 120, 85, 1200, 30, 214
Benjamin T. Guthrie, 80, 50, 1500, 20, 116

Wm. Robinson, 800, 500, 13000, 150, 1575

Igina S. Willis, 50, 46, 700, 23, 200

William Collins, 60, 48, 900, 27, 200

George F. Wedderbrism, 100, 65, 3000, 40, 291

Reubin Saunders, 100, 100, 3500, 50, 260

Charles H. Williams, 120, 130, 3500, 80, 430

John B. Duke, 40, 30, 1000, 20, 217

Robert T. Turner tenant, -, -, -, 20, 100

William H. Anderson, 200, 200, 10200, -, -

William H. Anderson, 50, 300, 4000, -, -

William H. Anderson, 85, 65, 1000, 300, 1241

John W. Seward, 50, 200, 1400, 30, 100

James H. Macomb, 40, 180, 1200, -, -

James M. Ben, 60, 70, 1000, 25, -

John B. Yarrington, 10, 80, 1500, 40, 324

Philip Estis, 75, 548, 1800, 25, 169

Hezekiah Ben, 25, 160, 925, 5, 54

James C. Foreman, 20, 120, 600, 30, 139

Laddy N. Moore, 175, 53, 2300, -, -

Laddy N. Moore, 100, 200, 1200, 40, 299

Peter Bray, 120, 140, 2000, 4, 325

Buckner E. Guthrie, 30, 73, 825, 10, 59

Francis A. Bland, 300, 200, 5000, 35, 260

Joseph W. Garrett, 30, 27, 1000, 25, 168

James C. Trice, 20, 55, 375, 125, 505

Richard F. Thurston, 50, 50, 500, 25, 55

James T. Seward, 40, 77, 500, 30, 148

William W. Brown, 20, 180, 1000, 25, 144

Thomas C. Bulman, 150, 175, 1800, 150, 420

Francis M. Birde, 65, 7, 400, 15, 50

Joseph B. Gatewood, 125, 75, 1500, 50, 280

James L. Seward, 200, 300, 4000, 35, 226

Mary Brown, 25, 5, 200, 20, 41

John E. Bray, 30, 145, 900, 30, 145

Lewis Boughton, 20, 8, 225, 10, 30

Maria Jackson, 300, 272, 2840, 50, 594

Alexander Atkins, 275, 129, 4000, 60, 620

Thomas Crittenden, 150, 150, 3000, 50, 358

Nancy Atkins, 300, 200, 5000, 45, 315

Alonzo Atkins, 200, 78, 2780, 25, 177

Robert Bowden, 225, 95, 3200, -, -

James A. Goalder, 210, 125, 3335, 100, 723

Beverley Anderson, 500, 325, 18000, 200, 1353

Thomas Edwards, 160, 107, 3000, 50, 444

John S. Leigh, 200, 159, 3700, 50, 400

Fanny Taylor, 100, 67, 1000, 25, 389

Miles C. Meridith, 250, 230, 3000, 50, 594

Absolum Bland, 150, 215, 2500, -, -

Martin Gouldman, 25, 15, 320, 8, 125

William S. Roane, 225, 75, 3000, 50, 692

William J. Ben, 50, 100, 750, 25, 127

Warner L. Wilson, 10, 10, 100, 25, 30

Samuel S. Crittenden, 50, 325, 1600, 50, 462

Richard Dungie, 25, 35, 240, 20, 70

Thomas Dungie, 30, 30, 240, 20, 101

Richard Walden, 140, 169, 1800, 75, 330

Edward Walden, 175, 175, 1700, 137, 973

William Eubank, 100, 50, 1200, 100, 298

Elizabeth R. Wright, 75, 75, 2000, 25, 257

Robert Driver, 40, 34, 700, 15, 215

Monroe D. Cooke, 60, 135, 1200, 50, 212

Ann L. Cooke, 150, 50, 2000, 100, 511

Robert T. Corr (Carr), 40, 37, 400, 15, 104

Bartholomew Y. Massie, 100, 136, 1000, 12, 230

William Y. Massie, 25, 215, 700, 8, 80

James A. Roane, 300, 140, 4500, 100, 860

George Brushwood, 100, 100, 3500, 9, 550

James C. Crittenden, 100, 50, 1500, 20, 281

Roderic Bland Sr., 700, 342, 8336,-, -

Roderic Bland Sr., 350, 254, 4432, 500, 1485

George P. Lively, 100, 60, 1000, 35, 328

Hogady Milby, 200, 218, 1800, 15, 205

James Guthrie, 30, 15, 200, 6, 45

Augustin Garrett, 100, 160, 1800, 25, 143

Samuel Thurston, 20, 20, 200, 5, 26

Henry C. Williams, 75, 57, 660, 25, 101

Allen Broach, 15, 10, 140, 10, 80

Bachelor Thurston, 275, 225, 4000, 50, 400

Peter Redd, 25, 7, 300, 10, 19

William Thurston, 40, 10, 500, 15, 40

John R. Breslow, 35, 11, 400, 10, 44

John R. Didlake, 150, 100, 1300, 30, 105

Benjamin Thurston, 122, 122, 1000, 10, 122

William Southern, 175, 175, 1400, 30, 109

Constantine Milby, 100, 100, 800, 15, 87

Katrina Williams, 22, 22, 180, 10, 85

Pondexter Corr, 45, 20, 900, 15, 185

Thomas C. Milby, 50, 150, 800, 10, 130

Frances T. Milby, 63, 63, 500, 10, 211

Fanny Jeffries, 200, 85, 4000, 20, 438

Andrew Wyatt, 150 250, 7000, 50, 299

James Brown, 25, 59, 350, 10, 87

Betsy Pollard, 400, 541, 7000, 150, 930

Absolum Muire, 45, 152, 1500, 50, 438

Ann R. Bland, 100, 70, 2500, 50, 326

John R. Cooke, 200, 200, 2500, 30, 460

Robert D. Didlake, 150, 30, 1500, 15, 100

James Corr, 225, 130, 4000, 100, 558

Samuel Milby, 100, 200, 600, 25, 162

Robert J. Milby, 60, 41, 250, 5, 132

James S. Yarrington, 100, 45, 600, 20, 119

Richard C. Benton, 80, 20, 600, 15, 144

John M. Benton, 80, 127, 2500, 25, 212

Richard Wayne, 125, 100, 3000, 50, 552

Irvin Taylor, 20, 5, 300, 10, 34

William Birch, 150, 175, 1500, 50, 377

Wiley Wright, 40, 26, 2800, 75, 474

Claiborne H. Bland, 250, 125, 3000, -, -

Claiborn H. Bland, 225, 287, 4000, 400, 1055

Marid N. Bland, 100, 25, 1500, 25, 165

John C. Daniel, 245, 75, 3500, 35, -

William C. Milby, 200, 150, 1000, 25, 289

John W. Street, 400, 200, 7000, 350, 975

Jenette Gresham, 120, 10, 780, 50, 369

John P. Williams, 400, 300, 7000, 100, 790

Richard Williams, 30, 20, 550, 20, 75

Charles Burgess, 30, 26, 1500, 30, 734

Leeland Cosby, 40, 35, 2000, 20, 110

Mortimer Jones, 100, 100, 3000,-, -

Mortimer Jones, 30, 90 1700, 25, 358

Elizabeth Hundley, 175, 80, 3000, 60, 569

William T. Fleet, 125, 125, 3200, 65, 446

Thomas Clevely, 16, -, -, -, 30

Samuel P. Ryland, 400, 200, 8000, -, -

Samuel P. Ryland, 400, 175, 7000, 500, 1826

Alfred Bagby, 60, 12, 2000, 50, 317

William G. Perkins, 160, 230, 5500, 125, 644

Robert Rose, 121, 121, 2550, 16, 297

Chaney Broach, 38, 38, 408, 25 176

Nicholas Dillard, 30, 85, 1200, 15, 112

Benjamin Boughton, 25, 20, 400, 12, 65

Vincent Coleman, 52, 5, 850, 15, 173

Moses Mitchell, 20, 8, 208, 5, 30

Larkin Williams, 30, 10, 800, 10, 75

Churchel Williams, 60, 30, 1000, 50, 360

John Williams, 30, 40, 1000, 20, 227

Thomas Deshazo, 20, 20, 500, 10, 85

William C. Milby, 95, 285, 1420, 10, 172

Thomas W. Redd, 80, 70, 500, 25, 378

Robert South, 100, 180, 1900, 25, 259

George B. Gibson, 40, 10, 400, 15, 75

William P. Courtney, 250, 200, 8000, -, -

William P. Courtney, 80, 10, 2000, 350, 975

Benoni Carlton, 500, 430, 12000, 250, 1360

Andrew Stone, 35, 15, 500, 7, 92

Samuel Doggins, 35, 17, 400, 10, 85

Joseph Pines, 10, 10, 200,-, 80

Sarah Pines, 140, 60, 2000, 25, 350

Mary Langham, 75, 60, 1400, 10, 160

Abraham Mahon, 20, 40, 500, 15, 80

John W. Deshazo, 120, 59, 5000, 60, 487

John Pruet, 50, 15, 600, 15, 166

Thomas Atkins, 100, 80, 1000, 25, 180

John W. Watkins, 140, 66, 2060, -, -

John W. Watkins, 225, 136, 3940, 10, 1000

William J. Clarkson, 600, 420, 12000, 200, 970

Robert Tate, 28, 10, 300, 10, 50

Alexander Fleet, 500, 125, 12000, 600, 1053

James R. Fleet, 400, 200, 12000, 800, 1190

John Faulkner Est., 600, 250, 10000, 150, 990

Richard Wright, 14, 12, 1500, 20, 250

Edward Gresham, 100, 226, 3000, -, -

Edward Gresham, 400, 582, 12000, -, -

Edward Gresham, 600, 341, 15000, -, -

Edward Gresham, 600, 340, 15000, 1000, 1430

Martha L. Harwood, 130, 56, 4000, 200, 650

Samuel P. Harwood, 350, 100, 10000, 330, 640

A. R. Harwood, 210, 10, 3500, 75, 325

Richard H. Bagly, 360, 250, 8000, 100, 1426

John A. Fleet, 125, 161, 4000, -, -

John A. Fleet, 350, 200, 7000, 350, 1249

Alfred T. Carlton, 118, 215, 4000,100, 650

Roderick Dew, 575, 225, 25000, 300, 1950

William A. Saunders, 325, 325, 111000, 395, 1550

Robert H. Spencer, 350, 250, 9000, -, -

Robert H. Spencer, 200, 236, 7000, 600, 15000

James Norman, 150, 300, 3500, 80, 400

John M. Garnett, 680, 365, 15000, 300, 1500

John N. Gresham, 300, 147, 11000, 250,700

Thomas Haynes, 500, 150, 35000, 300, 1000

James Mitchell, 350, 404, 15000, 100, 1222

William Walden, 400, 269, 12000, 250, 1343

Lewis C. Hart, 170, 164, 4000, 85, 245

Thomas M. Henley, 120, 84, 4000, 250, 600

Samuel G. Fauntleroy, 600, 450, 16000, -, -

Samuel G. Fauntleroy, 600, 460, 13000, 1050, 4350

William Boulware, 805, 805, 30000, 1000, 1880

Samuel S. Henley, 400, 300, 15000, 300, 1502

Charles Gresham, 400, 300, 10000, 200, 624

Robert Pollard, 370, 164, 15000, 200, 931

Joseph R. Garlick, 75, 75, 5000, 100, 500

James S. Bristow, 83, 100, 4000, 85, 300

Joseph Broumley, 200, 81, 3100, 100, 561

Carter B. Fogg, 185, 100, 2500, -, -

Carter B. Fogg, 500, 500, 20000, 300, 500

Ledford A. Vaughn, 200, 100, 3000, 50, 200

Robert Y. Henley, 600, 250, 17000, 1000, 1210

William B. Davis, 450, 250, 10000, 400, 1116

John R. Mann, 300, 200, 10000, 250, 931

Sarah M. Pendleton, 25, 15, 1200, 30, 285

Thomas D. Clegg, 100, 150, 3000, 75, 315

Alfred Gwaltney, 200, 200, 4000, 150, 695

James M. Jeffries, 550, 150, 17000, 300, 2000

Thomas P. Carlton, 52, 44, 500, 16, 103

King William County, Virginia
1860 Agricultural Census

The Agricultural Census for Virginia 1860 was microfilmed by the University of North Carolina Library under a grant from the National Science Foundation from original records at the Virginia Department of Archives and History in 1963.

There are forty-eight columns of information on each individual. Only the head of household is addressed. I have chosen to use only six columns of information because I feel that this information best illustrates the wealth of individuals. The columns are:

1. Name
2. Improved Acres of Land
3. Unimproved Acres of Land
4. Cash Value of Farm
5. Value of Farming Implements and Machinery
13. Value of Livestock

William Wallace, 342, 291, 12660, 550, 1916
Michael Wallace, 1250, 616, 37320, 613, 3000
Curtis Grymes, 150, 250, 6000, 150, 720
Jane Burchell, 125, 75, 3000, 75, 650
Benjamin Weaver, 150, 197, 4000, -, -
Edward T. Taylor, 885, 888, 50000, 600, 1800
P. P. Johnston tenant, 200, 38, 3000, -, -
Bladen T. Taylor, 285, -, 6000, 200, 600
W, Robman Taylor, 550, 200, 15000, 350, 2540
James C. Jones, 100, 35, 4000, 150, 630
James G. Taliaferro, 500, 588, 21760, 500, 419
George Anna L. Hore, 100, 50, 3000, 150, 375
James Hicks manager, 400, 200, 12000, 150, 225

Spotswood W. Corbin agent, 800, 350, 29400, 1000, 2776
Charles G. Jones, 300, 111, 5410, 100, 1480
Charles G. Jones tenant, 500, 130, 12000, 100, -
Susan W. Jones, 100, 100, 3100, -, 114
William A. Jones, 19, -, 225, 75, 410
William S. Brown, 250, 150, 5000, 125, 400
Frederick D. Davis, 80, 45, 1260, 25, 175
John Marmaduke, 11, -, 500, 10, 75
James Lee tenant, 70, 28, 686, 10, 130
Horace P. Ashton, 300, 100, 8000, 200, 761
Benjamin R. Grymes, 400, 450, 10000, 250, 1250
Mary W. Suttle, 30, 34, 385, 30, 238
John Dickman tenant, 100, 55, 1330, 25, 220
Susan Olive, 50, 53, 1000, 25, 255
James Arnold, 209, 100, 3000, 50, 325

Thomas Arnold tenant, 150, 300, 4500, - ,-

Cad. W. Crismond manager, 350, 200, 11342, 350, 1250

Robert Rollins, 40, 123, 1800, 12, 157

Milton Cunningham tenant, 63, 60, 1285, 25, 125

John T. Weedon, 53, 50, 830, 15, 125

Fielding Lewis, 600, 395, 25000, 1500, 3319

Mildred Allenworth, 150, 250, 4000, 25, 4

Thacker Rogers, 300, 300, 5000, 200, 800

Jno. R. Leigh, 150, 100, 2000, 50, 300

Jas. Woodward, 30, 20, 500, 30, 100

H. L. Abraham, 200, 150, 4500, 100, 400

Jas. Leftwich, 100, 140, 2000, -, 300

Jas. Leftwich, 200, 300, 6000, 200, 300

R. T. Mitchell, 30, 20, 1000, 20, 150

J. R. Alexander, 60, 60, 1300, 20, 150

Mary King, 150, 50, 2000, 100, 300

Jno. Vines, 200, 90, 5000, 100, 350

Wm. R. & P. H. Aylett, 1500, 600, 25000, 1000, 2000

Wm. R. Aylett trustee, 35, -, 3000, -, 250

Jos. A. Bond, 45, 30, 2500, 100, 300

Rd. Hillyard, 80, 60, 2000, 50, 100

P. H. Slaughter, 3, -, 1000, 20, 200

W. W. Jones, 10, -, 100, 20, 100

Junius Littlepage, 80, 84, 5000, 200, 1000

E. L. Powell, 150, 100, 5000, 200, 500

Wm. M. Willeroy, 150, 150, 5000, 200, 800

Wm. M. Willeroy, 25, -, 1000, -, -

J. D. Edwards, 120, 140, 4000, 200, 1000

Davy Straughn, 200, 100, 6000, 200, 700

Wm. B. Enos, 2, -, 1000, 50, 200

J. C. King, 110, 110, 2000, 200, 800

J. C. King, 50, 120, 2000, 100, 200

Wm. V. Cronton, 320, 250, 12000, 200, 1500

Wm. V. Cronton, 300, 200, 3000, -, -

M. C. Garlick, 160, 80, 4000, 200, 700

A. Brown, 400, 300, 14000, 500, 1700

J. M. Fauntleroy, 400, 150, 10000, 500, 1400

J. W. Taylor, 380, 100, 7000, 200, 1200

O. M. Winston, 500, 200, 12000, 300, 1500

Wm. Gwathmey, 900, 400, 20000, 1000, 2000

Thos. Eubank, 50, 30, 1000, 50, 100

Edwd. Hill, 500, 300, 10000, 700, 2000

Edwd. Hill, 300, 300, 4000, 200, 600

Dandy Sale, 800, 200, 12000, 500, 1200

John Eubank, 20, 10, 600, 30, 100

Wm. T. Eubank, 120, 60, 2000, 100, 300

Elvy Eubank, 175, 25, 2000, 100, 300

Carter Mahon, 220, 100, 3000, 200, 500

M. B. Trant (Trout), 350, 200 8000, 500, 1200

Wm. L. Harrison, 300, 300, 6000, 300, 1500

Wm. L. Harrison, 30, 80, 1000, -, 100

Wm. L. Harrison, 2, -, 800, -, -

A. W. Atkins, 15, -, 400, 50, 200

Silas Mahon, 170, 30, 3000, 200, 300

Silas Mahon, 40, 30, 1000, -, 200

R. A. Fox, 20, 40, 1000, 40, 150

James Trice, 100, 120, 2500, 100, 300

Wm. R. Berkeley, 1200, 400, 25000, 500, 3000

Mary Fox, 500, 200, 10000, 400, 1000

Wm. S. Ryland, 500, 300, 15000, 1000, 2000

Wm. S. Ryland, 27, 40, 3000, -, -

Wm. S. Ryland, 10, 2, 2000, -, -

R. W. Fox, 350, 200, 6000, 200, 1000

J. R. Alexander, 30, 20, 500, 50, 200

Bent. Tuck, 100, 80, 1500, 50, 300

Geo. Tuck, 100, 70, 1500, 50, 300

Jno. Tarrant, 80, 50, 2000, 50, 150

Ben White, 5, -, 200, -, -

Mary E. Dabney, 300, 100, 5000, 500, 1500

A. F. Dabney, 84, -, 1000, -, -

A. F. Dabney, 27, 100, 2000, -, -

Jno. B. Patterson, 80, 80, 1000, -, -

Augtn. Atkins, 40, 25, 500, -, 100

G. Floyd's Est, 50, 50, 1000, 30, 100

Richd. Tuck, 100, 100, 1500, 50, 200

Harris Tuck, 100, 50, 2000, 100, 300

Elizth Cocke, 300, 200, 6000, 500, 1000

Jno. Cocke, 60, 40, 1000, -, 400

Rd. Eubank, 70, 70, 1500, 200, 400

Rd. Eubank, 50, 50, 1000, -, -

B. J. Nicholson, 200, 200, 4000, 200, 500

Wm. T. Downer, 200, 120, 4000, 100, 400

G. W. Thompson, 20, -, 1000, 50, 200

Delphly Powell, 200, 200, 4000, 200, 500

Geo. Woollard, 30, 4, 300, 10, 50

W. W. Garrett, 100, 80, 3000, 60, 300

Jas. Figg, 20, 10, 500, 20, 100

D. K. Gregg, 600, 500, 15000, 100, 2000

D. K. Gregg, 100, 60, 1500, -, -

D. K. Gregg, 150, 200, 500, -, -

D. K. Gregg, 100, 50, 2000, -, -

D. K. Gregg, 120, 50, 2000, -, -

John Madison, 90, 10, 1000, 20, 2000

Wm. Wheeley, 10, 5, 2300, 20, 100

Miles Adams, 5, 90, 1000, -, 20

Jno. Madison, 6, -, 100, -, -

A. Walker, 200, 160, 3000, 50, 150

Peter Hormes, 15, -, 150, -, -

Ham Wheeley, 10, -, 150, -, -

Molly Adams, 8, - 100, 25, 100

L. J. Chappell, 80, 40, 2000, 100, 300

R. A. Dunstan, 1, -, 300, -, 100

W. L. Harrison, 2, -, 800, -, -

Jno. W. Walker, 300, 200, 4000, 300, 800

Jno. W. Walker, 100, 200, 2500, -, -

Wm. T. Slaughter, 80, 80, 1000, 50, 200

Jno. F. McGeorge, 200, 350, 6000, 500, 1000

Jno. F. McGeorge, 150, 50, 2000, -, -

Jno. F. McGeorge, 100, 50, 2000, -, -

J.K. Fisher, 300, 200, 5000, 200, 1000

B. B. Douglas, 400, 300, 12000, 600, 1000

Lucy Q Wyatt, 40, 50, 1000, 50, 200

Lucy Clements, 40, -, 400, -, -

Wm. K. Clements, 30, -, 500, 25, 150

Wm. K. Clements guardian, 50, 50, 800, -, -

A. Scott, 500, 500, 10000, 400, 1200

Lucy Minor, 7, -, 300, -, -

Ellin Smith, 80, 100, 1500, 50, 200

P. H. & W. R. Aylett, 2, -, 600, -, -

A. G. Sale, 400, 100, 6000, -, -

Jno. Tolly, 2, 1, 1200, -, -

C. H. Lee, 2, -, 800, -, 400

O. A. Gresham, 2, -, 2500, -, 150

Betty Pollard, 50, -, 1000, 25, 150

Wm. C. Pollard, 100, 100, 2000, 100, 500

Temple (Pollard), 200, 150, 3500, 100, 500

E. T. Powell, 250, 250, 6000, 200, 500

C. H. Boggs, 200, 140, 6000, 200, 700

J. B. Fox, 75, 75, 1000, -, -

D. K. Gregg, 1, -, 500, -, -

Elizth. Johnson, 600, 500, 12000, 500, 2000

J. C. Johnson, 300, 200, 6000, 200, 1200

J. C. Johnson, 300, 730, 6000, -, -

Ed. C. Pollard, 200, 150, 4000, 100, 600

Jno. T. Jones, 90, 35, 2000, 50, 300

Wm. Pollard, 200, 130, 3000, 100, 600

Jas. Fox, 30, 30, 600, -, 50

R. Read, 100, 160, 2500, 100, 400

Mary Ancarrow, 50, 50, 1000, 50, 200

Thos. Ancarrow, 60, 40, 1000, 25, 200

Wiley Atkins, 20, -, 2000, 20, 50

Jos. White, 40, 16, 500, 20, 60

J. J. Jones, 150, 275, 5000, 50, 300

W. Garnett, 30, 40, 1000, 50, 50

Temple Allen, 30, 30, 600, 25, 100

Tom Allen, 20, -, 300, -, 75

A. Campbell, 300, 300, 6000, 300, 1000

Lucy Atkinson, 100, 100, 2000, 100, 300

Wm. C. Floyd, 50, 40, 600, -, -

Richd. Garnett, 25, 25, 500, 50, 300

Mary Garnett, 50, 50, 1000, 50, -

Thos. Garnett, 30, 30, 1000, 50, 100

Th. Tucks est., 200, 100, 4000, 100, 400

Th. C. Sweet, 2, -, 1000, 50, 300

Giles Tignor, 40, 40, 1000, 25, 100

E. T. Wooddy, 60, 40, 1000, -, 50

Wm. Redd, 250, 100, 4000, 100, 600

Sam Terry est., 40, 10, 300, 25, 60

R. T. Redford, 5, 25, 300, -, 50

David Rider, 200, 100, 4000, 100, 400

Anna Terry, 100, 100, 2000, 50, 300

Geo. W. Adams, 100, 100, 1000, 25, 60

J. T. Moran, 80, 40, 1000, 50, 200

Mary F. Noel, 75, 75, 1000, 50, 100

Jno. Morrison, 30, 30, 500, 25, 75

T. C. Morrison, 10, 15, 200, -, -

Randal Twopence, 40, 10, 300, 25, 50

J. Beadles, 40, 40, 500, -, 10

Elston Edwards, 50, 50, 500, 25, 200

R. Abrahams, 20, -, 200, -, 25

John Spurlock, 20, -, 200, -, 20

John Anderson, 50, 100, 500, 25, 75

Richd. Anderson, 50, 100, 500, 25, 150

Ned Mills, 20, -, 200, 20, 50

Margt. Burruss, 100, 100, 2000, 50, 300

E. M. King, 200, 200, 300, 50, 300

Mildred Blake, 40, 40, 400, -, -

Th. B. Caltell, 50, 100, 600, 25, -

Lee A. Dunn, 160, 160, 2000, 20, 60

R. C. Pemberton, 250, 100, 5000, 100, 500

R. M. Tuck, 150, 100, 2000, 100, 200

R. Burke, 250, 150, 8000, 500, 1000

Wm. A. Spiller, 300, 40, 8000, 200, 1200

Elk. Clements, 150, 150, 3000, 200, 600

Martha Dudley, 30, 20, 800, 25, -

J. K. Birdsall, 40, 180, 1300, 300, 1000

J. H. Burch, 125, 75, 2000, 100, 600

John Robins, 230, 100, 5000, 500, 1200

John Robins, 100, 120, 2000, -, -

Jno. Guthrow, 30, -, 300, 20, 20

C. C. Davis, 150, 50, 200, 100, 440

J. T. Caldwell, 125, 130, 5000, 500, 1000

J. T. Caldwell, 1, -, 3000, -, -

J. T. Caldwell, 1, -, 1000, -, -
J. T. Caldwell, 1, -, 2000, -, -
B. Samuel, 4, -, 2500, 25, 200
B. Samuel, 1, -, 800, -, -
B. Samuel, 1, 400, -, -
C. C. Gary, 2, -, 1500, 35, 150
J. L. Latane, 600, 500, 12000, 500, 100
W. C. Satane(Latane), 200, 200, 5000, 100, -
R. W. Haynes, 200, 120, 3000, 300, 500
J. Alexander, 100, -, 500, -, -
Jno. Seger, 400, 160, 6000, 500, 800
C. Minor, 6, -, 200, 25, -
B. P. Wyatt, 200, 172, 3000, 50, -
Wm. E. Taliaferro, 400, 400, 6000, 300, 600
Th. P. Jackson, 160, 60, 3000, 200, 400
Geo. Taylor, 1200, 200, 60000, 500, 3000
Geo. Taylor, 1600, 400, 40000, 100, 2000
S. T. Norment, 400, 50, 15000, 500, 1000
S. T. Norment, 500, 200, 15000, 100, 800
D. W. Norment, 70, 90, 1500, 50, 250
D. W. Norment, 10, 30, 4000, 30, 80
Anna Blake, 40, 40, 500, 20, 300
E. M. Sutton, 400, 100, 6000, 500, 2000
M. E. Sutton, 100, 30, 1500, -, -
David Sutton, 100, 30, 1500, -, -
S. C. Sutton, 250, 100, 4000, -, -
H. Nelson, 700, 400, 22000, 300, 1500
L. A. Stevens, 50, 400, 4000, 100, 300
L. A. Stevens, 300, 200, 7000, 200, 800
L. A. Stevens, 150, 200, 2000, -, -
W. W. Hutchinson, 250, 250, 5000, 300, 1000

W. W. Hutchinson, 300, 200, 5000, 100, -
Rebecca Eubank, 250, 250, 5000, 300, 1000
Rebecca Eubank, 300, 200, 5000, 100, -
Ann Hay (Kay), 40, 60, 600, 20, -
A. Sozier, 100, 50, 5000, 50, 400
John Allen, 100, 60, 2000, 100, 300
T. O. Dabney, 230, 200, 4000, 200, 400
D. Atkins, 125, 125, 1500, 100, 200
Sam. McDonald, 1, -, 3000, -, -
Sam. McDonald, 1, -, 500, -, -
Sam. McDonald, 1, -, 800, -, -
H. Timberlake, 2, -, 2000, -, -
H. Timberlake, 1, -, 500, -, -
Th. Robinson, 160, 60, 5000, 300, 600
Han. Seal, 40, 20, 500, 20, 50
E. W. Satterwhite, 30, 12, 400, -, -
W. H. Beckeley, 1, -, 1000, -, -
G. W. Powell, -, 180, 1000, -, -
Rd. S. Ricer est., 40, 10, 300, 25, 50
Rd. Vian, 50, 60, 600, 30, 100
Phil. Eubank, 16, -, 300, 20, -
Amanda Adams, 40, 20, 500, 30, -
Nancy Beadles, 50, 75, 1200, 50, 300
Albr. Jackson, 150, 150, 3000, 100, 300
Otway Pollard, 200, 180, 3500, 100, 400
Harvey Pollard, 100, 25, 2000, 100, 300
L.C. Adams, 40, 20, 300, 25, 100
Cath. Adams, 70, 30, 600, 20, 150
Geo. Mitchell, 40, 10, 500, 20, 120
A. B. Walker, 40, -, 300, 30, 80
Richeson Cocke, 100, 160, 1500, 40, 150
J. R. Friendley, 100, 60, 1500, 50, 200
Mary Hawes, 120, 70, 5000, 1000, 2000
Richd. Hawes, 300, 100, 5000, -, -

W. A. Hawes, 100, 40, 3000, -, -
Alice Hawes, 100, 40, 3000, -, -
W. W. Dabney, 200, -, 5000, 50, -
J. N. Cocke, 40, -, 500, -, -
Ed. Cocke, 60, -, 600, -, -
John Sweet, 20, 10, 1000, 500, 250
Burch & Sweet, 30, 300, 1500, 100,
500
Mildred Huxster, 15, 15, 300, -, 50
Wm. Sweet, 50, 10, 600, 50, -
D. Atkinson, 250, 150, 4000, 200,
400
A. B. Puller, 50, 170, 2000, 100, 400
Lucy Puller, 170, 50, 3000, 100, 100
H. B. Tomlin, 330, 70, 20000, 1000,
1500
Wm. H. Mitchell, 50, 75, 600, -, -
F. Gregory, 300, 230, 12000, 500,
1500
F. Gregory, 100, 25, 2000, -, -
F. Gregory, 100, 50, 1500, -, -
Mary Braxton, 540, 100, 35000, 500,
2000
Wm. P. Braxton, 400, 300, 30000,
1000, 2500
R. A. Munday, 1800, 600, 50000,
2000, 4000
Wm. A. Braxton, 137, 100, 3000,
500, 2000
Wm. A. Braxton, 540, 300, 10000, -,
-
Wm. N. Turner, 350, 100, 10000,
500, 1000
Geo. Tremyer, 50, 30, 1000, 25, -
Jno. Edmund est., 20, -, 400, -, -
Tho. Houching, 20, -, 400,-, -
Gid. Trimmers est., 25, 15, 500, -, -
C. R. Rice, 50, 37, 500, -, -
Wm. Ware, 40, 10, 600, 30, 40
Lem. Pollard, 300, 250, 5000, 500,
600
Mary Powell, 130, 130, 2000, 200,
300
A. J. Fox, 120, 25, 2000, 200, 300
H. K. Foster, 100, 100, 1500, 100,
200

R. Neale, 150, 350, 5000, 200, 800
Cath. Madison, 50, 50, 500, 20, -
Sarah M. Lipscomb, 36, -, -, -, -
Sam. Robinson, 200, 150, 3500, 200,
500
Evelina Brooks, 400, 140, 10000,
100, 50
Nancy S. Gregory, 300, 250, 5000, -,
-
Wm. Ellett, 150, 150, 3000, 100, 400
Sarah King, 50, 100, 1000, 20, 100
R. C. Rice, 100, 100, 2000, -, 200
Junius Gregory, 100, 50, 2000, 50,
200
Alice Jackson, 20, 20, 400, 20, 20
Wm. B. Spencer, 100, 175, 2500, 40,
200
Lewis Harris, 20, 15, 300, -, -
Frank Martin, 400, 200, 8000, 200,
1000
A. B. Dangerfield, 140, 140, 3000,
100, 300
Jos. E. Ball ¼, -, 300, -, -
Emily Taylor, ¼, -, 1000, -, -
Edwd. Fleet, ¼, -, 2000, -, -,
G. W. Cooke, ¼, -, 500, -, -
Preston Lipscomb, 175, 200, 6000,-,
600
Wm. H. T. Lee, 1000, 1800, 350000,
300, 1200
Heely Lipscomb, 350, 50, 8000, 200,
1000
Heely Lipscomb, 50, 200, 3000, -, -
Wm. Hill, 250, 250, 6000, 1000,
1500
Wm. Hill, 250, 270, 5000, -, -
Wm. Hill, 50, 70, 4000, -, -
Hen. Corr, 200, 50, 2000, 500, -
Nancy Major, 100, 50, 2000, 50, 200
Wm. M. Ellett, 200, 20, 5000, 100,
400
B. C. Nelson, 260, 60, 4000, 100,
400
Th. Powers, 36, 12, 500, 30, 100
Caty Powers, 25, 25, 500, 20, 100
Henry Abert, 30, 30, 500, -, 20

Geo. Hance, 23, -, 300, -, 50
C. Wagner, 20, 30, 500, -, 100
Henry Corr, 600, 210, 20000, 500, 1200
Wm. Smith, 200, 150, 7500, 300, 500
Hen. Howard, 25, 25, 500, 25, -
Thos. Bew, 100, 36, 1000, 50, -
Elizth. Lipscomb, 200, 326, 5000, 75, 400
Bernd. Lipscomb, 100, 156, 3000, 50, -
Wm. J. Lipscomb, 6, 31, 2000, 20, 250
J. C. Houchings, 3, -, 1000, -, 100
S. S. Thornton, 40, 16, 2000, 200, 400
S. S. Thornton, 100, 70, 2000, -, -
R. Hargrove, 20, 5, 400, 30, 200
Sarah Davis, 112, 100, 3000, 100, 300
Bailey Davis, 50, 63, 1500, -, -
Robt. Davis, 50, -, 500, -, -
Ed. & W.A. Davis, 250, 140, 4000, 300, 800
Smith Davis, 150 75, 3000, 20, -
Geo. B. Mill, 300, 390, 7000, 150, 400
Geo. B. Mill, 10, -, -, -, -
Jas. H. Johnson, 200, 250, 10000, 300, 1000
Jas. H. Johnson, 50, 70, 1000, 25, 100
J. B. Green, 125, 126, 3000, 200, 4000
W. Dew, 40, 60, 500, 25, -
Steph. Howard, 30, 20, 500, 15, -
Jno. P. Johnson, 100, 110, 3500, 500, 500
Jno. P. Johnson, 100, 106, 2500, -, -
Hen. C. Johnson, 120, 100, 3000, -, -
J. H. Waring, 200, 150, 2000, -, 200
Ths. Stark, 250, 150, 4000, 800, 1200
Ths. Stark, 12, 75, 1000, -, -
Ths. Stark, 15, 65, 1000, -, -

Sarah G. Neale, 7, -, 1000, 25, 100
Lem. Edwards, 100, 123, 5000, 200, 600
E. C. Dillard, 100, 165, 3000, 25, 150
James Johnson, 300, 400, 6000, 500, 600
B. Stark, 125, 178, 4000, 100, 600
Cornel Davis, 50, 16, 600, 50, -
W. S. Baylor, 300, 225, -, -, -
Wm. Johnson, 400, 150, 10000, 1000, -
Sam. B. Lipscomb, 160, -, 3000, 250, -
Sam. B. Lipscomb, 40, 90, 2000, -, -
W. W. Fontaine, 1/8, -, 1500, -, -
J. J. Littlepage, ¼, -, 600, -, -
Wm. P. Taylor, 6, -, 2000, -, -
J. M. Hall, ½, -, 500, -, -
A. Dudley, 1, -, 1000, -, -
A. Grame, ¼, -, 250, -, -
Jas. Grame, ¼,-, 250, -, -
Jas. Greenwood, ¼, -, 600, -, -
Jas. Gwyn, ¼, -, 2000, -, -
Thos. Stark, ½, -, 3000, -, -
Dudly & Co., ¼, -, 250, -, -
Jas. Johnson, ¼, -, 250, -, -
Geo. B. Mill, ¼, -, 250, -, -
Nannie Gregory 2, -, 2000, -, -
R. M. Hord, ¼, -, 250, -, -
P. P. Duval, ¼, -, 2500, -, -
West. Pt. Land Co., 250, 100, 50000, -, -
P. P. Duval, 10, 20, 1000, -, 300
P. P. Duval, 200, 620, 3000, -, 150
Chas. Walker, 45, 45, 1000, 200, 150
Jessee Dungee, 40, 80, 1000, 25, 150
Wm. P. Taylor, 3000, 500, 70000, 300, 2000
Wm. P. Taylor, 400, -, 1000, 100, 1000
Wm. P. Taylor, 2, -, 3000, -, -
Ann F. Moore, 500, 350, 10000, 150, 1000
Amon Johnson, 300, 120, 7000, 100, 300

J. H. Purcell, 200, 130, 50, -, 500
S. D. Pilcher, 250, 80, 6000, 80, 1000
Anna Blake, 7, -, 2000, -, 200
P. H. Slaughter, 400, 150, 10000, 500, 1500
P. H. Slaughter, 350, 150, 8000, 100, -
P. H. Slaughter, 200, 250, -, -, -
P. H. Slaughter, 75, 125, -, -, -
P. H. Slaughter, 25, 75, -, -, -
J. P. Caldwell, 4, -, 4000, -, 200
Hen. Mitchell, 50, 50, 1000, 50, 10
L. C. Timberlake, 250, 200, 5000, 100, 300
Jas. H. Powell, 230, 100, 5000, 300, 1000
John W. Page, 150, 80, 2500, 20, 150
J. E. W. Toombs, 100, 50, 3000, 80, 500
S. Prince, 120, 20, 2000, 75, 500
Th. S. Jones, 400, 420, 20000, 500, 3000
Th. S. Jones, 250, 250, 10000, -, -
Th. S. Jones, 250, 350, 6000, -, -
Wm. Robinson, 80, 20, 2000, 100, 400
J. M. Davis, 25, 25, 1000, 100, 200
Caty Davis, 75, 100, 2000, 50, 200
Wm. A. Sweet, 75, 75, 2000, 50, 250
Wm. S. Fontaine, 800, 300, 20000, 500, 2000
Martn. Drewry, 350, 280, 15000, 500, 1500
Phil Sale, 100, 200, 3000, 50, 250
Th. Pollard, 200, 100, 4000, 100, 600
Henry Morrisson, 50, 50, 800, 15, 200
Tom Abraham, 50, 50, 800, 25, 150
Bent. Tuck Jr., 140, 140, 3000, 50, 400
Wm. Morrisson, 25, 15, 400, 20, 100
Lou Acree, 100, 37, 1600, 20, 200
Jno. O. Turpin, 200, 50, 3000, 100, 400

Ann Pollard, 75, 40, 1000, 50, 300
Ann Pollard, 15, -, 500, -, -
Lucy R. Eubank, 30, 24, 500, 20, -
R. Mitchell Sr., 26, 24, 600, 50, 150
Joshua Ellett, 10, -, 200, 25, 50
Ryland Blake, 12, 12, 300, 20, 50
R. E. Tompkins, 100, 200, 4000, 400, 400
W. W. Dabney, 300, 100, 5000, 500, 1500
W. W. Dabney, 50, -, 5000, -, -
Carter Wormley, 700, 300, 25000, 1000, 2000
Henry Timberlake, 150, 134, 5000, 700, -
Henry Timberlake, 100, 20, 1200, -, -
Henry Timberlake, 400, 400, 15000, -, -
Henry Timberlake, 200, 430, 8000, -, -
R. B. Garnet, 1, -, 800, 20, 150
Wm. F. Gardner, 100, 125, 1500, 50, 300
Aylett Moren, 150, 154, 3000, 25, 400
R. S. Crow, 60, 75, 1500, 50, 200
Jno. Quarles, 150, 100, 1500, 20, -
Jas. Roane, 500, 150, 15000, 500, 2000
Harfd. Timberlake, 100, 88, 4000, 50, 500
Isaac Davenport, 25, 62, 1000, -, -
P. Davenport, 100, 25, 3000, 500, 300
Wm. Taylor, 100, 56, 5000, 200, 400
S. C. Dabney, 200, 140, 3000, 25, 100
Y. J. Clements, 43, -, 3000, -, -
Y. J. Clements, 60, 90, 5000, -, -
Y. J. Clements, 500, 250, 10000, 300, 1500
Nancy Laundsun(Laundrun), 40, 60, -, -, -
Wm. Bosher, 200, 320, 10000, 1200, 1000

Wm. Bosher, 500, 250, 12000, -, 1000

Hen. T. Coalter, 425, 125, 28000, 400, 1500

Hart. Cocke, 200, 230, 6000,100, 500

Clarisa Robins, 200, 100, 5000, 100, 500

R. R. Turner, 100, 20, 4000, 100, 300

Law Trant, 100, 260, 4000, 50, 500

H. D. Burruss, 300, 500, 10000, 150, 1200

R. H. Lipscomb, 80, 100, 2500, 40, 400

Wm. B. Pointer, 200, 80, 4000, 50, 400

John Smith, 80, 60, 1500, 50, 300

C. S. Garrett, 125, 300, 8000, 125, 650

Jas. Pollard, 100, 70, 2000, 100, 400

L. S. Garrett, 125, 60, 5000, 300, 660

Egbert Lipscomb, 300, 240, 5000, 75, 500

Wm. J. Trimyer, 10, 30, 500, 50, 200

Wanon Lipscomb, 32, -, 300, 100, 400

Martha Littlepage, 50, 20, 1500, 50, 400

J. S. Lewis, 350, 140, 12000, 500, 1600

J. S. Lewis, 150, 250, 500, -, -

J. J. Newman, 200, 250, 7000, 100, 500

H. H. Hill, 300, 100, 8000, 300, 1200

A. White, 1000, 600, 30000, 500, 1500

Selim Slaughter, 200, 40, 4000, 100, 600

Selim Slaughter, 100, 240, 3000, -, -

S. B. Lipscomb, 50, 60, 1000, 50, 100

C. J. Hill, 400, 135, 6000, 150, 1000

J. Deffarges, 400, 150, 7000, -, -

C. J. Hill, 100, 67, 1000, -, -

H. Littlepage, 300, 113, 6000, 300, -

John Cardwell, 1200, 500, 15000, 1000, 2500

John Cardwell, 300, 35, 7000, -, 500

J. J. Waring, 175, 25, 3000, 200, 400

M. W. Pemberton, 300, 160, 8000, 500, 1000

Nat. Wiltshire, 13, -, 200, 25, 100

Thos. Whitlock, 32, 10, 500, 50, 100

M. G. Gregory, 212, 80, 8000, 500, 1500

Roger Gregory, 350, 160, 6000, -, -

Due Gregory, 200, 100, 3000, -, -

R. Hill, 300, 340, 12000, 300, 500

Nancy Gregory, 400, 260, 5000, 150, 500

M. Slaughter, 300, 250, 6000, 250, 800

R. King, 600, 410, 12000, 300, 1000

James G. Mill, 300, 100, 5000, -, -

Fanny Martin, 75, 37, 1500, 50, -

J. K. Custaloe, 16, -, 200, 20, -

B. Richards, 250, 250, 10000, 500, -

Wm. L. Baylor, 1, -, 10000, -, -

And. Eastwood, 50, 250, 3000, 50, -

Saml. Rice, 100, 100, 1500, 30, 100

A. T. Moshlar, 20, 17, 2000, 20, 300

T. Allman, 50, 67, 800, 1, 100

R. A. Hill, 100, 150, 2000, -, -

Sterlg. Lipscomb est., 200, 90, 4000, -, -

W. H. Sale, 200, 232, 3000, -, -

Thos. Tignor, 1, -, 800, 1, 150

P. H. Alexander, 100, 60, 1000, -, 100

Hen. Viars, 30, 16, 500, -, -

Jas. F. New, 2, -, 2500, 30, 400

Jno. Hoopper, 283, 100, 8000, 200, 600

Aren Garrett, 40, 20, 1200, 150, 200

M. E. King, 300, 200, 10000, 250, 700

W. P. Hogan, 275, 150, 10000, 400, 800

Wm. Gary, 300, 250, 1100, 750, 2000

Wm. Gary, 50, 470, 5000, -, -

Bagby & Gary, 700, 310, 12000, 300, 1000

Jno. Lewis, 230, 80, 8000, 300, -

John P. Lacy, 200, 100, 4000, 150, -

Reubin Hillyard, 1000, 500, 20000, 1000, -

R. A. Hillyard, 100, 120, 2000, -, 600

Geo. W. Lipscomb, 160, 60, 5000, 400, 1000

Dicey King, 80, 20, 2000, -, -

J. B. Edwards, 250, 170, 9000, 300, 1500

Geo. Edwards, 400, 220, 12000, 500, 1500

Warner Edwards, 300, 145, 8000, 300, 1500

Warner Edwards, 120, 50, -, -, -

Josiah Burruss, 600, 300, 30000, 400, 1500

Josiah Burruss, 150, 50, 2000, -, -

S. E. Blake, 300, 250, 10000, 300, 1000

Wm. Lakhard, 60, 60, 1000, -, -

H. K. Carter, 1300, 1060, 50000, 1000, 3000

Jas. H. King, 300, 125, 12000, 500, 1500

John Cooke, 900, 600, 25000, 500, 2500

Lew Litlepage, 450, 350, 20000, 500, 1500

Lew Littlepage, 37, -, -, -, -

Sarah A. George, 25, 6, 1000, 40, 120

R. M. Tebbs, 130, 40, 5000, 300, 500

R. King Sr., 30, 30, 700, 20, 40

Jno. Lakhard, 50, 40, 1000, 30, 80

M. Cobb, 12, 2, 300, -, 10

Jno. Cobb, 16, 6, 300, -, 50

Marsh Lipscomb, 14, -, 200, -, 10

Mary A. Madison, 15, 22, 300, -, -

Wm. A. Gresham, 100, 17, 2000, 50, 300

Ag. Slaughter, 40, 30, 500, -, 50

Herd. Lakhard, 150, 150, 3000, 25, 80

Corn Lakhard, 200, 100, 3000, 25, 150

Jas. K. Henry, 400, 264, 6000, 150, 600

Mary E. Roane, 450, 125, 12000, 400, 1000

Jno. H. Pitts, 70, 80, 3600, 150, 500

Jno. H. Pitts, 150, 150, 2000, -, -

R. W. Tomlin, 300, 50 20000, 500, 2500

Lancaster County, Virginia
1860 Agricultural Census

The Agricultural Census for Virginia 1860 was microfilmed by the University of North Carolina Library under a grant from the National Science Foundation from original records at the Virginia Department of Archives and History in 1963.

There are forty-eight columns of information on each individual. Only the head of household is addressed. I have chosen to use only six columns of information because I feel that this information best illustrates the wealth of individuals. The columns are:

1. Name
2. Improved Acres of Land
3. Unimproved Acres of Land
4. Cash Value of Farm
5. Value of Farming Implements and Machinery
13. Value of Livestock

Samuel Gresham, 250, 146, 4000, 500, 1538
Maria A. Robertson, 4, -, 40, -, 10
James Robertson, 150, 100, 1750, 25, 385
Jos. Tapscott, 20, 105, 750, 25, 60
Samuel Downing, 350, 450, 10000, 500, 1628
Thos. D. Eubank, 150, 185, 5000, 200, 844
John M. McKenny, 45, 15, 600, 40, 245
Nancy Tally, 90, 40, 800, 10, 187
Jos. H. Fallin, 90, 216, 2000, 100, 449
R. W. Eubank, 50, 220, 1500, 100, 365
W. O. Eubank, 200, 230, 3500, 250, 614
Wm. Boyd, 12, 77, 500, 20, 90
Thos. C. Callahan, -, -, -, 20, 158
Warner H. Haynie, 100, 125, 2000, 320, 1080
Winney A. Sebree, 30, 12, 220, -, 15
Ann Bush, 50, 200, 2000, 100, 185
D. W. Cockral, 60, 75, 1000, 50, 258

J. A. Holt, 60, 40, 1200, 130, 348
T. Christopher, 60, 198, 800, 80, 312
David Dorson, 10, 40, 400, 20, 62
Wm. Lumsford, 50, 50, 400, 35, 114
J. A. Headley, 75, 115, 1000, 25, 104
C. H. L. Jeffries, 15, 10, 150, 25, 133
James Bush, 75, 290, 1600, 60, 299
Thos. Edwards,-, -, -, 5, 21
M. D. Dunaway, 191, 110, 3000, 170, 842
S. G. Eubank, 200, 380, 3000, 75, 368
Thos. Rice, 40, 50, 900, 75, 278
J. M. Denny, 30, 20, 350, 20, 98
E. W. Bates, -, -, -, 50, 125
R. D. Rout, -, -, -, 75, 320
J. E. Elmore, -, -, -, 30, 178
T. Haynie, -, -, -, 100, 159
F. Thrift, -, -, -, 20, 105
E. E. Dunaway, 150, 170, 5000, 390, 831
Thos. Stott, -, -, -, 1, 28
W. H. Edwards, -, -, -, 1, 4
Martin Shay, 60, 73, 1350, 100, 160
James Marsh, 40, 48, 440, 10, 44
Fanny Jones, -, -, -, -, 5

Jos. Marsh, -, -, -, 10, 74
Indy Norris, 60, 78, 1104, 30, 277
W. Haynie, 35, 35, 1000, 75, 585
Luke Ball, 70, 30, 2000, 100, 410
W. H. Haynie, 10, 8, 200, 10, 120
Wm. Robertson, 75, 13, 1000, 80, 219
Wm. R. Knight, 75, 125, 1000, 35, 280
Wm. R. Knight, 20, 8, 150, -, -
James Smith, -, -, -, 100, 350
Ron Sullivan by J. Smith, 70, 200, 2500, -, -
Jos. S. Mitchell, -, -, -, 50, 68
Wm. G. Mitchell heirs given by J. S. Ritchum, 50, 150, 2000, -, -
Abner Revere, 40, 33, 1300, 80, 217
B. G. Burgess, 10, 60, 2800, 35, 129
F. Forester, 100, 125, 4000, 600, 675
J. R. Mitchell, 25, 113, 1200, 60, 88
R. B. Mitchell, 80, 170, 2000, 200, 471
R. B. Mitchell, 70, 280, 2500, -, -
E. E. Brent, 75, 100, 2500, 150, 501
J. V. Sullivan, 60, 120, 3000, 160, 365
Wm. Berrick,-, -, -, -, -
J. D. Gulie, -, -, -, 300, 420
John Hathaway, 110, 170, 3000, 330, 734
H. C. Norris, 4, 35, 600, 60, 142
O. Norris, 150, 153, 6000, 325, 848
Travers Sebree, 70, 40, 1800, 100, 257
A. Haggard, 15, 60, 500, 30, 72
Richd. Bloxum, 60, 40, 1000, 160, 280
J. G. Roberts, 40, 40, 800, 180, 473
J. G. Roberts, 35, 35, 800, -, -
Betty Mitchell, 40, 40, 800, -, -
Wm. Clarke, 75, 135, 1800, 20, 140
E. Rich, -, -, -, 20, 75
Nancy Rains, 100, 90, 1800, 80, 205
Thos. Taff, 30, 10, 500, 100, 139
Thos. Taff, 19, 19, 400, -, -
T. W. George, -, -, -, -, 384

James Webb, -, -, -, 200, 358
Polly Warnick, 30, 10, 500, 30, 95
J. Danson, 200, 200, 3500, 230, 485
P. R. Lampkin, -, -, -, 40, 241
M. Lewis, 100, 100, 3500, 375, 572
M. Lewis, 20, 32, 500, -, -
H. L. Biscoe by M. Biscoe, 100, 100, 3000, 100, 425
H. L. Biscoe by M. Biscoe, 95, 100, 2000, -, -
James Ransone, -, -, -, 225, 350
L. D. Mitchell, 21, 244, 17000, 780, 1345
J. P. Saunders, 250, 300, 11000, 275, 992
Wm. H. Kirk, 175, 175, 6300, 250, 1126
R. S. Rains, 25, 10, 513, 10, 65
Wm. H. Rains, 5, 2, 50, -, -
A. C. Tapscott, 130, 130, 5000, 300, 666
R. H. Chilton, 360, 180, 12000, 690, 2022
R. H. Chilton, 300, 150, 12000, -, -
J. D. Pursley, 16, 14, 600, 25, 70
J. Wm. Sebree, -, -, -, -, 15
RF. R. Kenn__, 50, 140, 900, 40, 295
J. H. Chowning, 300, 338, 7000, 600, 976
J. L. Sullivan, 180, 301, 7000, 315, 952
F. R. M. Oliver, 90, 90, 1800, 180, 298
L. Beane, 40, 30, 1400, 75, 140
J. B. McCarty, 180, 190, 8500, 215, 783
S. E. Mitchell, 180, 34, 4000, 90, 287
J. A. Rogers, 250, 350, 11000, 450, 1256
T. C. Callahan, 150, 214, 4000, 300, 692
T. C. Callahan, 100, 154, -, -, -
W. T. Dalby, 240, 302, 10000,750, 930

W. T. Jones by E. Shackleford, 400, 550, 20000, 759, 1810

E. M. Pursel, 60, 140, 3000, 225, 390

W. S. Doggett, 30, 22, 1000, 20, 101

R. W. Doggett, 160, 57, 4000, 110, 458

J. E. Blackmore, 130, 62, 4000, 335, 460

J. E. Blackmore, 200, 45, 6000, -, -

T. R. Clarke, 44, 44, 900, 50, 166

Warner Beane, 100, 155, 3000, 455, 794

Warner Beane, 20, 8, 100, -, -

Wm. Brent, 24, -, 3000, 310, 480

M. M. Gresham, 60, 105, 2500, 200, 415

M. P. Diggs, 40, 30, 800, -, 135

James W. Warwick, 100, 123, 3800, 500, 609

Thos. Daviee, 150, 173, 6000, 155, 660

M. N. George, 130, 100, 3400, 10, 340

Saml. Blackwell, 100, 60, 2000, 405, 346

J. R. Stephens, -, -, -, 225, 300

J. S. Chowning, 80, 50, 6000, 600, 481

John Norris, -, -, -, -, 184

James Simmonds, 150, 153, 6000, 200, 600

N. Robertson, 100, 93, 3000, 160, 378

A. C. Chilton, 210, 140, 7180, 350, 787

J. R. Chilton, 250, 140, 6000, 150, 409

Cyrus Hazzard, 50, 62, 2000, 75, 88

Elias Fendla, -, -, -, 3, 109

James Miller, 150, 300, 4000, 300, 121

J. S. Williams, -, -, -, 50, 150

R. Dunaway, 165, 62, 10000, 400, 920

R. Dunaway, 70, 80, 3000, -, -

W. Brown, -, -, -, -, 40

W. L. Downman by J. H. Davenport, 390, 340, 12000, 1000, 1662

A. L. Carter, 120, 200, 10000, 900, 2100

A. L. Carter, 170, 160, 8000, -, -

A. L. Carter, 165, 110, 11000, -, -

A. L. Carter, 350, 135, 15000, -, -

R. W. Brown manager, 900, 720, 40000, 780, 3025

B. R. Gains 520, 800, 27000, 300, 130

W. T. Jesse, 350, 328, 12000, 750, 1527

B. H. Robinson, 45, 71, 3000, 80, 255

Octavius George, 100, 100, 4000, 422, 484

W. W. C. George, -, -, -, 30, 285

A. H. Curric, 200, 150, 8000, -, -

B. George, 200, 160, 3600, 125, 575

E. A. Curric, 334, 560, 16200, 429, 1175

E. A. Curric, 180, 220, 4800, -, -

J. D. Glascock, 50, 50, 1000, 20, 122

J. H. Pearce, 90, 48, 1400, 275, 298

Jos. Webb, 30, 45, 1300, 75, 94

M. P. Webb, 8, 2, 500, 10, 140

A. V. Wiatt, 130, 338, 15000, 435, 2266

F. D. Jones, 250, 225, 15000,-, -

F. D. Jones, 80, 20, 2000, -, -

Wm. Doggett, 20, 12, 325, 20, 54

E. L. Dobyns, 60, 51, 888, 90, 305

B.M. Oliver, 60, 40, 1000, 30, 106

G. R. Waddy, 70, 90, 2000, 200, 410

James Ewell, 300, 300, 12000, 500, 1199

T. M. Wiatt, 100, 100, 8000, 771, 960

Walter Shay, 90, 112, 3000, 140, 380

Methodist Parsonage, 2, -, 1500, -, -

S. P. Gresham, 150, 200, 2400, 220, 715

Fanny Mitchell, 60, 340, 5000, 50, 307

Thos. Brown, 60, 98, 4000, 650, 1575

Thos. Brown, 120, 80, 3000, -, -

Thos. Brown, 10, 190, 1000,-, -

A. D. Watts, 50, 80, 1700, 15, 118

L. L. Lumsford est owned by W. C. Lumsford renter, 50, 65, 1500, 40, 178

J. E. Kenn__, 80, 90, 3500, 100, 450

Peter Williams, 60, 48, 1080, 300, 409

Peter Williams, 35, 92, 1016, -, -

E. Withers Episcopal Parsonage, 8, -, 2000, 5, 130

C. H. Leland, 140, 260, 8000, 475, 740

George H. Oliver, 75, 56, 1260, 30, 125

George H. Oliver, 28, 28, 560, -, -

A. N. Cundiff, 100, 230, 4000, 450, 770

A. N. Cundiff, 15, 155, 1500, -, -

A. Hall, 100, 75, 8000, 500, 574

Guard W.Eustace Heirs, 200, 133, 6000, -, -

James Anderson, -, -, -, 75, 449

T. Coppedge, 160, 45, 6000, 200, 347

T. Coppedge, 8, 72, 800, -, -

G. B. Cox, -, -, -, 100, 400

R. E. Beane, 80, 172, 2000, 210, 522

C. Flowers, 75, 10, 3500, 180, 278

W. E. Flowers, 70, 15, 3500, 100, 156

W. Tolman, 20, -, 1000, 10, 42

Jos. P. Flippo, 300, 60, 14400, 895, 987

E. T. Jefferson, 12, 3, 600, 17, 55

A. Anderson, 36, 16, 2500, 35, 165

Warren Hubbard, 90, 10, 5000, 440, 567

Charles Hubbard, 80, 100, 5000, 30, 393

Elisabeth Walker, 7, -, 280, 40, 20

Jos. Palmer, 200, 400, 6000, -, -

Elizabeth Lunsford, 40, 60, 800, 15, 113

E. C. Carter, 200, 392, 5500, 530, 753

W. W. Haydin, 60 50, 1000, 90, 165

Wm. Barnett, 100, 30, 1500, 300, 572

James George, 30, 8, 500, 100, 188

James West, 50, 40, 900, 75, 226

John W. George, 12, 8, 400, 10, 84

Jos. W. Brent, 87, 95, 2000, 200, 328

R. D. Carter, 350, 250, 6000, 200, 224

James Connelle, 20, 50, 700, 50, 209

W. Y. Downman, 560, 412, 40000, 1055, 1917

John Beane, 80, 120, 1200, 45, 150

Sally Marsh, 30, 30, 300, 15, 62

John Boatman, 20, 60, 250, -, -

George Talley, 60, 40, 600, 30, 145

Chit Davies, 60, 60, 800, 25, 210

R. R. Dunaway, 120, 223, 2900, 225, 275

Richd. Douglass, 50, 161, 2500, 150, 415

Thos. S. Dunaway, 500, 350, 20000, 1000, 1465

Wayland Dunaway, 70, 30, 2200, -, -

T. H. Pinkard, 450, 200, 15000, 100, 1325

T. H. Pinkard, -, 396, 4000, -, -

Fanny Brent, 80, 20, 4000, -, -

Digma Brent, 15, 12, 135, -, -

Celia Brent, 15, 12, 135, -, -

M. F. Brent, 100, 100, 3000, 200, 575

George Brent, 100, 50, 2500, 300, 750

George Brent, 100, 50, 2500, -, -

George Brent, 50, 150, 2000,-, -

George Brent, 50, 50, 500, -, -

George Brent, 30, 70, 500, -, -

Thos. T. Lampkin renter, 60, 100, 1500, 100, 168

J. H. Coleman, 90, 30, 2000, 250, 410

J. W. Moore, 25, 15, 800, 50, 175

G. L. Smith, 20, 5, 200, 20, 71

Wm. Doggett, 60, 40, 1000, 131, 516

Wm. Doggett, 60, 40, 1000, -, -

Wm. Doggett, 60, 40, 1000, -, -

Thos. B. Payne, 100, 100, 3000, 40, 840

J. D. Oliver, 150, 100, 3000, 300, 494

Wm. Fendla, 40, 15, 1000, 50, 160

Thos. Pitman, 100, 150, 2500, 275, 453

J. W. Stiffy manager for R. Carter's Heirs, 300, 442, 6000, 200, 500

A. G. Gipson, 80, 170, 3500, 60, 126

B. B. McKenney, 750, 723, 15000, 650, 2098

Richd. Callia, 20, -, 400, 10, 12

John Winder, 5, -, 125, 10, 50

N. Spriggs, 199, 63, 5240, 400, 710

Aaron Pinns, 10, 3, 300, 10, 27

T. Taylor, 275, 146, 9000, 325, 962

G. G. Lee, 40, 32, 2500, 190, 304

G. H. Landon, 25, 60, 1200, 200, 130

J. F. Christopher, 24, -, 500, 15, 105

J. F. Christopher, 20, 10, 500, -, -

J. Y. Currell, 18, 7, 500, -, 75

Josiah Doggett, 6, 85, 1500, 20, 50

E. L. Doggett, 18, 2, 300, 10, 30

A. L. James, 40, 20, 1200, 20, 182

J. W. Gresham, 240, 60, 12000, 300, 794

J. W. Gresham, 400, 550, 14500, 520, 596

J. W. G. Guard. S. Mitchell, 50, 250, 1500, -, -

J. W. G. Guard. S. Mitchell, 50, 250, 1500, -, -

J. W. G. Guard. S. Mitchell, 20, 280, 1800, -, -

S. W. G. Guard Palmer's Heirs, -, 18, 100, -, -

S. W. G. Guard Palmer's Heirs, 20, 50, 2000, -, -

J. W. G. Guard: S. Stott, 60, 60, 600, -, -

J. W., G. Guard: S. Stott, -, 73, 800, -, -

Maria Currell, 20, 10, 300, -, -

Maria Currell, 20, -, 400, -, -

Wm. M. Saunders, 80, 91, 2000, 175, 410

A. M. Saunders, 40, 20, 1200, 388, 462

Wm. N. Kirk, 300, 250, 12000, 435, 676

Wm. N. Kirk, 100, 450, 4000,-, -

Wm. N. Kirk, 150, 50, 6000, -, -

W. E. Jones, 6, 11, 400, -, -

J. Currell, 130, 120, 4000, 50, 312

W. Henderson, 500, 1000, 18000, 850, 1505

L. Chase, 20, -, 3000, 290, 356

J. C. Thorn, 225, 547, 11000, 200, 670

R. B. Mitchell, 200, 377, 8000, 200, 671

B. C. Chinn est., 40, 200, 1000, -, -

B. C. Chinn est., 40, 132, 1000, -, -

P. H. Beane, 75, 12, 1600, 300, 365

W. C. Jones, 40, 335, 3350, 60, 150

Wm. Kesterson, 32, 100, 1000, 125, 373

Wm. Doggett, 20, 68, 800, -, -

G. W. Flowers, 150, 50, 4000, 250, 206

W. T. Chase, 70, 75, 3000, 500, 350

J. B. James, 90, 75, 5000, 300, 480

P. Towles, 130, 70, 6000, 350, 772

Wm. H. Robins, 80, 20, 3500, 500, 244

A. McNamara, 7, -, 400, -, 27

O. Lanson's(Lawson's) Heirs by G. Lunsford, 170, 30, 4000, 150, 208

O. Lanson's Heirs by C. C. Cundiff, 200, 100, 6000, 100, 433

Jasper Stott, 75, 25, 2600, 75, 150

B. P. Chilton, 46, 6, 2000, 200, 170

S. H. Robertson, 11, 11, 600, 300, 165

Thos. Armstrong, 250, 150, 8000, 200, 860

Thos. Armstrong, 3, 568, 8000, -, -

Warring Kenner, 45, 29, 2000, 50, 155

Rich. H. Ingram, 25, 35, 900, 45, 160

Dru Cox, 60, 51, 2200, 125, 267

J. W. Saunders, 60, 70, 3000, 130, 306

J. M. George, 40, 90, 2000, 30, 182

Judson James, 25, 58, 1500, 90, 215

J. Robins, 20, 10, 600, 150, 195

William George by Robt. Harris, 10, 2, 400, -, -

E. Currell, 38, 11, 2500, 50, 390

A. H. Lock, 60, 45, 2500, 190, 534

O. Lawson's Heirs by Sam Ingram, 15, 15, 600, 20, 85

Francis Harcum, 115, 36, 5000, 400, 465

Eugene Chase, 30, 45, 2000, 275, 162

Nancy Wilder, 25, 12, 800, 48, 25

James Hayden, 15, 14, 1000, 50, 68

Jesse George, 50, 30, 3000, 120, 110

Lucy D. Davenport, 75, 19, 800, 8, 110

Isaac Cundiff, 150, 95, 3000, 130, 347

James K. Ball, 213, 399, 12000, 500, 1350

R. W. Doggett, 40, 40, 800, 125, 203

R. W. Doggett, 20, 10, 400, -, -

R. J. Mitchell, 140, 320, 16000, 230, 938

R. F. Peirce, 150, 50, 3000, 600, 480

A. M. Peirce, 50, 25, 2250, 350, 540

H. C. Peirce, 30, 48, 1000, -, -

W. R. Peirce, 20, 89, 1000, -, -

A. C. Peirce, 30, 48, 1000, -, -

Ella C. Peirce, 10, 68, 1000, -, -

W. E. Rogers, 12, 2, 800, 95, 114

R. H. Hayden, 40, 100, 2000, 95, 268

J. S. Frankle, 70, 5, 3000, 300, 460

J. S. Frankle, 30, 20, 1500, -, -

J. S. Frankle, 35, 35, 500, -, -

J. S. Currell, 135, 19, 7000, 410, 864

W. C. Currell, 250, 50, 9000, 400, 1382

W. C. Currell, 125, 75, 5000, -, -

W. C. Currell, 80, 50, 2060, -, -

C. N. Lawson, 150, 85, 10000, 1200, 1136

C. N. Lawson, 130, 270, 8000, -, -

S. S. Buckon, 175, 150, 12000, 505, 550

Polly Buckon, 35, -, 1750, 20, 165

Lucy Williams, 170, 50, 7000, 500, 864

Jane Edmonds, 300, 63, 10000, 700, 1120

K. Towles, 160, 140, 12000, 250, 614

Hugh Brent, 230, -, 10000, 1000, 770

H. H. Hill, 192, 133, 10000, 300, 572

H. H. Hill, 100, 100, 2000, -, -

D. George, 20, -, 200, 10, 205

James Cornelias, 25, 5, 600, 5, 58

Wm. Cornelias, -, -, -, -, 64

Wm. K. Lewis, 180, 100, 7500, 300, 1021

Nancy C. George, 200, 50, 6000, 300, 674

A. H. Lee, 50, 22, 1200, 75, 328

Wm. H. George, 140, 130, 6000, 250, 819

Griffin Williams, 130, 26, 4000, 75, 202

John Freakle, 30, 6, 2000, 50, 98

H. S. Hathaway, 250, 100, 8000, 500, 825

Jos. F. Yerby, 150, 25, 1500, 100, 359

D. D. Berrick, 160, 170, 3000, 200, 198

Thos. Pritchett, 80, 170, 2500, 25, 265

E. O. Robinson, 70, 90, 3000, 300, 361

C. S. Dunton, 100, 50, 4500, 235, 384

Wm. L. G. Mitchell, 130, 430, 8000, 200, 836

Lee County, Virginia
1860 Agricultural Census

The Agricultural Census for Virginia 1860 was microfilmed by the University of North Carolina Library under a grant from the National Science Foundation from original records at the Virginia Department of Archives and History in 1963.

There are forty-eight columns of information on each individual. Only the head of household is addressed. I have chosen to use only six columns of information because I feel that this information best illustrates the wealth of individuals. The columns are:

1. Name
2. Improved Acres of Land
3. Unimproved Acres of Land
4. Cash Value of Farm
5. Value of Farming Implements and Machinery
13. Value of Livestock

Charles Daugherty, 200, 400, 15000, 200, 2290
Wm. Oaks, -, -, -, 200, 30
Jamiel Shelburn, 65, 764, 5803, 50, 243
Wm. Baswell, -, -, -, -, 343
Andrew Edmonson, 30, 125, 1500, 120, 472
J. H. Fulkerson, 470, 420, 15520, 200, 8470
Jency Fulkerson, -, -, -, -, 1580
Jacob Bumgardner, 20, 5, 300, 35, 396
Joseph A. Davis, -, -, -, -, 65
Wm. Jonson, 45, 100, 1010, 70, 259
Edmon Pace, 70, 150, 1500, 30, 285
John A. Lynch, -, -, -, 100, -
Isabell Robison, 100 170, 1620, 200, 177
Wm. Robison, -, -, -, 800, 100
Edwin Pace, 20, 30, 300, 400, 115
John Hyet, -, -, -, 110, 218
George W. Powel, 25, 5, 300, 10, 202
Wm. A. Cook, 150, 170, 3000, 10, 344

Anthony Rutledge, 45, 55, 1500, 300, 183
Sarah Edds, -, 15, 116, -, 19
Lavina Edds, -, -, -, - 28
Robbert E. Clauson, -, -, 200, -, 66
Fideler Balee, 85, 90, 1275, 65, 764
John C. H. Ely,-, -, -, -, 35
Wm. Reeves, -, -, -, 2, 60
George R. Lindsey,-, -, -, 2, 2
John C. Hamblin, -, -, -, 1, -
John Speak, 75, 105, 1080, 15, 201
John Patten Jr., -, -, -, 3, 10
Nathan Morgan, 200, 280, 7000, 100, 683
Willis Harris,-, -, -, -, -
Wm. P. Balis, 100, 350, 8000, 100, 997
James Brim, -, -, -, 100, 112
George Seale, -, -, -, 5, 83
Archilaus Balis -, -, -, -, 1540
John H. Balis, -, -, -, -, 150
Absalom Robberson, -, -, -, 20, 616
John Yearry, 50, 85, 810, 4, 161
George Upton, -, -, -, -, -
David Eldridge, -, -, -, -, -

Hyram L. Ely, 200, 300, 8000, 150, 1681

Moses S. Ely, 75, 75, 2000, -, -

Robbert B. Ely, -, -, - 15, 18

Lavishe Cheek, -, -, -, -, -

Wm. Motley, -, -, -, 5, 63

Thos. Vandeventer, -, -, -, 3, 37

Harison B. Crockett, -, -, -, 92, 304

China McAphe, -, -, -, -, 50

Wm. Thompson, 60, 45, 3000, 130, 587

Silvester E. Thompson, -, -, -, -, 112

Henry Motley, 40, 90, 1500, 71, 221

Wm. F. Gibson, 450, 800, 14725, 75, 1905

Wm. Mintin agt., -, -, -, -, -

Huey Eldridge agt., -, -, -, -, 15

Stokley Dagley, -, -, -, 5, 133

Jas. Woodard agt., -, -, -, 1, 44

Constantine Smith agt., -, -, -, 5, 115

Wm. J. Monveau agt., -, -, -, 3, 42

Jos. Eldridge, -, -, -, -, -

Jos. Mink, -, -, -, -, 26

Wm. Marcum, -, -, -, 116, 468

Nathaniel Ewing, 250, 275, 10000, 140, 1255

Elizabeth Mink agt,-, -, -, 3, 111

Joseph Mink agt., -, -, -, -, 40

Edwin Mink agt., -, -, -, -, 50

Margaret Mink, -, -, -, -, 15

Jessee Taylor, -, -, -, -, 53

Lefore A. Marcum, -, -, -, -, 393

John J. Ewing, -, -, -, 5, 345

Andrew J. Smith agt., -, -, -, 5, 100

John Martin agt., -, -, -, -, 46

Pavira Ayseo agt., -, -, -, -, 33

Jas. M. Pica, -, -, -, -, 17

Thos. Ball, 225, 375, 10000, 90, 483

Moses Ball, -, -, -, 5, 380

Wm. Ball, 600, 2000, 20000, 200, 4315

Moses McAphe, 105, 151, 3000, 30, 295

George W. Ball, 180, 420, 8000, 60, 786

M. J. Moss agt for J. H. Gibson, 225, 800, 10000,-, 1500

Robbert Woodard, -, -, -, 1, -

Absalum Z. Lankford agt., -, -, -, 80, 108

David C. Chadwell agt., -, -, -, 3, 203

George W. Ball Jr. agt., -, -, -, 100, 389

Samuel Chadwell, 35, 75, 800, 4, 266

Andrew M. Ely, 115, 47, 4000, 115, 960

George W. Cloud Sr., 175, 225, 10000, 50, 917

Daniel Ayres, -, -, -, 3, 3

Wm. K. Brittain agt., -, -, -, 75, 893

Peter Migeso, -, -, -, 55, 286

Wm. S. Collinsworth agt., -, -, -, 150, 682

Wm. F. Collinsworth agt., -, -, -, 5, 51

Wm. Robberts agt., -, -, -, 5, 3

Alexander Chadwell Sr., 300, 400, 12000, 50, 907

Cass. Chadwell agt., -, -, -, 1, 192

Golvin Chadwell agt., -, -, -, 1, 90

Duff Chadwell agt., -, -, -, 1, 145

John Chatman, -, -, -, -, 209

David Chadwell Sr., 75, 70, 2000, 90, 331

Vandearman Minten, -, -, -, 2, 26

Wm. McNeel agt., -, -, -, 2, 305

Susan Willis, 500, 1400, 12000, 80, 1765

David Willis, -, -, -, 44, 775

Jane Willis, -, -, -, 10, 420

Hyram Hoskins, 200, 230, 5500, 90, 973

David Chadwell Jr. agt., -, -, -, 2, 308

Peter Rolen, 150, 100, 5162, 5, 185

Michael Rolin, -, -, -, 5, 125

Wm. H. Pittwell, -, -, -, 3, 250

Daniel Rolen, -, -, -, 3, 157

Chadwell Brittan, 350, 1200, 16700, 80, 3050

George W. Mintin agt., -, -, -, 3, 50

Balis Litteral, 75, 85, 3000, 30, 507
Samuel Lock, 170, 130, 1700, 25, 38
Jas.M. Pridemore agt., -, -, -, 3, 256
Sarah Mintin, -, -, -, 3, 367
Jacob Susong, 75, 90, 3000, 20, 498
John Ball, 25, 5, 500, 3, 341
Andrew _. Susong, 50, 50, 1200, 5, 282
Amos Snaveley, -, -, -, 5, 197
Wilbourn F. Catson, -, -, -, -, 332
Sarah Rollins agt., -, -, -, -, 269
Jonas Snavely agt., -, -, -, 6, 49
Ezecheal Cox agt., -, -, -, 3, 45
Wilbane Robinson, -, -, -, -, 37
Jas. M. Wheeler, 40, 20, 1200, 3, 115
Robbert M. Ely, 1600, 7000, 50000, 400, 3550
Moses S. Ball, 200, 235, 7050, 300, 2360
Wm. H. Novell, 25, 30, 400, 10, 128
John Jackson agt., -, -, -, 5, 66
Wm. Ridings agt., 10, 340, 500, 20, 204
Mary Baylor agt., -, -, -, -, 44
Jobe Baglary agt., -, -, -, 3, 50
James Kibert agt., -, -, -, 5, 495
Wm. Robbert(Kibert) agt., -, -, -, 2, 101
Jas. S. Bush agt., -, -, -, 5, 144
Henry Colson, 50, 50, 1000, 10, 760
Charles Baley, 40, 100, 500, 5, 219
Sarah Colson, 85, 93, 3560, 5, 109
Henry F. Colson, -, -, -, 3, 126
Wiley Carmack agt., -, -, -, -, 27
Jas. B. Colson, 90, 10, 2000, 50, 560
John Colson Sr., 75, 62, 3000, 20, 630
Daniel Litterell(Littwell), 200, 100, 5000, 100, 662
William M. Dilman,-, -, -, -, 46
Alexander K. Brent, 200, 230, 10000, 170, 1430
John Brent, -, -, -, 3, 400
John Colson Jr., 60, 20, 5150, 65, 639
Andrew M. Comer, -, -, -, 80, 131

Eliza Colson, 10, 80, 600, 3, 100
Richard Crabtree, 175, 350, 5700, 35, 170
Henry Yonce, 10, 100, 600, 3, 65
John Sprawls, 30, 120, 600, 5, 125
Patrick Lenard, 25, 25, 400, 3, 70
George H. Hoskins, 283, 125, 7550, -, 1575
Elisabeth Shumake, 30, 20, 600, 3, -
Steaven Arrial, -, -, -, 3, 315
Mary H. Pearman, -, -, -, 3, 122
Wm. Sanderfer, -, -, -, 3, 47
Robbert E. Crockett, -, -, -, -, 139
Joseph Renfro, 300, 450, 16000, 133, 2460
Thos. J. Pridemore agt., -, -, -, 3, 176
George W. Ball, -, -, -, -, 30
Moses McGipson, 130, 115, 5000, -, -
Easter J. Brown, 100, 50, 4000, 50, 375
Wm. Sawyers Sr., 200, 200, 8800, -, 305
Wm. G. Sawyers, -, -, -, -, 233
Thos. S. Ensor, 80, 37, 300, -, 381
Daniel McPhearson, 280, 150, 9680, 100, 1113
Levi Carmack agt., -, -, -, -, 117
Hyram C. Wiseman, 116, 205, 4000, 200, 833
Wm. Prince agt., -, -, -, -, 138
George Hoskins agt., -, -, -, -, -
John Provence, agt., -, -, -, -, -
John Lattin agt., -, -, -, -, 394
Jerry Elison agt., -, -, -, -, 73
A. J. Crockett, -, -, -, -, 155
C. Childers agt., -, -, -, -, 45
Hyram Heller, -, -, -, -, 123
Sarah Jones agt., -, -, -, -, 15
Wm, Woodson, 300, 200, 18000, 300, 2565
Daniel J. Cross agt., -, -, -, 75, 291
Jos. Herrin agt., -, -, -, -, 37
Robbert Crockett, 140, 460, 9000, 20, 453
Isam Goins, -, -, -, -, 230

Thos. J. Jackson, -, -, -, 3, 49
Noel Stanly agt., 50, 50, 1600, -, 55
John Hubbard, -, -, -, -, 23
Daniel Killian, -, -, -, 5, 124
Alexander Chadwell, -, 200, 6, 3, 357
Catharine Chadwell, 60, 70, -, -, -
Bacob Brown, 20, 24, 175, 15, 240
Thos. J. Sandson agt., 100, 255, 1520, 3, 35
J. B. Rowlett, 30, 67, 200, 5, 184
Thos. Pillian agt., -, -, -, 3, 112
Henry F. Farris, -, -, -, 5, 158
B. G. Short, 275, 425, 5600, 25, 895
John Pillian agt., -, -, -, 3, 34
John Brown, 30, 50, 500, 30, 573
Calvin Fay agt., -, -, -, 3, 90
Wm. Hinton agt., -, -, -, -, 135
Michael Brown, 225, 600, 4500, 10, 389
C. W. Garrett agt., -, -, - 55, 144
Levi Chance, 100, 168, 2500, 62, 486
John T. Chance, -, -, -, 5, 233
Joseph Brown, 40, 40, 500, 5, 196
Arther Ball agt., -, -, -, 15, 397
Jas. Marcum, -, -, -, 3, 126
Nancy King, 85, 75, 1500, -, 195
Stephen Chance, -, -, -, 5, 179
Luther Rowlett, -, -, -, 3, 29
John Yearry, 75, 55, 1500, 5, 242
Wm. Short, 90, 120, 1500, 10, 489
Jacob Mondy, -, -, -, 5, 169
Jas. Combs agt., 300, 400, 4000, 10, 227
Wm. Combs,-, -, -, -, 120
Henry Yearry, -, -, -, 5, 203
Josiah Chance, 45, 71, 800, 5, 116
Joseph Ely, 40, 140, 1500, 3, 213
Wm. W. Neal, 80, 120, 2000, 5, 171
George Marcum, 50, 80, 400, 5, 259
Wm. Dan, 40, 60, 1000, 5, 233
Davis Moore, 150, 150, 2500, 135, 658
Abner Moore, 30, 40, 1000, 3, 84
Jas. Brooks, -, -, -, -, 155

Michael Herd, -, -, -, -, 136
Charles W. Noe, 200, 600, 6000, 10, 419
Wm. C. Parrott agt., -, -, -, 3, 95
Randolph Noe agt., -, -, -, 20, 268
Eligah Edens agt., -, -, -, 3, 123
Ann Muncy, -, 7, 34,-, 31
John Bartley, 30, 63, 800, 3,334
Joseph Thomas, -, -, -, 3, 123
David Moore, 80, 185, 2400, 60, 272
Rebecca Bray, -, -, -, 3, 133
Hesakiah Parsons agt., -, -, -, 3, 60
George Parsons agt., -, -, -, -, 40
Alexander Carter, 30, 30, 500, 3, 140
John Sebolt agt., -, -, -, 3, 33
David P. Blevens, -, -, - -, 91
Daniel Balis, -, -, -, 10, 333
Vincen Balis, 270, 730, 7000, 120 860
Thos. Sladarraway, -, -, -, 3, 140
Stephen Balis, 150, 270, 3000, 120, 984
Pleasant Smith agt., -, -, -, -, 47
Solomon Hobo agt., -, -, -, -, 29
Robbert Marcum agt., -, -, -, 3, 28
Boen E. Polly agt., -, -, -, 3, 110
Ely Eldridge agt., -, -, -, -, 27
Elin Campbell, 60 80, 1200, 4, 175
Elish. Hobs, -, -, -, -, 295
David C. Campbell, -, -, -, 105, 139
Jesse R. Edds, 100, 50, 1300, 20, 628
Harison Edds, 150, 201, 3510, 15, 410
Albert F. Hartgrove, -, -, -, 78, 210
E. B. Balis, 75, 125, 1575, 5, 384
Daniel Widner agt., -, -, -, 10, 121
John Martin agt., -, -, -, 2, 45
John Marcum, -, -, -, -, 30
John M. Hamblin, 60, 90, 1200, -, 85
Lucresey Polly agt.,-, -, -, -, 70
Hue C. Pauly, 30, 37, 450, 3, 71
Wesley Hampton, -, -, -, 40, 140
John Snodgrass, 40, 60, 1000, 5, 149
Elisabeth J. Edds, 100, 140, 3000, 30, 506
R. H. Edds, -, -, -, -, 194

R. M. Balis, 500, 1800, 40000, 200, 3000

Joshiah Ewing, 200, 200, 10000, 50, 1833

John R. Baldwin, -, -, -, 3, 107

Joseph Jones agt., -, -, -, -, 34

John B. Linsey, -, -, -, 43, 180

Elaner L. Fulkerson, 105, 165, 4350, 90, 1046

John F. Howard, 125, 875, 8000, 3, 110

Smith Crabtree, -, 150, 300, 3, 263

Christley Brant, 65, 77, 2300, 60, 641

Charles G. Deeds agt., -, -, -, 3, 150

Fedrick McDaniel, -, -, -, 5, 217

John Sims, 150, 335, 5820, 15, 370

Lewis Trass agt., -, -, -, 30, 255

Jessee Perry agt., -, -, -, -, 76

Ervin H. Russel, 60, 103 1200, 70, 430

F. A. Muncy, -, -, -, 280, 461

Jas Shuper, 45, 110, 1550, 30, 278

Jacob Grabille, 300, 1000, 10000, 5, 285

July A. Wells, 60, 85, 2150, 5, 150

John W. M. Ely, 10, 27, 150, 75, 544

Isam H. Scott agt., -, -, -, 63, 88

George Grabille, 45, 85, 1000, 23, 145

David Ely, 135, 113, 5500, 5, 180

John W. Bailey, 12, 25, 925, 15, 136

John W. Millen, 50, 100, 1200, 3, 32

John England, 20, 5, 250, 3, 185

Anderson England agt., -, -, -, -, 44

John England, 40, 35, 500, 3, 145

Josephus Grabille, 22, 11, 650, 20, 776

Wm. H. Linn agt., -, -, -, 3, 98

George P. McDanel, 140, 102, 3500, 15, 151

Andrew W. Jonson agt., -, -, -, 5, 35

I_y F. Deen agt.,, -, -, -, -, 20

Jas. Thompson agt., -, -, -, 1, 20

Henry England, -, -, -, -, 46

Herad Loftus, 5, 108, 500, -, 72

Wm. King, -, -, -, 3, 25

David Daugherty, agt., -, -, -, 5, 256

Andrew Grabill, -, -, -, 5, 40

A. J. McDanell, 45, 529, 1500, 50, 101

Eli Hubbard, 75, 1000, 10000, 75, 766

Wm. Thompson, 10, 50, 300, 3, 191

Jas. Burgan, 15, 35, 150, 3, 158

Jeramiah Bergan, 15, 35, 150, 3, 154

Peter Bayze, 3, 400, 500, 3, 108

John Carrol agt., -, -, -, 2, 78

Sylvester Bond, 25, 200, 800, 5, 189

John Hartsock, 20, 230, 600, 3, 196

John Penington, 55, 237, 1000, 5, 483

David Penington, 166, 1200, 1935, 150, 569

Wm. Hughes agt., -, -, -, -, 35

Budine Jonson agt., -, -, -, 2, 100

Ephrame Jonson, 40, 75, 1000, 3, 126

Jas. D. Jonson, -, -, -, -, 40

Corden Jonson, 35, 40, 400, 5, 137

Jas. M. Parson, 75, 325, 1200, 55, 559

Edward Garrett agt., -, -, -, 3, 89

John Garrett agt., -, -, -, 3, 32

Wm. M. Penington, 45, 175, 1500, 10, 475

Wm. Smith, -, -, -, 3, 228

Abraham Penington agt., 300, 2210, 20000, 125, 1425

John Penington, 90, 120, 3000, 10, 609

Abraham D. Zion, 65, 109, 1350, 130, 961

Wm. Cordel agt, -, -, -, -, 35

John Zion, 50, 117, 5000, 80, 570

Patterson Zion, 71, 209, 8000, 57, 704

Edward Penington agt., -, -, -, 5, 236

Thos. H. Smith agt., -, -, -, -, 35

Wm. T. Graham, -, -, -, 5, 445

John E. Burk, 175, 115, 12150, 115, 1034

Rachael Myres, 167, 115, 10150, 115, 1034

Robbert Witt agt., -, -, -, 3, 69

Henry Graham, 145, 196, 6820, 120, 551

Joseph Myres, 30, 28, 700, 90, 937

John D. S. Russell, 140, 765, 6350, 45, 671

Joseph Ely, 100, 125, 5000, 85, 492

John P. Graham, 65, 75, 3600, 25, 568

Alford K. Russel, 200, 420, 10000, 121, 1021

Wm. Zion, 60, 76, 3300, 80, 428

Jefferson Hedrick, -, -, -, 3, 59

Frankling Sprinkle, 40, 160, 800, 5, 90

Lorenzo D. Russel, 35, 58, 1500, 12, 547

John Hedrick, 100, 200, 2800, 35, 447

Wm. Stapleton, 25, 175, 100, 3, 185

Daniel Thompson, 13, 51, 300, 3, 109

Henry Thompson, 50, 185, 800, 40, 178

Stephen S. Crocket, 120, 560, 4000, 80, 700

Marida Hurt, 100, 158, 3000, 90, 236

Wm. Davis, 230, 200, 8000, 40, 620

Wm. Hamblin, 300, 500, 16000, 50, 755

Jas. Retherford agt., -, -, -, 5, 315

David Orr, -, -, -, 5, 315

Lewis Grady agt., -, -, -, 3, 107

James Penington, 100, 86, 3000, 60, 1022

John Woodard, -, -, -, 5, 226

Wm. W. Woodard, 90, 118, 3020, 30, 623

Jas. R. Shepard, 80, 190, 1200, 10, 554

_. A. Woodard, 200, 300, 6000, 150, 1206

Gasuaway Carrel, 100, 200, 3000, 5, 258

Henry Woodard, 100, 219, 4000, 90, 813

Thos. P. Heartsock agt., -, -, -, 3, -

Washington Clark, 20, 25, 2000, 30, 219

Elisabeth Harper, 50, 60, 1000, 5, 143

Andrew J. Williams agt., -, -, -, 3, 106

Henly C. Hall agt., 50, 50, 2000, 7, 367

Arthur Blankenship, 100, 400, 3000, 75, 470

Wm. Robberson, 80, 370, 3000, 47, 208

John K. Farris, -, -, -, 35, 248

Charles J. Taylor, 30, 75, 400, 3, 180

Geo. F. C. Burgan, 50, 30, 500, 4, 75

Elihu K. Howard, 75, 350, 3000, 10, 300

Henry H. Yearry, 100, 100, 4000, 100, 572

Henry M. Yearry, -, -, -, 3, 27

Wm. D. Yearry, -, -, -, -, 103

Elisabeth Yearry, 50, 100, 600, 5, 299

Wm. M. Taylor, 80, 120, 4500, 110, 546

Patterson Deeds agt., -, -, -, 5, 253

Hyram Hubbard, 55, 95, 2500,-, 257

John C. Webb agt., -, -, -, 35, 156

Silas H. Wolf agt., -, -, -, -, 23

Edward H. Snodgrass agt., -, -, -, -, 110

Joseph H. Bundy, 125, 189, 4000, 40, 445

Wm. E. N. Mark, 300, 325, 12000, 100, 1460

John M. P. Ely, 45, 80, 2000, 5, 406

Jas. D. Morgan, 100, 50, 5250, 85, 1075

George Scott, 200, 165, 8000, 70, 330

Elisabeth Looker agt., -, -, -, -, 69

Wm. McDonell, 200, 248, 5376, 80, 722

Geo. W. McDonell, -, -, -, 3, 196
Andrew M. C. Speak, 30, 65, 1000, 10, 275
Andrew Wright, 12, 25, 250, 3, 120
Wilson Lowe, 280, 101, 3000, 125, 515
Almon Oaks, 60, 75, 800, 8, 272
Edward Parson, -, -, -, 8, 173
Wm. C. Thompson, 40, 4, 1000, 116, 208
Henry J. Arnold agt., -, -, -, 5, 150
I. J. Arnold, -, -, -, 65, 240
Wm. Hughes, 75, 125, 4000, 10, 405
Wm. L. Alen, -, -, -, 5, 265
Ivy Lucas, 60, 90, 1600, 5, 538
David J. Yearry, 30, 10, 800, -, -
John Jayne, 100, 100, 4000, 80, 930
Wm. Patten agt., -, -, -, 5, 271
Jas. H. Clouson agt., -, -, -, 10, 710
John Marion, 50, 50, 2000, 10, 376
Michael Cecil, 200, 200, 8000, 200, 1135
Robbert W. Winn, 150, 150, 600, 150,740
George W. Wilson agt., -, -, -, -, 76
Jobe Hobbs, 80, 20, 2000, 75, 818
N. G. Baley, 140, 100, 6000, 32, 1083
Jas. M. Flaunary, 52, 98, 2500, 10, 530
Lcaner Flamney, 300, 950, 16000, 215, 2956
G. W. Young, 150, 100, 5000, 110, 1020
Aaron Hobbs, 130, 470, 6000, 25, 790
Jeramiah Scaggs, 175, 225, 6000, 40, 759
John Brooks agt., -, -, -, -, 495
Carr Baley Jr., 100, 260, 5000, 20, 358
George McKinon, 40, 10, 1500, 40, 517
Williamson Coomer, 40, 60, 2000, 5, 495

John Scaggs, 200, 500, 12000, 60, 225
Jonathan Ritchmon, 500, 1400, 15000, 100, 6318
John Rasor, 180, 620, 10000, 105, 1574
Jas. F. Jones, 350, 150, 18000, 104, 970
George W. H. Haburn, 250, 3250, 16000, 25, 735
Margret Stump, 300, 3000, 14000, 20, 333
Aaron Collier, 100, 200, 3500, 150, 654
Aaron J. Collier, 35, 465, 500, 10, 253
Samuel Dingas, -, -, -, -, 41
John Robberts, -, -, -, -, 85
Solomon Collier, -, -, -, -, 28
Jas. M Bevins, -, -, -, 5, 353
Wm. T. Leg, 40, 100, 1000, -, -
Dale C. Leg, 35, 100, 1000, 5, 70
Russel W. Leg, 50, 90, 500, 5, 157
Mary J. Aroy(Avoy), 50, 75, 2200, 25, 230
Jos. Barret, 25, 50, 500, 5, 295
Jas. Worley, 40, 100, 1700, 10, 262
John Milbern, 125, 675, 6000, 50, 590
A. C. Loyd, 125, 105, 4000, 135, 1108
Archable McElroy, 130, 95, 4000, 100, 592
John Bickley, -, -, -, 100, 887
Noe. Daugherty, 70, 90, 5000, 118, 630
Robbert Perry(Qerry), 4,75, 600, 5, 188
John Laramore, 200, 300, 12000, 200, 2010
Robberson Daugherty, 80, 80, 5000, 35, 205
John McElroy, 450, 550, 20000, 290, 1230
Cowan McElroy, 125, 125, 5000, 110, 723

Arthur J. B. McElroy, -, -, -, 75, 963
Andrew J. Wilson, 200, 301, 8000, 125, 1248
Jas. T. Loyd, 200, 200, 10000, 200, 1705
Henry Beasly, -, -, -, -, 90
Marcus Bell, 75, 65, 3000, 105, 295
George W. Stout, -, -, -, 70, 893
Wm. Arnal, -, -, -, 5, 266
Willy O. Andess, -, -, -, 80, 270
Henry Barker, 250, 4000, 16000, 10, 1225
Edward Smith, 25, 25, 600, 5, 200
John Gobble, 30, 120, 300, -, -
Fluner Musick agt., 25, 75, 200, -, -
Alexander Stout, 150, 250, 4000, 75, 870
Jas. McNeely, agt., -, -, -, 80, 409
Jas. M. Venerable, 60, 130, 2500, 90, 507
John M. Andish agt., -, -, -, 10, 227
John Jessee, 275, 165, 10000, 150, 1368
Ivy Warmer, 100, 140, 3000, 60, 748
George A. Crabtree, 100, 89, 4000, 100, 593
Wm .Carns, 900, 3015, 37779, 280, 2717
Wm. Carns agt., 80, 10, 4000, -, -
Ira G. Sprinkle, 200, 300, 5000, 215, 1935
George Hughes, -, -, -, -, 40
Samiel Trett, 200, 300, 8000,185, 770
Alexander Litton, 150, 150, 5000, 80, 1340
Wilson V. Litton, 120, 180, 3000, 20, 500
Godfrey D. Stout agt., -, -, -, -, 156
Wm. Parott, 85, 120, 5000, 185, 1336
Solomon Vanhuss, -, -, -, 10, 63
Thos. Moore, 115, 120, 3000, 120, 693
Benedick Yeary, 50, 80, 2000, 50, 619

John W. Debusk, 3, 38, 300, 5, 134
Pulser Debusk, 60, 140, 4000, 60, 300
C. N. Robinson, 300, 521, 15000, 125, 1246
Wm. J. Canter agt., -, -, -, -, 100
Henley Robinson, 40, 60, 1500, 3, 140
Jas. H. Greever (Gruver) agt., -, -, -, -, 217
Daniel Tuck agt., -, -, -, 5, 181
Alex. M. Robinson agt., -, -, -, -, 438
Rebecca H. Smith, 12, 38, 600, 3, 194
Stephen S. Parrott, 35, 80, 600, 3, 158
Charles Wilson, 100, 190, 3000, 5, 298
James M. Drake,-, -, -, 25, 157
James Lathem, 7, 115, 500, 5, 131
Henry D. Hamblen, 180, 469, 7000, 80, 868
Wm. Adams, 200, 180, 600, 75, 537
Robert S. Ewing, 100, 120, 5000, 75, 794
Jas. Thomas, -, -, -, 120, 580
Alexander Ewing, 270, 520, 12000, 160, 1130
John Fips agt., -, -, -, 1, 32
Francis Wood agt., -, -, -, 5, 38
Adaire Quinley agt., -, -, -, 3, 116
Wm. Clifton agt., -, -, -, -, 34
Goldman Davis, 35, 45, 1000, 5, 40
Austin Clifton, 100, 200, 4000, 85, 304
Peter Miller, 80, 120, 1400, 75, 350
David Wheeler agt., -, -, -, 5, 175
Mary Deeds, 40, 47, 1000, 5, 175
Jas. B. Burgan agt., -, -, -, -, 100
Alfred Covey, 50, 106, 1000, 5, 90
Henry Rock, 75, 117, 2500, 75, 260
John Worley, 18, 18, 345, 5, 92
Mary Wadle agt., -, -, -, -, 33
John Bolin agt., -, -, -, 115, 111
Andrew J. Smith agt., -, -, -, 6, 28

Edward Snodgrass, 100, 120, 1760, 80, 437

Thos. Givens, 65, 60, 1000, 5, 103

Wm. Albert, 50, 75, 1500, 6, 115

Helen(Helm) S. Wood agt., -, -, -, -, 101

Pearce Neapres agt., -, -, -, -, 249

Joseph Thomas agt., -, -, -, -, 25

Arthur C. Ely, 30, 70, 1000, 5, 88

Mathew Grissom, -, -, -, -, 63

Jas. Rock, -, -, -, -, 30

Henry Welch, 11, 22, 500, 3, 65

George R. Ely, 70, 74, 1500, 43, 110

W. S. Hall agt., -, -, -, 5, 58

Elisha Smith, 45, 55, 1000, 10, 403

Jas. K. Burnett, 10, 40, 500, 5, 152

Nathaniel Scott, -, -, -, 3, 144

Isaac Phips, -, -, -, 5, -

George F. Ely, 75, 125, 2200, 125, 511

Charles Milbourn, 46, 32, 500, 2, 124

Jacob H. Stemp, 45, 70, 1600, 75, 465

Jacob Shupe, -, -, -, 3, 62

Charley Burten, 30, 19, 700, 5, 429

Charly C. Blankenship agt., -, -, -, 50, 187

Jas. Wilburn, 135, 165, 2000, 110, 371

Louis M. Mise agt., -, -, -, 5, 120

James Ramy, -, -, -, 5, 474

Wm. J. Hutton, 10, 33, -, 5, 40

Lenard Hutton, 15, 28, 500, 6, 125

John W. Howard agt., -, -, -, 3, 133

Gabrel Jackson agt., -, -, -, 5, 164

Ezekiel Callaham, 60, 115, 1500, 5, 82

Nicholas Speak, 15, 15, 300, 5, 50

Wm. Thompson agt., -, -, -, 5, 125

Joseph Thomas agt., -, -, -, -, 40

Jacob Spangler, 100, 100, 2000, 3, 89

Nicholas Heneger, 140, 280, 4000, 60, 1297

David Lester agt., -, -, -, 60, 92

John Snodgrass, 150, 508, 4000, 70, 190

Wm. Elliot, 40, 106, 1000, 75, 419

Wm. Snodgrass agt., -, -, -, 5, 154

Joel Leady, 150, 240, 3000, 175, 914

Caleb Quinly agt., -, -, -, 3, 66

Rufus Leady agt., -, -, -, 5, 71

Henry Thomas, 80, 116, 700, 5, 82

Francis P. Thompson agt., -, -, -, -, 61

David B. Redwine agt., -, -, -, 5, 170

Jacob Thomas agt., -, -, -, 5, 44

John M. Thomas, -, -, -, -, 6

Eli Flunor, -, -, -, -, 175

Drury Flunor, 100, 550, 5000, 35, 271

Berry H. Flunor, 40, 132, 2500, 80, 350

Simon P. Andis, 70, 65, 1000, 4, 144

Samuel Roop, 80, 220, 1500, 200, 380

Christopher Woodard, 110, 120, 2000, 5, 245

Wm. J. Woodard, 50, 250, 1000, 5, 246

Franklin Garrott, -, 82, 500, 5, 166

Thos. Neaper, 100, 175, 1200, 80, 390

Wm. Garrett, 50, 100, 500, 5, 127

Andrew Gott, agt.,-, -, -, 25, 66

John Moore agt., -, -, -, -, 30

Jacob Vickars agt., -, -, -, 5, 148

Elias Thompson, 100, 200, 1000, 5, 162

Baxter Simpkins agt., -, -, -, -, 90

John Crabtree, 100, 200, 1800, -, 15

Henry Johnson, 40, 45, 1000, 3, 134

Wm. Johnson agt., -, -, -, -, 73

Henly C. Collins, -, -, -, 5, 100

Nancy Garrett, 17, 2, 150, 10, 75

Wm. A. Jones, 202, 308, 7000, 40, 475

John Green,-, -, -, 5, 265

John H. Alen, 100, 128, 3480, 90, 576

Andrew Bumgardner, 280, 720, 10000, 100, 576

Henry Bumgardner agt., -, -, -, 5, 180

John A. Bumgardner agt., -, -, -, 5, 192

John W. Moncy, 58, 95, 2000, 15, 310

Jas. Garison agt., -, -, -, 5, 140

John Parsons, 75, 80, 2000, 10, 413

David Cole agt., -, -, -, 5, 83

Elizabeth Bolin, 45, 55, 600, 5, 141

David Smith agt., -, -, -, 10, 277

John W. Gilley, 100, 200, 6000, 60, 825

John Smith, 200, 240, 5000, 75, 483

Isaac Russel, 50, 80, 1200, 8, 233

Rebecca Parsons, 80, 76, 1200, 10, 598

John D. Gage, 130, 80, 3000, 70, 616

Wm. W. France, 50, 61, 1500, 165, 486

Alexander Orr, 100, 110, 2500, 50, 582

Wm. R. Orr, 20, 45, 600, 8, 129

Leroy Venable agt.,-, -, -, 5, 70

Robert W. Moncy, 75, 55, 2000, 8, 407

Jas. Shelburn, 200, 200, 4800, 75, 515

Henry Milburn, 80, 100, 6000, 165, 469

George B. Milburn, 100, 380, 4000, 170, 496

Josiah Winn, 100, 310, 5000, 50, 320

Martin Collier, 200, 400, 4000, 100, 474

Elijah Turner, 100, 40, 1000, 100, 345

Jas. Russel, 4, 80, 428, 5, 195

Benjamin Terry agt., -, -, -, 5, 118

Zion Pinington, 100, 214, 4000, 50, 375

Jas. Folin agt., -, -, -, 5, 93

Joseph Barnes, 150, 300, 4000, 80, 648

Thos. Dalton, -, -, -, 5, 120

Jos. Green, 60, 400, 1000, 10, 234

Wm. A. Stapleton agt., -, -, -, 7, 42

John Graham, 175, 275, 6000, 50, 636

George Wooliver agt., -, -, -, 8, 218

Joseph Wooliver, 40, 84, 1000, 8, 130

Alexander R. Russel agt., -, -, -, -, 140

Robert Travis, 225, 564, 3000, 30, 510

Jas. R. Neaper agt., -, -, -, 5, 20

Henry Hoover, 12, 48, 700, 5, 127

David C. Clinger, 30, 80, 700, 10, 435

Joseph Miller, 30, 45, 1500, 10, 187

John Miller agt.,-, -, -, 5, 80

Wm. Pottiller, 130, 246, 2800, 15, 235

Andrew Johnson agt., -, -, -, 80, 565

John Herrell, 200, 100, 4000, 30, 587

John Phillips, -, -, -, 5, 80

Benjamin L. Poteet agt., -, -, -, -, 101

Jos. C. Yearry agt., -, -, -, -, 146

Samuel Oxford agt., -, -, -, 5, 45

Wm. G. Standifer, -, -, -, 5, 150

Jas. Cheek, 165, 120, 3000, 60, 839

John Thompson, 75, 64, 2000, 30, 909

Lorinda Jonas, 107, 143, 3000, 90, 1010

Wm F. Whitt, 65, 95, 2000, 20, 678

Jas. G. Whitt, 75, 125, 2000, -, -

G. G. Henderson, 70, 130, 2000, 15, 531

Mary Herd, 100, 170, 3200, 20 500

Cornelias Fetts, 350, 650, 20000, 400, 2349

George Rudy, -, -, -, 5, 137

Isaac Hues, 60, 148, 1500, 5, 633

Harison Warren, 40, 160, 2000, 5, 290

Joseph Stanaford, -, -, -, 5, 66

Jas. H. Davis, 240, 383, 7000, 100, 855

Joseph P. Bishop agt., 50, 186, 1100, -, 437

Joseph C. Adams, 40, 260, 1200, 5, 190

Samuel A. Burchet, 100, 250, 3000, 60, 458

Benjamin A. Burchett, 34, 117, 1600, 5, 125

Jas. Bishop, 180, 220, 7000, 150, 967

Eldridge F. Devault, 50, 100, 3000, 15, 600

Wm. Young, -, -, -, -, 15

Wm. Brown, -, -, -, -, 40

Francis Muncy agt., -, -, -, 7, 211

Wilaby Muncy, 150, 150, 6000, 60, 812

Jas. Jayne, 57, 57, 3000, 100, 627

Jefferson Naff, 80, 20, 2500, 25, 697

Samuel B. Muncy, -, -, -, 20, 699

Francis Muncy, 70, 35, 2000, -, 235

George W. Rolins, -, -, -, -, 33

Henry Towel, 100, 100, 1500, 60, 624

Wm. Jayne, 65, 115, 3500, 90, 594

Wm. Muncy, 90, 210, 4000, 50, 521

Andrew M. Fetts, 150, 350, 3000, 100, 676

Wm. F. Naff, 40, 40, 800, 10, 157

Claiborn T. Anderson, 150, 500, 12000, 46, 638

Jas. G. Browning, 100, 180, 4000, 25, 437

Wm. Blackamon, 40, 60, 600, 10, 115

Burdine Waggle, 150, 484, 7000, 200, 1210

John Hamblen, 100, 200, 6000, 50, 675

Jourdan Louis, -, -, -, 5, 71

Stephen Kimberlain, 25, 50, 600, 3, 121

Mary Kimberlain, 50, 75, 1000, 50, 155

Eli B. Crockett, 15, 2, 200, 5, 40

Joel Taylor, 150, 152, 2500, 15, 521

Squire Blackamon, 50, 150, 2000, 60, 260

Wm. Noe, 40, 38, 600, 10, 365

Louis Smith, 45, 60, 1000, 50, 160

Jonathan Smith, -, -, -, 5, 125

Nymrod Noe, 50, 76, 1000, 10, 195

Elizabeth S. Beaty, -, -, -, 10, 366

Joseph Anderson, -, -, -, 5, 52

Thos. M. Hamblin, 110, 110, 4000, 40, 870

John P. Orr, 30, 70, 1000, 10, 579

Samuel Edsall agt., 55, 145, 2000, 5, 352

John Sprinkle, 50, 80, 2000, 70, 500

George Poff, -, -, -, -, 90

Jefferson P. Cox, -, -, -, 5, 40

Francis Mills, 40, 124, 1000, 5, 235

Thos. S. Ely, 75, 75, 2000, 20, 504

Alexander Ely, 80, 70, 3000, 85, 684

John M. Beaty, 500, 1000, 24000, 500, 1610

Wm. Moncy, -, -, -, 60, 325

Richard B. Beaty, -, -, -, -, 200

John M. Beaty agt., 160, 250, 6000, -, -

Thompson Mink, 75, 25, 1200, 8, 427

E. P. Spencer, 200, 133, 10000, 150, 1452

Charles C. Barker, -, -, -, 10, -

Robbert P. Spencer, 100, 71, 5000, 130, 1181

Elijah Bishop, 150, 170, 5000, 150, 798

Jas. F. Moncy, 150, 250, 5500, 60, 1127

Wm. B. Spencer, 100, 50, 3500, 10, 305

L. K. Tyler, 200, 100, 3500, 25, 592

Lipscomb Parrot, 400, 600, 7000, 200, 2713

Wm. Mink, -, -, -, -, 29

John Snodgrass, 70, 130, 2500, 100, 534

Jas. N. Thompson, 50, 100, 2000, 35, 420

John Thompson, 75, 225, 3000, 90, 599

Caleb Thompson, 50, 100, 2000, 100, 255

John Goliham, 60, 110, 2500, 5, 596

John Fletcher, -, -, -, 5, 85

John G. Fletcher, 200, 200, 3000, 75, 912

Jas. N. Thompson, 80, 120, 3000, -, 260

Jacob A. Runbeskir, 50, 75, 1000, -, 130

John W. Lambert, 175, 125, 2000, 10, 246

Richard Noel, 60, 40, 600, 10, 144

Wm. Marshal, 100, 217, 3000, 50, 515

Salina Sumers, 50, 87, 500, -, -

James Edwards, 90, 10, 1500, 10, 407

Jesse Stapleton, -, -, -, 5, 173

Wm. Graham, 75, 100, 1500, 35, 500

D. S. Dickinson, 450, 950, 13400, 200, 3140

George Price, 40, 60, 600, 8, 70

Wm. Howe, 30, 170, 600, 8, 237

John D. Sims, 375, 725, 15000, 300, 1761

Jas. Retherford, -, -, -, -, 130

John G. Barnes, 100, 146, 4500, 125, 1163

Russel B. Davault, 130, 170, 7000, 175, 1025

M. S. Jaynes, 175, 225, 6000, 75, 668

Wm. Colins, -, -, -, - 137

Henry Daugherty, 143, 257, 6000, 110, 621

Cheral Warner, -, -, -, 325, 1144

Jas. W. Jaynes, 100, 200, 5000, 75, 1004

C. S. Jaynes, 100, 150, 4000, 70, 833

Briar Bowls, -, -, -, -, 185

Martin Sims, 135, 225, 6000, 85, 628

Hannah Davault, 75, 127, 2500, 20, 280

Robert Clark, 150, 125, 8000, 100, 818

Jacob Davault, 80, 100, 4000, 110, 1307

Sarah Boles, -, -, -, 5, 296

Nancy Worly, 90, 75, 200, 5, 127

G. W. M. Furgerson, 100, 718, 3000, 115, 839

John M. Whitehead, 50, 155, 3000, -, 175

Nelson Marion, -, -, -, 5, 123

Wm. Vandeventer, -, -, -, 135, 1016

Burrell Burchet, 75, 225, 4000, 35, 295

John Burchett, -, -, -, 10, 192

James J. Money, 48, 87, 2000, 5, 389

John Jaynes, -, -, -, 15, 301

Wm. Jaynes, 10, 230, 4000, 70, 415

Alexander Janus, -, -, -, 3, 106

Sarah Ewing, 100, 75, 5000, 50, 374

Joseph Ewing, 175, 185, 8000, 10, 521

Wm. Marion, 30, 70, 2000, 5, 55

A. R. Burgmer(Surgmer), 50, 150, 1200, 30,491

W. G. Chandler, 100, 680, 2000, 10, 364

Daniel B. Newbery, 125, 575, 4000, 110, 309

Richard Maxy, 25, 25, 300, 5, 120

Abraham J. Levingo tenant, 200, 560, 4000, 90, 847

John A. Moore, 150, 507, 8000, 120, 714

Charles M. Hill, 80, 220, 1500, 10, 468

Wm. Flanian, -, -, -, 5, 193

Allen Chandler, 50, 50, 1000, 5, 189

Sanders Levasy, -, -, -, 5, 107

Andy Levasy, 150, 215, 6000, 40, 437

Robert Levesy, -, -, -, 5, 117

Martin Anderson, 60, 140, 2500, 35, 356

Aaron Anderson, 50, 150, 2500, 115, 403

Joseph Johnson, 40, 600, 600, 5, 140
George Waller, 100, 100, 1000, 10, 411
Henry Fisher, 50, 200, 1000, 5, 115
Amos Roller, 30, 70, 1000, 5, 145
Dickenson Goble, -, -, -, 5, 234
Jas. Robinson, -, -, -, 5, 385
Wm. Robinett, -, -, -, 5, 170
Louis Stapleton, 40, 60, 800, 5, 115
Isaac H. Robinett, 20, 800, 5000, 60, 755
Jefferson Chandler, 45, 35, 600, 37, 256
Enoch Ausburn, 50, 197, 700, 5, 130
David Herd, 30, 30, 200, 3, 30
Hiram Moore, 50, 250, 1000, 10, 75
Sterling S. Maness, 200, 300, 2500, 10, 780
Joseph Willis, 50, 250, 1500, 80, 138
Samuel Moore, 50, 40, 500, 3, 125
Dire Lawson, 100, 200, 2000, 10, 350
Arthur Rogers, 20, 180, 400, 3, 83
Jas. M. Young, 60, 290, 2000, 10, 263
Ira Lawson, 50, 179, 1500, 5, 525
Joseph Bledsaw, 30, 45, 300, 3, 52
John Stapleton, 40, 100, 500, 5, 103
Normessa Bledsaw, 75, 75, 1500, 5, 140
Jesse Roberds, 15, 60, 300, -, 115
Elias Bledsaw,-, -, -, 3, 172
Nancy Wolfenbarger, -, -, -, -, 147
Jacob Wolfenbarger, 150, 170, 4000, 50, 491
John Meek, 30, 85, 1000, 10, 403
Joseph A. Hardy, 150, 100, 4000, 90, 410
George Bales, 18, 55, 600, 5, 306
Archable Bales, 36, 172, 1450, 40, 484
Joseph Haunshell, 21, 274, 3000, 5, 270
Jas. Bartley, 243, 393, 4400, 60, 640
Jas. A. Speak, 75, 105, 1080, 10, 240

Samuel Speak, 100, 81, 1080, 10, 120
Wm. H. Speak, -, -, -, 5, 202
Wm. H. Rosenbum, 75, 200, 2500, 65, 420
Joseph A. Speak, -, -, -, 5, 147
Fielden Lial, 82, 118, 2100, 20, 446
Marqis Yearry, -, -, -, 165, 158
Michael Grubb, 16, 134, 150, -, -
Andrew Milburn, 300, 600, 14000, 300, 2252
Leaner Winn, 200, 400, 10000, 45, 608
Jas. Hobbs agt.,-, -, -, -, 22
David A. Martin, 120, 380, 1000, 55, 464
George W. McColly, -, -, -, -, 207
John Miller, 300, 250, 10000, 230, 1034
Wm. Woods agt., -, -, -, -, 30
Wm. Hoston, 140, 160, 7000, 140, 1013
Joseph Napper agt., -, -, -, 7, 208
George Napper agt., -, -, -, 6, 276
Peter Fips agt., -, -, -, 3, 75
Henry Watts, 90, 110, 3000, 78, 544
Wm. Kinsor agt., -, -, -, 5, 123
Wm. Shelton agt., -, -, -, 5, 210
Robert Shelton, 40, 75, 1200, 85, 475
Wm. Kinsor, 130, 470, 5000, 40, 385
John Kinsor agt.,-, -, -, 3, 163
Jacob Kinsor agt., -, -, -, -, 28
Joseph R. Kinsor agt., -, -, -, -, 39
Mary Caliham, 80, 420, 2000, 5, 345
Felix B. Mink agt., -, -, -, 3, 164
Jas. Huten agt., -, -, -, -, 12
Jas. Crace, 55, 70, 1200, 5, 259
D. B. Tacket, -, -, -, -, 42
Jas. M. Jackson, -, -, -, -, 101
Wm. Sims, 100, 180, 5000, 20, 794
Robbert Sims, 90, 110, 5000, 50, 365
John A. Long, 100, 300, 6000, 100, 343
Jas. S. Long, 110, 260, 5000, 15, 656

Hary M. Hall, 55, 180, 2000, 105, 429

Alen Milham, 20, 650, 8000, 80, 377

Wm. H. Long, -, -, -, 5, 43

David Poteet, 200, 229, 6000, 70, 881

Martin Warrick, 50, 700, 3000, 45, 108

Wayman Warrick, 65, 300, 2500, 5, 114

Joseph R. Thomas, 40, 150, 1500, 3, 172

Autry King agt., -, -, -, -, 116

Stephen Medlock, 8, 42, 300, 3, 36

Joel Warrick agt., -, -, -, -, 13

Jas. Woodard, 13, 121, 1500, 6, 15

Jas. Thomas, 50, 50, 1000, 7, 382

John C. Debush, 85, 97, 1275, 10, 123

Jobe McDaniel agt., -, -, -, -, 200

Jas. Snodgrass agt., -, -, -, -, 293

Alexander Snodgrass, 40, 82, 700, -, -

Abraham Whitman, 75, 85, 1280, 125, 419

Jas. Stone, 40, 116, 1500, 5, 155

Eli Davis, 50, 100, 2000, 5, 272

Thos. Thompson, 45, 315, 2500, 60, 302

Jesse James, 11, 89, 1000, 30, 50

Thos. Martin, -, -, -, -, 150

D. C. Martin, 50, 100, 3000, 20, 425

Charles Heneger, -, -, -, -, 52

George R. Fletcher, 85, 35, 2500, 135, 600

Elias Debush, 40, 90, 1200, 5, 195

Absalum Robberson, 110, 110, 3000, 10, 648

Daniel Vanhuss, 20, 38, 800, 5, 125

Michael Vanhuss, 50, 65, 2000, 70, 64

David Wolf agt., 60, 50, 1000, 70, 398

Tilman Ball, 25, 75, 1000, 5, 244

Alexander Wolf agt., -, -, -, -, 61

Ephram M. Wolf agt., -, -, -, -, 87

Jonathan Hanes, 30, 30, 500, 5, 100

John Botner agt., -, -, -, -, 100

Hiram H. Bashier, -, -, -, 15, 528

Wm. Sage, 200, 400, 6000, 300, 1192

Jacob Simkins, -, -, -, 50, 244

Jas. Hamelton, -, -, -, 5, 138

Lany Flitcher, 150, 150, 5000, 100, 500

David Banner, 200, 300, 6000, 100, 995

Jacob Rasnick, 100, 400, 4000, 63, 402

John C. Rasnick, 30, 120, 1000, 20, 159

Peter Lambert, 30, 70, 500, 10, 378

John Mebllin, 120, 205, 5000, 95, 529

Frances Coblen, 110, 151, 5000, 275, 415

Sampson Sage, 110, 50, 2000, 100, 721

Andrew J. Litton agt., 30, 70, 2000, 110, 345

Abraham S. Young agt., -, -, -, 5, 290

Jas. B. Ward agt., -, -, -, -, 80

Thos. D. Duff, 140, 260, 8000, 200, 915

Claibourn Young, 200, 300, 7000, 200, 1098

Delila Young, 250, 700, 6000, 100, 1000

Joseph Petters agt., -, -, -, 40, 290

Martin Drake, 150, 150, 3000, 150, 874

David Vickers agt., 11, 89, 600, 2, 15

Thos. Vickers, -, -, -, -, 25

Samuel Word, 450, 600, 7000, 160, 3083

Hiram Goble agt., -, -, -, 70, 338

Granderson Bush, -, -, -, 5, 199

Samuel H. Duff, 110, 3040, 2500, 105, 1344

Henry D. Duff, 40, 60, 1500, -, -

Wm. Robins, -, -, -, 4, 70

Claibourn Stanley, -, -, -, 5, 160
John Hobbs, -, -, -, 7, 24
Margrett Duff, 100, 100, 6000, 70, 755
Joseph Duff, 150, 159, 4000, 60, 853
James M. Adams, -, -, -, 3, 239
Simrard Adams,-, -, -, 120, 877
David Young, 100, 200, 4500, 70, 588
Ezra Countress, -, -, -, 100, 305
Elizabeth Duff, 300, 800, 16000, 175, 1316
Brant Fannon, 100, 138, 3000, 70, 378
A. C. McNile, 100, 100, 5000, 100, 816
George D. Gibson, 180, 155, 10000, 200, 438
Johnathan G. Oxford,-, -, -, -, 22
Francis Sleggle, -, -, -, -, 400
John Ausburn, 75, 225, 3000, 25, 265
George Ausburn, 125, 875, 3000, 78, 219
Golaman Ausburn, 3/20, 130, 500, 5, 181
Enoch Ausburn, 50, 250 800, 5, 300
Thos. Moneyhon, 28, 32, 600, 5, 374
Alfred Hall, 10, 50, 300, 3, 126
Joseph Kinsor, -, -, -, -, 120
Andrew J. Roller, 25, 75, 700, 30, 353
Phillip Roller, 200, 600, 8000, 40, 348
Harvey Lambert, -, -, -, 3, 106
Wm. T. Cheek, -, -, -, -, 182
Daniel Lockheart, 45, 55, 300, 5, 40
Claibourn Duff, -, 6, 500, 110, 414
Jas. Marcum, 9, 16, 400, 5, 309
Samuel Price, 20 165, 500, 3, 80
Marion Price, -, -, -, -, 270
Thos. R. Burk, 100, 100, 3000, 5, 73
Andrew Lockheart, 150, 150, 5000, 85, 1048
Jesse Ball, 25, 25, 100, 5, 165
Thos. G. Burk, 75, 225, 1800, 5, 411

John H. Gword, 35, 105, 1200, 5, 149
Joshua McClancy, 50, 150, 1200, 10, 138
Abraham B. Cason,-, -, -, 10, 116
Edmon Livsay, -, -, -, 8, 245
Jas. Chandler, 140, 170, 5000, 40, 624
Wm. Gilbert, 30, 70, 700, 5, 224
Riel Lawson, 300, 600, 11000, 100, 927
Joel M. Furgurson, 125, 295, 1200, 50, 532
Dicy Furgurson, 75, 300, 2500, 40, 339
Thos. Lermer, 240, 360, 10000, 140, 1804
Alexander Hamelton, 40, 167, 1000, 5, 299
A. G. Hickman, 125, 175, 3500, 50, 787
Squire Edwards, -, -, -, 3, 15
Nelson Preston, 100, 200, 3000, 30, 460
Benjamin Hill, 150, 150, 5000, 95, 1188
Edimelolas Glass, 15, 45, 700, 5, 273
Wm. G. Glass, 25, 40, 700, 5, 222
Steaven McLearson, 100, 100, 4000, 85, 622
John Roach, 60, 320, 3000, 125, 840
John Riddle, -, -, -, 100, 575
Jahiel Collier, 70, 20, 2000, 10, 756
Isaac Collier, 70, 20, 2000, 15, 650
Wm N. G. Barron, 300, 300, 1600, 200, 3428
Carelton Waid, 150, 150, 6000, 100, 487
James H. Reason, 100, 300, 6000, 40, 530
Peter Reason, 125, 175, 4000, 75, 1070
John W. Slemp, 150, 700, 8000, 90 1725
Isaac Reece,-, -, -, -, 30
M. D. L. Wilson, -, -, -, 5, 50

110

Alpha Slemp, -, -, -, -, 205
Wm. Ward, 100, 400, 4000, 80, 630
Wm. Wilson, 200, 250, 8000, 165, 1690
Joseph A. Jones, -, -, -, 5, 77
Jacob Olinger, 70, 94, 1200, 50, 452
David Olinger, 40, 160, 200, 10, 460
Hue L. Slemp, 160, 350, 7000, 20, 617
John C. Olinger, -, -, -, 80, 1173
John C. Olinger Sr., 200, 2500, 20000, 82, 1625
Carr Bailey, 75, 2000, 3000, 20, 562
Richard Robinson, 20, 10, 300, -, -
Joseph K. Olinger, 100, 300, 2000, 20, 853
David Cox, 60, 90, 2000, 5, 230
Wm. DeMinter, 80, 300, 2000, 165, 845
Wm. Ewing, -, -, -, -, 1197
Dickson G. Litton, 1100, 969, 43500, 100, 3565
Henry C. Slemp, 100, 150, 2000, 20, 1323
Jobe K. Yearry, 100, 50, 2000, 105, 112
John W. Scott, 150, 266, 3000, 25, 360
Elisha Cox, 35, 115, 800, 25, 360
Jobe C. Cox, -, -, -, 5, 61
Hiram Ely, 200, 800, 8000, 85, 650
Nimrod C. Ely, 150, 350, 10000, 80, 769
France Zion, 200, 400, 10000, 170, 970
Joseph Ely, 50, 100, 500, 10, 175
Mouman M. Penington, 50, 50, 2000, 10, 376
Andrew J. Ely, 200, 110, 10000, 75, 866
Tobias Penington, 100, 500, 8000, 35, 624
Joseph N. Ely, 90, 310, 2000, 30, 880
Wm. Stuart, 60, 40, 1000, 10, 484
Jos. H. Stuart, 75, 85, 1500, 25, 601

David Spangler, 75, 65, 1200, 100, 495
Vincen. Hobbs, 100, 340, 2100, 5, 406
John Lucas, 40, 120, 700, -, 300
Larkin Herndon, 75, 62, 2000, 15, 448
John Lucas, 130, 260, 4000, 120, 820
Wm. Rivers, -, -, -, 3, 20
Nancy Schritch, 100, 50, 1500, 80, 362
Caleb W. Hobbs, 60, 40, 1500, 20, 395
Mary Hobbs, 150, 250, 4000, 20, 570
Doctor F. Wells, 120, 50, 4000, 35, 781
John Penington, 150, 200, 10000, 220, 1421
Samuel C. Stallone, 50, 50, 2000, -, 425
Silas Flanery, 100, 120, 7000, 148, 1030
James Flanery, 140, 120, 8000, 30, 1079
Thos. Collinsworth, 80, 115, 6000, 80, 872
Shelby Hobbs, 200, 125, 10000, 20, 524
Levi Penington, 200, 200, 8000, 100, 1365
Timothy Dalton, 100, 175, 1800, 5, 466
Charley Barker, 75, 275, 1500, 5, 543
Wm. Bailey, 75, 160, 1200, 5, 268
John Morris, 60, 400, 1500, -, 526
James M. Clark, 30, 200, 800, 3, 351
Thos. Clark, 25, 35, 300, -, 125
John V. Reese, 40, 148, 500, 5, 87
Hiram W. Russel, 90, 390, 2000, 130, 449
Thos. M. Leg, 50, 450, 1800, 5, 281
Thos. Morris, 50, 310, 1000, 3, 435
Jacob Morris, 45, 193, 1000, 5, 253

Jonas M. Whisman, 20, 55, 400, -, 103

Albin Whisman, -, 150, 300, -, -

Archable Witt, 30, 100, 500, -, -

Edmon Witt, 50, 350, 1200, -, -

John Payne, 100, 1200, 1500, -, 327

Thos. Falcum, -, 80, 100, -, -

David Garrison, 120, 680, 3000, 10, 365

Anasias Robbles, 60, 440, 2500, 10, 375

Wm. Neaper, 5, 14, 300, -, -

John S. Bailey, 60, 600, 3000, 10, 335

Larkin H. Robins, 30, 270, 1000, 5, 92

Martin Witt, 30, 170, 500, 3, 86

Green B. Pennington, 50, 250, 1500, 10, 250

Wm. Short, 8, 62, 300, -, -

Obadiah Hoover, 30, 70, 400, -, -

Zion Robbins, 100, 200, 2000, 10, 430

Thos. Parsons, 50, 250, 1500, 10, 330

Thos. Robbins, 25, 475, 1000, -, -

Wm. Parsons, 20, 130, 250, -, -

Charles Pennington 80, 86, 1500, 5, 372

Drury F. Carter, 100, 400, 1500, 10, 383

Wm. Parsons, 50, 550, 1500, 5, 240

Antony Witt, -, -, -, 10, 132

Wm. A. Parson,-, -, -, 5, 200

James Johnson, 20, 35, 600, 5, 310

Henry M. Harber, 19, 189, 500, 5, 153

Charley Harber, 12, 207, 500, -, -

Daniel Woodard, 30, 100, 250, -, -

Arthur Harber, 60, 40, 800, 10, 464

George Neaper, 30, 70, 500, -, -

Stephen Neaper, 60, 44, 500, -, -

Richard Kirk, 50, 50, 500, -, -

John Spurlock, 25, 175, 500, -, -

John Birgan, 60, 240, 1500, 10, 314

Robert Stapleton, 10, 290, 300, 5, -

A. J. Smith, -, -, -, 3, 169

Joseph Stapleton, 40, 110, 500, 3, 155

Preston Kirk, 30, 20, 200, 3, 145

Henry C. Shop, 300, 1400, 17000, 90, 549

Frances Bishop, 200, 440, 8000, 100, 878

Moses Skelton, -, -, -, -, 200

Aley Winn, 150, 330, 5000, 70, 593

Timothy Sisk, 1, -, 1500, -, 100

Edwin Dickinson, 100, 200, 300, -, 300

Harrison Bailor, 75, 125, 3000, 175, 585

B. H. M. Cole, 1, -, 500, -, -

Samuel L. Saul, 6, -, 1250, -, 210

John D. Wood, 1, -, 1200, -, -

Marian D. Richmond, ½, -, 1200,-, -

James Miles, ½, 25, 5000, -, -

Elijah Hill, 65, 25, 5000, 90, 515

John M. Couk, ¾, -, 800, -, 150

Elbert S. Martin, 1, -, 1000, -, -

Wm. Davidson, 2, -, 1500, -, -

Andrew M. Hankins, 90, 28, 2500, 85, 150

M. B. D. Lane, 25, 201, 2200, 14, 101

Boyd Dickinson, 100, 200, 5000, 260, 850

H. S. Miles, 4, -, 2300, -, -

Henry J. Morgan, 110, 61, 4000, -, -

Thos. Whitton, 200, 100, 800, 100, 810

James B. Bolin, -, -, -, 5, 262

Joseph Asberry, -, -, -, 5, 85

L. S. Fulkerson, 100, 200, 6000, 75, 825

Thos. S. Gibson, 200, 150, 1200, 200, 2020

Loudoun County, Virginia
1860 Agricultural Census

The Agricultural Census for Virginia 1860 was microfilmed by the University of North Carolina Library under a grant from the National Science Foundation from original records at the Virginia Department of Archives and History in 1963.

There are forty-eight columns of information on each individual. Only the head of household is addressed. I have chosen to use only six columns of information because I feel that this information best illustrates the wealth of individuals. The columns are:

1. Name
2. Improved Acres of Land
3. Unimproved Acres of Land
4. Cash Value of Farm
5. Value of Farming Implements and Machinery
13. Value of Livestock

Jas. H. Wilson, 50, 94, 1440, 50, 440
Wm. W. Tallmer, 80, 50, 1860, 100, 300
Wm. P. Hutchison, 200, 115, 315, 40, 652
Thomas M. Wrem, -, -, -, -, 160
John T. Presgraves, 225, 40, 2000, 125, 700
Francis Lambert, -, -, -, -, 180
Legohev of Jno. Moxley Dcd., 90, 248, 3380, 50, 570
Lovel H. Middleton, 120, 50, 3000, 125, 350
Rice Croson, -, -, -, -, 160
Robison Croson, 80, 50, 650, -, 50
Liddia Whaley, 5, 20, 500, -, 75
Elizabeth Rose, 10, 15, 500, -, 85
Nelson B. Wilson, 50, 200, 2450, 40, 245
Jas. M. Holtzclaw, 420, 135, 5000, 150, 1590
Jas. W. Mankins, 300, 400, 2800, 100, 735
Jas. W. Mankins, 300, 200, 2000, -, -
Jas. Smith, 180, 60, 2400, 100, 350

E. F. Hutchison, 200, 100, 3000, 100, 370
John W. C. Beavers, 125, 30, 1240, 50, 66
George W. Fling, 200, -, 2400, 100, 568
John I. Coleman, 300, 100, 6000, 450, 1806
Jonathan C. Coleman, 225, 75, 6000, 200, 1045
John H. Tippett, 31, 7, 760, 100, 412
John R. Burr, 138, -, 2750, 200, 360
John Aukers, 200, 150, 4200, 100, 500
George W. Hummer, 74, 16, 1080, 75, 600
Albert T. Blincoe, 25, 20, 445, 15, 200
Eleaner Caylor, 50, 7, 500, -, 146
Lucinda Blincoe, 37, 28, 650, -, 40
Daniel W. Blundell, 35, 76, 1332, 125, 410
Sarah Offett, 149, 30, 1790, 10, 300
David Smith, 100, 75, 2625, 100, 340
Orville Hitchins, 30, 5, 600, 100, 147

Joseph Smith, 40, 135, 2625, 100, 315

Leah H. Tippett, 300, 33, 3330, 20, 176

Jonathan E. Edwards, 200, 100, 3000, 100, 850

Richd. H. Summers, 350, 70, 3160, 200, 1110

David McCullock, 100, 50, 1800, 150, 200

John W. Ellmore, 5, 5, 150, -, 76

John W. Moran, 250, 75, 3750, 25, 950

Garrett B. Walker, 100, 50, 1500, 20, 150

Benj. B. Heard, 185, 100, 4000, 70, 400

Mary A. Mills, 175, 50, 3375, 30, 250

Susan Ambler, 300, 30, 1650, 50, 194

Lewis F. Mankins, 325, 120, 4480, 150, 1293

Margarett French, 140, 40, 1800, 10, 120

Ruben Leyock, 80, 125, 2220, 8, 527

___ G. Skinner, 225, 165, 3950, 30 666

Joshua F. Lee, 156, 44, 4200, 50, 359

John Elmore, 120, 40, 1600, 200, 790

Mary W. Elmore, 100, 20, 1200, 50, 300

Lewis M. Millard, 280, 140, 8400, 205, 1380

Thomas J. Doughty, 90, 20, 1800, 200, 300

Alfred G. Damwood, 140, 100, 1400, 100, 385

A. C. Freeman, 140, 40, 3000, 90, 469

Kamer _. Millard, 400, 138, 7970, 250, 1140

Margaret A. Ryan, 390, 80, 5100, 100, 1031

Aldridge Bridges, 250, 50, 3000, 150, 686

John A. Rees, 350, 412, 7620, 150, 600

W. B. Paxson, 77, 30, 1070, 20, 310

Lydia J. Bradshaw, 125, 125, 5000, 150, 501

Jas. M. Orrison, 100, 115, 2150, 25, 435

George Paxson, 63, 20, 1660, 30, 327

Robert Mathews, 38, 9, 800, 40, 180

Alex. Lyons, 120, 12, 1310, 50, 321

Wm. M. Moran, 90, 10, 1010, 50, 223

Ellen Fouch, 65, 30, 950, -, -

John Layock, 200, 53, 3036, 150, 989

Thomas H. Lyere, 200, 53, 3825, 150, 933

Sandford Fling, 200, 105, 3050, 80, 250

Jas. F. Fling, 200, 105, 3050, 80, 250

Jas. H. Greenelease, 150, 96. 8060, 150, 898

Benj. F. Taylor, 180, 50, 4000, 20, -

George W. Presgrace, 450, 197, 6470, 325, 2451

Samuel Tippett, 35, 65, 900, 75, 275

William W. Presgrace, 35, 140, 1780, 50, 718

William M. Moran, 17, 122, 1780, 50, 525

Francis Keene, 100, 214, 4710, 20 175

Lester Loyd, 90, 35, 2600, 150, 303

W. _. Mandal, 75, -, 1500, 25, 150

Levi Hammer, 60, 10, 1400, 75, 287

Elizabeth Keene, 210, 140, 3600, 75, 1361

Benj. Bridges Jr., 200, 50, 2500, 150, 1172

Benj. Bridges Sr., 420, 60, 4800, 50, 610

Wright Travus, 80, 54, 1050, 125, 422

Brianda Russard, 46, 34, 1600, 100, 440

Benj. F. Waikell, 170, 60, 3000, 50, 360

Lewis Blossom, 100, 5, 1050, 35, 172

George M. Words, 180, 20, 6000, 225, 792

William S. Edwards, 80, 20, 1600, 25, 100

Elizabeth Edwards, 15, 5, 800, 25, 350

John W. Dayley, 52, 21, 2000, 40, 550

Samuel Aukers, 68, 12, 1000, 60, 423

Wm. Misskrell, 25, 15, 600, 20, 100

Benj. S. Morgan, 400, 100, 10000, 200, 710

Charles W. Green, 39, 30, 1725, 150, 315

Jas. W. Daynes, 190, 52, 4500, 150, 645

Jacob Ish, 200, 53, 2500, 50, 150

Daniel F. Palmer, 160, 40, 2400, 150, 692

Washington Hammer, 265, 50, 4725, 225, 835

Jonathan Lewis, 100, 50, 1500, 150, 767,

Wm. C. Hammer, 250, 68, 4770, 125, 839

Armstead M. Taylor, 150, 50, 5000, 300, 920

Cordine Bartlett, 100, 100, 5000, -, 60

Wm. R. Presgrace, 160, 200, 5400, 150, 1000

Richard H. Presgrace, 250, 100, 8750, 300, 1546

M. M. Jones, 240, 100, 3400, 150, 1270

W. S. Greenelease, 325, 75, 4000, 200, 1349

Robert G. Allnutt, 350, 50, 4000, 150, 1160

Richard Caleis, 60, 20, 3200, -, 130

Thomas Darius, 180, 20, 2000, 25, 400

Jonathan C. Paxson, 117, 30, 260, 200, 1900

Richard Houser, 1000, 700, 17000, 500, 2035

John January, 481, 200, 14000, 500, 200

John M. Orr, 280, 230, 3000, 100, 808

Chas. M. Mathew, 75, 5, 800, 25, 166

Jas. Grubb, 125, 81, 3600, 125, 725

Catherine Daner, 209, 120, 3290, 80, 395

John Price, 100, 88, 4500, 150, 812

John George JK, 135, -, 6750, 300, 1875

George J. Rust, 450, 279, 29160, 700, 2660

John Dailey, 500, 70, 14250, 300, 2640

Alfred Dulin, 100, 26, 1890, 150, 631

William Kendrick, 115, 20, 1800, 100, 288

Jasper C. Kepheart, 200, 230, 10000, 285, 1745

Joel Hunt, 40, 25, 1500, 75, 154

Alfred D. Keene, 70, -, 2000, 100, 300

Wm. J. Young, 500, 350, 17000, 100, 1624

John McCamerly, 150, 150, 3000, 75, 450

Jas. Arundle, 213, 100, 2130, 250, 836

John A. Darnes, 100, 90, 1900, 125, 470

Samuel M. Myers, 100, 220, 3200, 75, 175

Susan Dawson, 100, 98, 1980, 100, 500

Bayley D. Cockrell, 30, 30, 600, 75, 90

Alex. Moran, 27, 10, 600, 75, 200
Jas. Beavers, 75, 125, 2000, 125, 650
Henry Cowden, 75, 75, 1184, 80, 235
Isaiah Bedine, 100, 100, 2000, 20, 80
Wm. Alexander, 90, 10, 600, 50, 125
George H. Winkman, 60, 40, 800, 75, 425
Robert Power, 100, 75, 1750, 50, 475
Craven & Johnston, 221, 60, 3372, 225, 721
Jas. H. Gulick, 385, 40, 8500, 101, 1100
Harmon Ritzer, 180, 78, 3000, 300, 800
Jas. A. Moffett, 100, 50, 1700, 50, 570
John T. Iden, 550, 150, 28000, 400, 2035
Francis Thrift, 285, 40, 13505, 350, 1660
C.W. & P. T. Moffett, 175, 27, 8080, 150, 1120
George E. Leyock, 90, 42, 2000, 100, 842
Wm. Leyock, 80, 80, 2400, 30, 270
Henry Lefever, 150, 50, 2000, 15, 192
Samyuel Leyock, 37, -, 444, 25, 338
Thomson Cornel, 80, 32, 1120, 20, 200
Richard Y. Moran, 120, 32, 1824, 60, 682
Isaac Wortman, 62, 31, 900, 30, 88
Samuel of Jas. Lafever, 300, 156, 5340, 175, 922
Wm. M. Havenner, 40, 9, 1000, 40, 110
Jas .Whaley, 70, 30, 300, 300, 900
Saml. Jenkins, 90, 10, 3000, 250, 620
Geo. W. A. Hammer, 180, 32, 1800, 20, 365
Jas. F. Darnes (Dames), 175, 37, 2100, 25, 80

Wm. H. Russell, 200, 25, 5500, 450, 1220
Thos. J. Misskell, 550, 100, 26000, 300, 1500
Saml. L. Jenkins, 350, 125, 7125, 75, 1754
Frederick W. Pleasants, 530, 530, 15900, 350, 1160
Thos. Veal & Bros., 11034, 566, 34000, 500, 2725
P. M. Walker, 250, 250, 4000, 40, 400
John Auston, 300, 160, 4500, 200, 1420
Jas. Readford, 70, 33, 1500, 250, 477
William Dow, 100, 98, 3000, 150, 860
Ann M. Lowe, 150, 105, 3825, 60, 900
Enoch Lowe, 300, 93, 19650, 100, 725
Jas. Crokrill, 100, 35, 2700, 120, 1260
Alex. H. Rogers, 270, 30, 15000, 600, 1500
Sarah Craven, 180, 77, 20560, 150, 1000
Chas. F. Fadeley, 170, 30, 6000, 150, 1200
Smart & Bentley, 1350, 400, 51000, 1000, 6450
T. M. C. Paxson, 105, 30, 8100, 400, 1410
Robt. G. Bowie, 100, 45, 7000, -, 347
Burr W. Harrison, 15, -, 1800, -, 150
Thos. Claxton, 12, 30, 1250, 10, 150
Power & Bedine, 200, 80, 16600, 250, 1050
Henry A. Ball, 567, 190, 50000, 1000, 4515
Richd. E. Furr, 497, 89, 29300, 500, 2650
Robt. W. Gray, 400, 100, 20000, 500, 1800

Chas. Douglass, 400, 140, 21600, 300, 800

William Ivy(Sey), 180, 65, 1700, 75, 565

Wm. S. Gray, 144, 40, 6000, 400, 870

Eliza V. White, 230, 70, 12000, 500, 1320

Cumberlin G. Orison, 90, 75, 3000, 1500, 625

John McKinney, 20, 57, 6000, 200, 1000

Jas. Sey, 65, 100, 3300, 150, 300

Jesse Coleman, 250, 60, 4000, 100, 1000

Wash. C. Trammell, 100, 14, 5000, 150, 786

John Williams, 300, 50, 7000, 300, 1400

George W. Chick, 75, 25, 800, 75, 220

Peter J. Slayman, 60, 76, 1700, 100, 350

Randolph Barnhouse, 225, 16, 5000, 150, 853

Robert Ryan, 250, 250, 5400, 100, 676

David Shreve, 250, 20, 6000, 300, 740

John Thomas, 115, 17, 4620, 100, 800

George W. Ball, 556, 160, 43000, 500, 2470

John Hoffman, 200, 50, 17000, 500, 2020

Margarett Jackson, 20, -, 400, 20, 75

Wm. Bedily, 500, 150, 29250, 930, 4542

Arthur M. Chidester, 300, 58, 19650, 600, 1000

Arch. K. Sanders, 20, 7, 3900, 200, 520

Wm. Hough, 15, -, 375, 20, 100

Chas. Attwell, 60, 15, 2550, 80, 200

Jas. H. Whitmore, 250, 65, 18125, 350, 1700

Alfd. C. Belt, 380, -, 19000, 1300, 6900

Geo. M. Frey, 400, 300, 21000, 150, 600

C. W. Paxson, 110, 7, 3000, 101, 590

Cephas Hempston, 160, 23, 9150, 600, 1250

Martin Frey, 15, -, 560, 20, 150

Jas. Mathew, 20, -, 1000, 70, 288

Tilghman Gore, 466, 120, 23520, 300, 1766

Wm. Stocks, 140, 22, 6480, 100, 300

Jas. W. Minor, 100, 195, 8850, 700, 1160

Wm. C. Luckett, 24, -, 900, 40, 128

Wm. B. Jackson, 290, 96, 11000, 350, 1800

Alfd. Belt, 230, 50, 18000, 400, 1050

Jas. Mathew, 125, 50, 7000, 100, 260

Wilson C. Sanders, 282, 102, 15360, 540, 1200

Jas. V. Brown, 120, 80, 6000, 100, 770

Chas. Williams, 330, 105, 21750, 750, 1500

Daniel F. Shreve, 280, 45, 16000, 800, 1545

Chs. G. Giddings, 263, 40, 9000, 316, 830

Mathias Fletcher, 800, 500, 52000, 1300, 5150

Elizabeth C. Mason, 300, 300, 24000, 300, 250

Wm. Stream, 30, -, 900, 10, 150

Saml. Snoots, 10, -, 200, 10, 15

Sarah Dawson, 250, 50, 12000, 1200, 1750

Michael Mullen, 200, 400, 75000, 500, 1500

Saml. H. Houser, 45, 117, 4000, 100, 386

Tavis Titus, 130, 150, 9000, 300, 800

John Woodheart, 35, 9, 1000, 125, 320

Wm. Hickman, 110, 40, 6500, 350, 600

Wm. Ball, 575, 25, 30000, 700, 3250

Robt. Harper, 150, 40, 5790, 75, 380

Henry Sanders, 243, 100, 8575, 350, 1246

Jas. W. Downs, 500, 67, 8500, 100, 610

Jas. Jones, 200, 114, 9420, 500, 1388

Luther A. Thrasher, 300, 375, 13500, 150, 1385

Thos. W. Muse, 242, 50, 8000, 150, 860

Edward Hanes, 100, 50, 4500, 250, 460

Noah Downs, 40, 10, 1000, 15, 180

Jas. W. Soloman, 64, 25, 2000, 100, 270

Daniel Diedrick, 30, -, 900, 75, 190

Jochan Richlus, 20, 12, 800, 10, 128

Aquilla Bauckman, 27, 20, 1175, 200, 336

Jas. H. Daily, 25, -, 375, 75, 230

Cornelia Hanes, 50, 16, 900, 50, 110

Cornelius B. Wynkoss, 50, 50, 1500, 75, 108

Wm. Burgess, 50, 50, 1000, 150, 305

Jas. Ault, 182, 25, 2655, 150, 567

Jno. Compher, 60, 80, 3000, 200, 750

Edith Fanley, 16, -, 630, 40, 160

Adam Cordell, 20, 10, 900, 10, 130

Henry S. Williams, 50, 10, 3000, 250, 1200

Wm. Stocks, 130, 30, 6400, 100, 400

Albert Best, 23, 9, 1280, 100, 309

Sydney Williams, 300, 65, 16400, 1200, 1300

Saml.Wright, 115, 65, 8000, 150, 664

Jno. Heaton, 160, 65, 9000, 350, 800

Elizabeth Heefner, 15, 18, 700, 20, 120

Jos. Dixson, 130, 159, 8000, 100, 330

Wm. S. Schooley, 100, 15, 1180, 200, 325

Jas. L. Fullin, 20, 10, 900, 50, 100

Geo. Umbaugh, 50,-, 1750, 25, 288

Jas. Umbaugh, 310, 20, 12000, 200, 500

Jno. Williams, 7, 3, 3500, 20, 328

Chas. J. Harper, 60, 10, 2000, 150, 400

Joseph Fry, 60, -, 1800, 100, 780

John F. Money, 65, 30, 1175, 50, 220

Isaac M. Rice, 186, 40, 9000, 400, 711

Jas. Brown, 200, 50, 10000, 350, 1200

Wade Barrett, 200, 55, 11250, 200, 770

Elizabeth Vincel, 55, 10, 4000, 125, 280

Randolph White, 160, 140, 12000, 200, 1100

Jas. Segars, 60, 10, 2700, 35, 175

Jas. P. Wright, 380, 70, 14000, 150, 400

Isaac Warner, 225, 160, 13475, 50, 4350

Jno. Brown, 160, 70, 9200, 150, 750

Jas. Shugars, 250, 100, 14000, 75, 360

Geo. Aldee, 156, 50, 9270, 300, 1040

Joshua Pusey, 105, 59, 8200, 150, 2550

Jas. W. Orrison, 140, 48, 5600, 280, 580

Wm. H. Franklin, 240, 60, 15000, 300, 2330

Joseph Hough, 105, 30, 8000, 150, 650

Eli L. Schooley, 200, 99, 15000, 100, 1225

Geo. D. Smith, 230, 98, 12720, 350, 2000

Geo. D. Smith, 200, 50, 7000, -, -

Wm. H. Cassaday, 400, 100, 18000, 350, 1223

Saml. Baker, 208, 100, 13000, 200, 1230

Wm. Geddings, 420, 200, 30000, 1000, 2500

Lewin T. Jones, 180, 20, 11000, 500, 1251

Wm. Fauly, 91, -, 5800, 300, 790

Syner Bennith, 175, 30, 8200, 350, 1300

Danl. Trillipfoe, 170, 68, 9600, 150, 760

Jas. M. Downey, 53, 12, 12000, 300, 670

Jonas J. Compher, 120, 42, 7100, 250, 860

Saml. Slater, 168, 10, 8000, 350, 620

Chas. James, 300, 100, 20000, 250, 1170

Jno. Bumhouse, 130, 30, 12800, 100, 1200

Chas. Bayne, 70, -, 4500, 125, 599

Jas. W. Cockrill, 150, 30, 9000, 200, 1000

Saml. Compher, 87, 13, 4900, 150, 597

Jno. Louder, 118, 6, 7440, 500, 1340

Geo. Blamer, 400, 130, 26500, 700, 2500

Thos. M. Edwards, 125, 25, 7500, 150, 380

Geo. Ritchie, 85, 15, 5000, 100, 750

T. M. Paxson, 40, 25, 3000, 150, 110

Emanuel Stoutscubuger, 24, 8, 1600, -, -

Benj. J. Grubb, 385, 100, 24300, 75, 3040

Jno. Swank, 115, 35, 9000, 200, 1300

Jno. H. Craven, 150, 20, 8000, 150, 483

Adam Log, 150, 40, 1860, 100, 718

A. T. M. Filler, 280, 32, 21840, 500, 2900

Solomon Smith, 30, -, 1500, 60, 240

Thos. H. Hickman, 40, -, 2400, 175, 315

Saml. Fry, 56, 4, 3600, 250, 612

Wm. Smith, 100, 28, 6400, 150, 645

Saml. Smith, 100, 30, 6300, 100, 575

Susan Loudee, 170, 33, 10000, 200, 900

Henry Fauly, 350, 200, 1500, 250, 400

Edward Tromson, 95, 6, 5000, 250, 400

Chas. Wight, 80, 50, 5200, 50, 400

Wm. Spring, 200, 5, 8100, 150, 800

Jas. Fauley, 100, 60, 6400, 75, 575

Jas. Yakey, 170, 84, 13000, 600, 920

Jas. W. Rust, 85, 15, 4500, 125, 670

Casper Spring, 60, -, 3000, 150, 155

Saml. Orrison, 250, 78, 13000, 100, 950

Sarah Heaton, 80, 33, 4500, 200, 440

Geo. W. Wenner, 129, -, 7700, 100, 510

Catherine Hickman, 210, 63, 16000, 150, 800

Jno. Hickman, 45, 9, 3000, 10, 69

Saml. Ropp, 350, 118, 23400, 200, 2100

Andrew Seitz, 191, 67, 11600, 400, 1480

Jacob Wenner, 135, 18, 9000, 200, 1000

Mary Frazier, 180, 20, 12000, 300, 1500

Mary Vincel, 52, 9, 3600, 50, 100

Geo. H. Wenner, 110, 50, 8000, 150, 820

Elizabeth Short, 30, 4, 1800, 50, 200

Wm. U. Wenner, 75, 25, 7700, 1000, 1800

Jno. W. Fauborn, 130, 52, 10900, 350, 775

Jas. W. Nixson, 110, 79, 9000, 400, 1350

John Boyer, 100, 55, 10600, 150, 600

John V. Leslie, 375, 100, 28000, 1400, 3300

Emanuel Waltman, 200, 62, 14000, 100, 1300

Saml. C. E. Ramsey, 130, 30, 8900, 260, 836

Jas. & Jos. Heath, 200, 20, 13200, 150, 1125

Saml. W. George Jr., 135, 90, 13500, 100, 569

Michael Landbour, 100, 20, 6000, 300, 630

Jonas Harvey, 28, 4, 2100, 20, 125

Andrew Bockman, 60, -, 3000, 20, 78

Michael Carver, 150, 25, 8750, 150, 420

Jas. Schaver, 80, 3, 3780, 150, 500

Jas. Everheart, 200, 25, 13400, 250, 450

Jno. Mann, 103, 12, 6900, 150, 550

Joseph Mann, 140, 20, 9600, 1000, 425

Emanuel Wenner (Weimer), 285, 92, 16800, 400, 1500

Jos. Everheart, 80, 20, 4900, 50, 360

Israel Everheart, 100, 20, 7000, 250, 960

Jas. Booth, 8, -, 1500, 150, 225

John Long, 19, 10, 640, 100, 320

Daniel Cam, 60, 10, 3100, 20, 210

Jas. A. Washington, 150, 75, 8500, 400, 1200

Louisa Vincel, 100, 80, 5200, 50 190

Richd. C. Marlow, 410, 105, 20600, 650, 2300

Thos. Swan, 800, 401, 50000, 1000, 2800

J.T. M. Rust, 425, 110, 32000, 775, 4000

Chas. Williams, 320, 120, 25000, 500, 1125

Noble S. Braden, 330, 81, 24000, 300, 6000

Catherine E. Fox, 50, 10, 5000, 250, 900

Mary Clark, 160, 18, 11500, 600, 1400

Saml. Steen, 115, 25, 7000, 250, 550

Rachel Steen, 85, 15, 5000, 100, 200

Jas. W. Walker, 195, 42, 14200, 500, 2210

Wm. C. Sharven, 165, 75, 25500, 350, 1450

Jacob Myers, 69, 50, 7000, 250, 525

Israel Warner, 200, 49, 14000, 150, 890

Israel Warner, 200, 95, 12000, 100, 27

Oscar S. Braden, 225, 73, 17900, 400, 3600

Edward Y. Mathews, 135, 25, 9600, 500, 750

Bushrod L. Fox, 115, 35, 91000, 350, 775

Washington Myers, 65, 9, 4450, 200, 250

Rodney C. Braden, 350, 53, 20250, 700, 2150

Chas. E. Paxson, 80, 5, 8000, 150, 950

Amasa Hough, 190, 40, 14000, 600, 1501

Jno. McGouch, 21, -, 15000, 150, 360

Uriah Beans, 80, 32, 6160, 300, 580

George H. Wine, 140, 42, 11700, 300, 670

Jas. E. Walker, 135, 60, 14624, 300, 1158

Jno. A. Hope, 470, 130, 40500, 300, 2165

Wm. H. Miller, 130, 40, 8350, 250, 900

A. M. Vandeventer, 175, 40, 16000, 200, 2000

Jas. Wright, 80, 30, 770, 250, 720

Griffeld N. Paxson, 98, 6, 7300, 300, 788

Chas. L. Hollingsworth, 70, -, 5600, 100, 600

Joseph James, 150, 72, 13000, 210, 880

Jno. Compher, 150, 50, 12000, 400, 1624

Ebinesa Grubb, 173, 35, 16000, 350, 3000

Wm. Veits, 166, 33, 13000, 400, 1000

Wm. Veits, 168, 40, 12500, 200, 800

Jas. Curry, 140, 20, 12000, 400, 1775

Joseph Fry, 152, 2, 4800, 350, 760

Edward Hamilton, 170, 70, 13000, 300, 1500

Geo. W. Russle, 166, 50, 12950, 1300, 900

Thos. Philips, 210, 20, 13800, 300, 1350

Jno. W. Veits, 200, 50, 15000, 350, 1500

Curtis Grubb, 225, 47, 16400, 410, 1800

Jno. Woolford, 100, 25, 7500, 500, 600

Wm. H. Adams, 50, 15, 4600, 50, 400

Geo. L. Moore, 230, 20, 15000, 700, 1850

Joshua White, 180, 10, 14000, 600, 920

Jonathen Evens, 87, 10, 5800, 200, 360

Mary A. Smith, 140, 30, 10200, 300, 510

Isaac Camp, 150, 100, 12500, 500, 550

Josiah T. White, 220, 78, 17000, 200, 1500

Aquilla Butts, 35, 45, 2000, 100, 175

Malin Stalks, 120, 70, 6000, 100, 1200

Nathen White, 100, 71, 10000, 300, 710

Ezekeal Potts, 140, 84, 11000, 450, 1500

Jno. Jones of Jno., 140, 60, 10000, 200, 2000

Jno. Thomson, 320, 320, 21000, 501, 3400

Harmon Copeland, 310, 177, 14700, 150, 1700

Saml. P. Thomson, 100, 7, 5000, 100, 500

Ruth Clendenning, 186, 100, 14300, 400, 1700

Jas. H. Clendenning, 129, 60, 7500, 100, 500

Zedekia Kidwell, 130, 24, 6000, 150, 1000

Nathen Prints, 60, 24, 4000, 200, 700

Ebenezer Coward, 150, 104, 7300, 200, 1100

Edward Harding, 157, 72, 9000, 150, 800

Philip Derry, 90, 50, 4000, 100, 225

Jonas Potts, 40, 36, 2500, 75, 240

Jno. Potts, 25, 10, 1000, 75, 200

William Grubb, 200, 46, 9500, 350, 1500

Henry L. Wince, 65, 40, 4000, 150, 750

Elsy Chamblin, 73, 20, 3000, 75, 400

Jas. W. Conard, 120, 25, 5800, 250, 1201

Philip Derry, 100, 25, 5000, 150, 600

Geo. Able Lee, 110, 34, 4300, 150, 400

Geo. F. Able, 105, 30, 4000, 150, 500

Jas. Clendenning, 250, -, 4000, 150, 800

Saml. Edwards, 40, 16, 2250, 50, 260

Geo. Wear, 100, 34, 3500, 100, 700

Elizabeth Edwards, 100, 67, 6000, 150, 900

Wilson B. Everheart, 75, 56, 2600, 150, 700

Levi Waters, 150, 150, 7000, 150, 625

George P. Hunter, 100, 25, 3700, 100, 400

Jno. F. Waters, 44, 35, 1740, 100, 300

Malin Demery, 129 40 4000, 150, 700

David Shriver, 160, 45, 8000, 200, 700

Jos. Conard, 180, 100, 11000, 150, 800

Jos. Conard, 100, 64, 4000, 50, 550

Jno. Nisewaner, 120, 230, 9000, 150, 850

Wm. Bagent, 50, 34, 3000, 100, 550

Abner Conard, 267, 125, 15600, 200, 1300

Benj. Leslie, 130, 50, 9000, 150, 1200

Saml. Crince, 70, 15, 4250, 500, 810

Ann Russle, 135, 15, 3750, 50, 275

Wm. Denovey, 90, 40, 3250, 150, 360

Saml. N. Grubb, 105, 71, 3000, 125, 450

Geo. W. Derry, 80, 60, 3500, 100, 725

Jno. Miller, 27, 3, 900, 100, 270

Armstead Miller, 76, 42, 3820, 150, 620

Jno. P. Derry, 175, -, 4700, 100, 600

Mary J. Smith, 120, 25, 5050, 150, 560

Geo. H. Donel, 87, 25, 3920, 150, 825

Benj. Grubb, 150, 63, 7455, 100, 450

Jno. Grubb, 327, 185, 18350, 400, 1600

Jno. Fritz, 78, 23, 4000, 100, 450

Jonathen Painter, 49, 11, 2000, 100, 315

Jno. W. Nees, 130, 60, 7600, 400, 825

Hugh S. Thomson, 140, 120, 6800, 100, 850

Nathen Nees, 100, 12, 5600, 600, 1850

Hiram Faire, 20, -, 800, 50, 275

Edwin Potts, 200, 35, 11225, 250, 1150

Wm. Potts, 200, 30, 11225, 200, 1125

Wm. Clendenning, 375, 100, 14250, 1300, 2875

Wm. W. Gray, 400, 100, 35000, 1500, 3000

Wm. E. Vins, 150, 60, 10000, 150, 800

Chas. Y. Veits, 195, 35, 10000, 250, 1000

Jno. Veits, 160, 40, 12000, 400, 1000

Jas. Reese, 175, -, 9000, 300, 1360

Jacob Shoemaker, 65, 15, 3200, 150, 550

Henry Ruse(Reese), 147, 33, 9000, 150, 650

Jacob Veits, 180, 20, 10000, 150, 1350

Jonasl P. Schooley, 130, 20, 10500, 75, 2300

Robt. Camtell, 50, 10, 900, 100, 250

Alex. Johnson, 200, 116, 6320, 200, 820

Ebinesa L. Grubb, 132, 20, 6080, 50, 225

Green Baker, 90, 25, 5750, 200, 460

Jesse Mann, 87, 20, 5350, 125, 690

Saml. Ball, 80, 20, 5000, 150, 537

Edward Thomson, 94, 7, 5050, 125, 392

Saml. C. Luckett, 190, 22, 10600, 500, 1280

Geo. Baker, 125, 25, 5500, 150, 850

Albert Auldeer, 200, 20, 13000, 300, 1700

Geo. Slater, 140, 25, 8800, 30, 1000

Jacob Crusen, 50, 26, 4000, 150, 450

Wm. Veits, 100, 30, 7200, 150, 665

Asa R. Woldford, 82, 3, 500, 100, 650

Wm. Beaty, 135, 20, 6240, 400, 770

Jno. Bartlett, 37, 5, 1680, 100, 475

Peter Stonebuner, 112, 15, 5750, 200, 760

Saml. Frey, 55, 5, 3000, 125, 650

Wm. Came, 75, 5, 3600, 125, 985

Jno. Wise, 160, 10, 6800, 125, 490

Daniel Boland, 400, 48, 24600, 1000, 2000

Daniel Boland, 100, 40, 7700, 200, 600

Peter Wise, 95, 14, 5000, 150, 600

Benj. Grubb of Jno., 300, 100, 18000, 200, 1900

Jerard Lockes, 30, -, 2000, 100, 360

Geo. Cooper, 106, 45, 7000, 350, 800

Wm. Cooper, 100, -, 6000, 200, 700

Jos. Hickman, 156, 31, 10300, 150, 1000

Jacob Householder, 271, 46, 15900, 400, 1200

Peter Weeb, 249, 30, 11000, 300, 1000

Jacob Smith, 100, 45, 7000, 300, 1200

Adam Cooper, 90, 10, 5000, 125, 570

Jno. P. Fry, 19, 1, 1000, 50, 150

Frederick Miller, 8, -, 500, 50, 200

Jno. Bushie, 30, 1, 1200, 75, 200

Gideon Householder, 154, 8, 8900, 125, 750

Saml. H. Crumbacker, 90, 14, 4150, 150, 450

Joshua Everheart, 30, -, 22000, 150, 700

Jacob Potterfield, 18, 12, 1000, 50, 65

Geo. Vincel, 50, 45, 3500, 100, 500

Saml. Porterfield, 120, 36, 6200, 150, 725

Chas. Johnston, 250, 75, 10000, 500, 1000

Michael Arnold, 75, 25, 5000, 200, 500

Michael Cooper, 50, 20, 3200, 150, 300

Christian Wisewaner, 122, 18, 7000, 350, 900

Simon Arnold, 120, 49, 10000, 100, 950

Michael Wiard, 45, 20, 2450, 100, 350

Jonas Potterfield, 35, -, 1750, 150, 720

Jas. G. B. Kolb, 94, 23, 8930, 150, 960

Elias D. Kolb, 45, -, 3130, 100, 260

Saml. J. Kolb, 250, 30, 13280, 200, 1193

Elizabeth Conard(Coward), 51, 2, 2640, 100, 200

David Akline, 131, 30, 6400, 250, 1067

Thos. Kolb, 80, 15, 5250, 150, 500

Jas. N. Conard (Coward), 100, 10, 5500, 150, 500

Mary C. Conard, 98, 10, 5400, 100, 400

Jas. & Geo Cooper of P, 87, 13, 5850, 150, 640

Jas. Cooper of Geo., 110, 26, 7000, 200, 1300

Susaniah Cruise, 80, 12, 5500, 100, 550

Saml. L. Stone, 150, 50, 12000, 250, 800

Geo. Shoemaker, 178, 50, 14300, 500, 1600

Calhoon Hough, 60, 5, 60, 5, 3900, 300, 500

Joseph Compher, 136, 35, 8500, 300, 900

Elias Cooper, 25, -, 1500, 100, 500

David Fry, 150, 100, 12000, 250, 1100

Wm. Vickers, 140, 8, 8800, 200, 700

Noah Fry, 109, 10, 5900, 200, 800

Wm. Crim, 170, 20, 11000, 200, 850

Jas. & Jos. Heart, 100, 30, 17000, 200, 850

Saml. Shoemaker, 20, -, 1000, 50, 250

Jas. T. Merchant, 100, 40, 8000, 150, 450

Jonathen Shoemaker, 89, 10, 8450, 100, 400

George Shumaker, 47, 5, 3200, 110, 425

Chas. W. Myres, 60, -, 2400, 200, 600

Michael Beamer, 100, 34, 8050, 250, 1450

Jno. E. Stoth, 146, 35, 2000, 150, 300

Philip Fry, 65, 15, 1300, 100, 675

Jno. G. Slater, 120, 80, 6000, 150, 1650

Saml. W. George, 300, 100, 20000, 275, 2100

Jacob Arnold, 57, 68, 2400, 100, 250

Peter Hickman, 160, 20, 8000, 250, 1000

Richd. James, 100, 20, 5650, 200, 825

Philip Vincel, 150, 47, 9650, 400, 1100

Aaron Cooper, 100, 16, 5700, 400, 750

Edward Simons, 37, -, 1750, 100, 350

Nicholus Lynn, 44, 11, 3300, 150, 650

Dabriel Householder, 100, 60, 10200, 500, 1000

Jos. H. Fry, 200, 97, 14850, 450, 1150

Emeline Cooper, 30, 5, 2000, 100, 350

Renard Jacobs, 83, 20, 6240, 150, 510

Thos. B. Marche , 171, 25, 12360, 350, 1000

A. Jas. Wightman, 100, 25, 7000, 150, 550

Geo. F. Crimm, 44, 13, 2900, 100, 250

Jno. Winnce, 148, 35, 10900, 200, 1000

Jno. Fry, 136, 20, 2800, 250, 750

Smith Reed, 140, 118, 7800, 200, 1000

Jobe Smith, 135, 20, 6200, 200, 750

Joseph White, 180, 60, 9400, 350, 1000

Jno. Stone, 200, 40, 16000, 250, 900

Saml. Clendening, 225, 75, 10500, 400, 700

Saml. Clendening, 100, 20, 4800, 100, 400

Robt. L. Wright, 725, 177, 48950, 900, 5950

Horatia Trundle, 1350, 357, 69000, 1200, 10000

Richard M. Bently, 350, 200, 16000, 200, 1700

Mathew P. Lee, 400, 100, 10000, 200, 1600

James H. Swain, 40, 160, 800, 50, 40

Lucinda Skillman, 136, 50, 1500, 20, 280

Daniel Sloper, 135, 22, 1500, 30, 200

Daniel Summers, 150, 75, 1500, 50, 1000

John Hutchison, 58, 4, 500, 20, 100

James H. Palmer, 50, 40, 1100, -, 40

James S. Hutchison, 90, 25, 1500, 100, 460

Redding Hutchison, 50, 10, 1000, 20, 300

Andrew Hutchison, 200, 80, 1400, 100, 1000

Lewis Bradshaw, 70, 10, 800, 20, 300

William A. Hutchsion 145, 30, 1000, 100, 600

Pickering Hutchison, 60, 20, 800, 20, 120

Henry J. Obannon, 145, 40, 1800, 100, 800

Alonzo M. Obannon, 75, 25, 1000, -, 100

David D. Lee, 25, 35, 1000, 40, 300

Hezekiah Clemmings, 55, 25, 650, 20, 100
Loyd Lowe, 120, 30, 1300, 300, 500
Charles Manley, 37, 10, 400, 10, 25
Newton Foley, 92, 50, 800, 30, 400
Smith James, 200, 124, 1500, 50, 600
David B. James, 600, 200, 5000, 150, 1500
Tomas Hutchison, 100, 33, 800, 35, 300
Nelson Suttle, 224, 50, 1400, 50, 600
Hamilton R. Bruer, 120, 13, 700, 50, 200
George W. Rose, 200, 60, 1000, 20, 30
John S. Wilson, 180, 10, 1000, 30, 550
James Boumonest, 200, 50, 1000, 30, 100
Robert Cunningham, 200, 100, 2500, 20, 200
Wilmoth Cunningham, 60, 40, 500, 20, 100
Elias Ayrs, 300, 380, 7000, 50, 650
Edward L. Carter, 250, 60, 6000, 100, 700
Robert Lewis, 100, 100, 3000,-, -
Benj. S. Chinn, 50, 30, 1500, -, -
William H. Thornton, 275, 50, 5000, 50, 1100
George W. Byrne, 500, 150, 8000, 100, 800
John W. Hutchison, 250, 50, 3000, 50, 3000
S. Arniss Buckner, 1100, 400, 30000, 500, 2600
Sidney L. Hodgson, 700, 500, 16000, 400, 1400
Henry S. Swart, 120, 10, 1600, 100, 300
William Foley, 300, 100, 3000, 150, 400
John W. Skillman, 150, 50, 2000, 75, 200

Alex. D. Lee, 1130, 600, 20000, 250, 2400
Jonathan T. Wyckoff, 60, 10, 600, 20, 300
Nicholas H. Wyckoff, 60, 10, 600, 10, 430
Thomas B. Lewis, 700, 400, 13000, 600, 2100
Charles W. Lane, 135, 15, 1200, 10, 150
Benj. Mershon, 175, 55, 1800, 75, 600
Mary A. Whaley, 182, 108, 3500, 50, 5500
Susan Laffer, 200, 50, 4000, 150, 1490
John Ratrie, 150, 75, 3000, 25, 125
Martain H. Swart, 175, 45, 3000, 300, 700
Harrison Cross, 230, 30, 3000, 200, 800
Nimrod Cross, 200, 178, 2000, 50, 300
Adren L. Swart, 170, 30, 3000, 50, 600
John C. Swart, 120, 30, 3000, 50, 200
William H. Stephenson, 200, 33, 2300, 50, 700
Robert Poland, 125, 25, 3000, 100, 500
Leah Hutchison, 160, 40, 2000, 100, 1100
Bevely Hutchison, 350, 175, 9450, 200, 1390
Orien Lewis, 135, 20, 2100, 50, 440
Charles B. Adam, 100, 50, 2700, 50, 600
Mary Oden, 450, 150, 15000, 500, 1700
John R. Skinner, 200, 20, 5000, 175, 1160
Jonah Hood, 280, 50, 9900, 100, 700
William N. Berkley, 500, 200, 31000, 700, 4650

William F. Adam, 170, 80, 1170, 100, 1100

William Wilson, 175, 35, 8500, 50, 300

Anna Swart, 210, 30, 9000, 35, 450

Norbourne Berkley, 350, 150, 30000, 500, 3150

John J. Curle, 200, 50, 6000, 75, 330

Francis Chinn, 250, 80, 12000, 100, 700

John Hixson, 125, 63, 5500, 50, 650

Sandford Rogers, 270, 100, 11000, 200, 1070

Richard Turner, 453, 40, 6000, 40, 690

Rufus Smith, 100, 23, 6200, 350, 680

William Rogers, 250, 80, 18000, 300, 1165

Hamilton Rogers, 900, 300, 50000, 700, 5760

Asa Rogers, 260, 80, 18000, 400, 1500

Asa Rogers, 16, 3, 8000, 50, 300

Asa Rogers, 590, 150, 30000, 400, 3000

Richard Rogers, 100, 20, 7200, 115, 700

T. McVeigh, 445, 75, 31200, 500, 4250

W. Dodd, 135, 35, 6800, 150, 1450

Wm. Sullivan, 125, 29, 6000, 200, 790

B. D. Bartlett, 40, 2, 2500, 40, 375

Jno. A. English, 200, 25, 15000, 200, 2810

Burr P. Noland, 180, 30, 10000, 500, 1300

Burr P. Noland, 300, 100, 8000, 12, 3270

L. Chancelor, 200, 60, 11000, 250, 2250

E. A. Smith, 170, 20, 9500, 300, 980

A. G. Smith, 45, 15, 2400, -, 1600

Benj. Skinner, 275, 50, 12800, 150, 1400

Henry F. Davis, 190, 17, 6200, 500, 1400

Washington Garrison, 225, 16, 4400, 100, 920

Stephen W. McCarty, 180, 50, 11500, 250, 1130

Hugh Smith, 450, 100, 22000, 250, 2180

Joseph Meade, 200, 200, 15000, 400, 2290

Joseph Meade, 300, 100, 12000, 60, 200

Stephen Smith, 60, 17, 2300, 40, 300

Francis L. Fredd, 300, 50, 16500, 500, 2500

Wm. Fitzhugh, 300, 65, 15000, -, 250

Billington McCarty, 75, 28, 4280, 80, 630

Sarah Wilson, 200, 52, 10080, 25, 300

Joseph Eidson, 210, 40, 12500, 150, 1350

Melton McVeigh, 160, 50, 8000, 150, 630

Jackson Hogueland, 125, 25, 6000, 100, 400

Samuel Cox, 150, 63, 8500, 100, 860

Geo. L. Bitzer, 200, 55, 10200, 400, 770

Jonah Tavener, 118, 45, 5700, 150, 865

James Priest, 220, 70, 11150, 20, 375

Joseph Frey, 80, 80, 4400, 20, 400

Wm. F. Carter, 171, 23, 8730, 125, 590

R. S. Chinn, 380, 40, 16800, 100, 2000

Fanny Beatly, 97, 15, 6000, -, -

Jos. Campbell, 200, 40, 12500, 50, 325

Eblin Farr, 120, 12, 7200, 150, 1190

Wm. R. Moore, 100, 11, 4000, 100, 500

Jno. S. Carter, 181, 30, 10500, 200, 10500

John M. Rawlings, 223, 35, 12900, 150, 1475

John A. Carter, 775, 150, 37000, 450, 3135

Richd. H. Dulany, 1150, 350, 75000, 2000, 21450

Townshend Seaton, 180, 80, 11880, 100, 1475

William Seaton, 250, 65, 14500, 150, 800

Thomas Glasscock, 650, 120, 38000, 600, 6300

Joshua Fletcher, 1000, 150, 69000, 300, 9224

Hugh Rogers, 304, 1500, 15890, 200, 2840

Charles Blakeley, 245, 70, 12600, 200, 2020

Jesse Richards, 248, 75, 20900, 300, 2500

George Ayre, 525, 75, 30000, 600, 7550

William Wilkinson, 45, 10, 3300, 40, 470

John M. Scott, 253, 29, 20000, 600, 2350

M. W. Fitzhugh, 390, 14, 16160, 100, 1115

Bruce Gibson, 260, 55, 15000, 150, 800

Kimble G. Hicks, 400, 100, 20000, 150, 1830

Theod. Leath, 232, 10, 9600, 75, 1200

Francis W. Kendall, 100, 21, 2500, 50, 500

Chas. Kendall, 25, 25, 150, 25, 200

Jas. Slack, 80, 50, 1000, 20, 375

Nathaniel Thomas, 235, 40, 15000, 125, 970

Vincent Moss, 271, 40, 14000, 200, 1154

Charles Lucas, 200, 50, 10000, 108, 1100

Wm. Fleming, 100, 40, 4200, 80, 649

Saml. M. Trussel, 80, 20, 400, 20, 310

Travis Miley, 25, 20, 300, 5, 225

Herod Thomas, 300, 230, 15900, 200, 2256

Joseph G. Gray, 600, 190, 3200, 600, 3173

John M. Harrison, 400, 100, 22500, 400, 3570

John M. Harrison, 480, 70, 24750, 350, 1265

A. B. Carter, 225, 75, 10000, 200, 895

John Ross, 115, 110, 4600, 100, 700

John T. Ross Jr., 265, 107, 7950, 100, 1200

Jas. Alexander, 40, 56, 400, 10, 150

Kemp Farr, 50, 138, 1000, 25, 600

Clinton Farr, 60, 200, 1200, 15, 400

Jno. Moreland, 100, 50, 900, 15, 430

Townd. Fraizier, 210, 49, 11000, 200, 2322

Wm. G. Farr, 290, 100, 12640, 100, 2550

Milton V. B. Waltman, 120, 57, 6200, 200, 1325

Washington Beavers, 60, 50, 2750, 10, -

Jas. B. Throckmorton, 238, 80, 14310, 500, 2140

Frank. W. Moore, 228, 60, 11500, 200, 1565

Jas. W. Nicholls, 280, 81, 12635, 280, 2000

Robert James, 215, 130, 13600, 300, 1525

Hester Alder, 100, 50, 1000, 50, 400

M. Selcott, 200, 96, 9500, 175, 6450

S. R. Mount, 90, 100, 4500, 50, 350

A. M. Moore, 95, 32, 5000, 100, 300

Marcus Dishman, 108, 50, 5900, 100, 545

Levin P. Chamblin, 120, 70, 10000, 150, 1165

Frank. P. Grady, 225, 35, 14600, 250, 1500

Edwd. Powell, 150, 100, 8800, 350, 1000

Thomas Marshall, 255, 107, 17500, 150, 1000

Sanford Leckey, 140, 48, 7500, 150, 800

John L. Powell, 175, 150, 16300, 350, 1930

Alpheus Gibson, 309, 80, 19500, 300, 3000

William Ewers, 300, 80, 19000, 300, 1500

James Haws, 120, 20, 6000, 100, 725

David Tavener, 95, 23, 3500, 100, 700

Geo. W. Bowman, 95, 16, 6000, 200, 825

Sydney Hawling, 180, 41, 13260, 200, 900

John W. Garrett, 90, 10, 5000, 150, 500

Enoch Garrett, 125, 35, 8000, 300, 850

Levi Shumaker, 66, 30, 5760, 60, 500

William Tate, 121, 50, 11000, 100, 1006

Geo. B. McCarty, 270, 100, 15000, 100, 746

John R. Carter, 350, 90, 22000, 400, 2764

Thomas Francis, 231, 75, 12240, 300, 1460

Joseph T. Rector, 177, 30, 8280, 125, 1775

Harvey Rector, 100, 23, 4800, 125, 800

Robert C. Bowman, 457 100, 22280, 200, 1600

Wm. L. Gochnauer, 100, 19, 4760, 100, 600

Howard Leith, 115, 25, 8000, 100, 475

Wm. Leith, 170, 12, 9100, 200, 2290

Wm. Leith, 125, 25, 7500, -, -

Nancy Anderson, 250, 25, 15000, 450, 1545

Alfred Anderson, 150, 65, 8600, 150, 1270

Joseph Carr, 243, 40, 14000, 150, 1580

William Chamblin, 300, 71, 14000, 200, 1650

Wm. A. Reeder, 150, 60, 12000, 250, 2200

Joseph Baldwin, 77, 30, 5000, 50, 490

Isaac H. Lauck, 200, 32, 5800, 100, 300

Richard Carter, 113, 113, 9000, 100, 660

John Littleton, 100, 21, 5000, 150, 600

Alison Grason, 290, 60, 7000, 250, 1720

Hered Frazier, 280, 60, 12000, 150, 1520

Phebe Humphrey, 200, 55, 10200, 150, 1120

Francis Osburn, 200, 35, 7000, 150, 1500

Harriett Carpenter, 150, 35, 7500, 150, 1120

Geo. Keene, 470, 150, 18600, 300, 200

Joseph A. Hutchison, 240, 44, 11000, 125, 1670

A. G. Chamblin, 120, 20, 5600, 400, 755

John Butcher, 400, 100, 20000, 300, 2800

Wm. P. Thomas, 200, 60, 15600, 300, 2000

Benj. Stringfellow, 80, 20, 3600, 50, 300

D. P. Neill, 120, 19, 6000, 80, 250

Lucinda Palmer, 50, 10, 2500, 25, 150

Catherine Farr, 90, 10, 4000, 50, 400

Adison Osburn, 110, 40, 8250, 200, 790

Fenton Farr, 120, 28, 7000, 300, 1320

Paten Moore, 53, 15, 5000, 20, 325

Thomas Fred, 180, 50, 11500, 200, 940

Mary Hand, 80, 4, 4000, 20, 150

Thomas M. Humphrey, 150, 23, 7000, 100, 5000

Wm. Benson, 100, 35, 5360, 75, 600

John Bearson, 120, 60, 7000, 80, 500

Wm. Farr, 215, 45, 12000, 225, 1020

Turner Galleher, 80, 34, 4000, 100, 500

Frederick Leith, 200, 46, 12000, 400, 2000

John Keene, 300, 89, 20000, 250, 2000

Henry Plaster, 200, 100, 13000, 200, 700

John Plaster, 100, 30, 4500, 50, 300

Henry Plaster, 280 90, 12900, 150, 1280

Rossle Jacobs, 50, 20, 2550, 50, 300

Amos Johnson, 147, 40, 2000, 200, 570

Elizabeth Hutchison, 600, 100, 20000, 200, 2000

Joseph Garrett, 57, 6, 1500, 75, 600

Leven Richards, 150, 10, 5600, 125, 600

Wm. Atwell, 10, 50, 300, 15, 200

Francis Gulick, 200, 80, 6500, 50, 900

Bassill Ganes, 175, 25, 8000, 50, 300

Wiat Allen, 18, -, 1000, 25, 150

James Crage, 175, 50, 6600, 50, 900

Samuel Pillett, 170, 53, 5500, 50, 400

Squire Mathew, 192, 35, 8000, 150, 910

Hamelton R. Gulick, 340, 40, 13300, 50, 1820

Jos. H. Gulick, 88, 28, 6960, 150, 740

Joseph L. Hawling, 600, 208, 16000, 250, 5100

Thomas Moss, 30, 20, 500, 25, 300

Thomas Munday, 65, 60, 1000, 50, 300

Sarah Branaugh, 40, 10, 700, 20, 200

Patrick Hogen, 200, 50, 3740, 50, 1050

Henry Moffett, 150, 92, 4840, 175, 1364

William Dennest, 157, 157, 2900, 75, 500

Penelope Tyler, 300, 65, 4000, 100, 1000

Jonathan Beard, 190, 40, 6500, 125, 1175

John Ridecer, 220, 20, 5000, 75, 550

Chas. Ridecer, 275, 105, 5000, 75, 440

Thomas Ayres, 140, 100, 3600, 50, 680

Nathan Skinner, 150, 100, 2500, 40, 280

John B. Oden, 360, 180, 13500, 600, 2325

John B. Lever, 460, 150, 9500, 200, 1200

Moses Thomas, 500, 250, 22500, 200, 2200

George Barr, 110, 10, 3000, 25, 60

Samuel Simpson, 300, 42, 17000, 100, 2200

John Flinn, 72, 15, 1750, 15, 200

Sandford P. Rogers, 300, 100, 12000, 150, 1600

M. K. Baldwin, 36, 1, 5000, 20, 350

James R. Simpson, 160, 40, 8000, 200, 2650

Bushrod Skillman, 30, 2, 3000, 20, 150

Henson Simpson, 260, 75, 14000, 450, 1385

Philip Vansickler, 230, 70, 1500, 225, 1150

James T. Otley, 24, 6, 2500, 20, 230

John H. Simpson, 230, 92, 13000, 200, 985

John Cockrill, 226, 20, 10000, 200, 900

Wm. G. Young, 100, 39, 6000, 50, 300

John B. Young, 35, 12, 2500, 10, 190

Richard Tavener, 70, 12, 3300, 50, 350

George Gregg, 75, 15, 4200, 100, 650

Romulus Furgerson, 150, 50, 3000, 100, 860

Romulus Furgerson, 125, 43, 3000, -, -

Chas. Mount, 300, 140, 22000, 125, 3000

William Walker, 20, 8, 1680, 25, 300

Jonathan Ewers, 100, 70, 4000, 300, 1650

Jas. Albert Cox, 135, 30, 5800, 120, 700

Fielding Tavener, 43, 22, 4000, 60, 450

Samuel H. Nichols, 130, 53, 7500, 150, 1300

John F. Pirce, 56, 8, 3000, 100, 550

Abner G. Humphrey, 225, 90, 14600, 150, 1500

Harmon Lodge, 200, 50, 11500, 200, 1000

James Carlisle, 35, 12, 1800, 50, 400

Paten Davis, 20, 10, 1800, 20, 640

William Hampton, 136, 40, 6400,-, -

Thomas L. Humphrey, 250, 115, 16400, 200, 2820

James Alder, 140, 23, 7400, 100, 500

Flavius Lodge, 265, 100, 14600, 200, 1140

Moses Arnett, 100, 50, 4200, 100, 500

Rebecca Lodge, 130, 86, 9700, 200, 1090

John Beaty, 175, 41, 10800, 150, 1060

James W. Hill, 250, 150, 18000, 300, 1100

Phemius Osburn, 100, 42, 8200, 300, 750

Jonah Thomas, 150, 120, 7100, 150, 1000

George H. Allder, 100, 38, 4500, 80, 1000

Jonah Osburn, 265, 200, 20000, 200, 2160

Abraham Young, 30, 114, 7000, 100, 1300

James Howell, 72, 50, 1600, 50, 360

Jonah Purcell, 200, 80, 11500, 125, 1370

Crarew(Craven) Howell, 285, 150, 12500, 300 2060

Ashford Weadon, 125, 66, 6000, 150, 1260

Isaac B. Beans, 116, 15, 6500, 78, 780

Abner H. Beans, 140, 40, 3400, 200, 1500

Labin L. Hill, 270, 30, 15000, 200, 225

Harmon Gregg, 125, 48, 8300, 170, 860

James Heaton, 220, 100, 16000, 250, 1725

Thomson Osburn, 200, 79, 12500, 275, 1660

Joseph Worthington, 300, 60, 16500, 200, 1000

Chas. Hammerly, 80, 4, 3500, 50, 500

Washington Stone, 350, 50, 14000, 250, 1600

Bushrod Osburn, 226, 83, 13550, 200, 1750

F. M. Potts, 150, 50, 10000, 100, 600

N. R. Heaton, 230, 155, 16000, 100, 1000

Peter Compher, 90, 10, 4250, 100, 600

Thomas H. McCarter, 50, 12, 1500, 25, 200

Edward J. Potts, 225, 125, 14550, 500, 2100

Harrison Osburn, 250, 75, 19500, 300, 1700

Rheuben Jenkins, 30, 28, 1500, 150, 790

Wm. N. Hough, 160, 61, 9000, 425, 1860

Malen Beans, 100, 40, 5500, 40, 525

Joseph P. Grubb, 53, 48, 4500, 150, 737

Thomas Beans, 140, 46, 8400, 100, 775

Noble B. Peacock, 120, 30, 9000, 300, 670

Jonah Orrison, 90, 32, 6700, 200, 600

Lott Tavenner, 190, 50, 13200, 200, 1100

Wm. H. Frances, 160, 65, 9000, 250, 1100

John Frances, 120, 40, 11000, 50, 416

Hannah P. Wade, 150, 10, 65400, 300, 1000

Jacob H. Manning, 185, 16, 12000, 500, 1800

Joseph Pierpoint, 155, 30, 9000, 100, 750

George Warner, 120, 20, 8500, 100, 795

Samuel Pierpoint, 100, 20, 8000, 200, 500

Wm. Beans, 44, 4, 4000, 100, 500

Nathan Gregg, 120, 30, 9500, 250, 780

Joseph Virts, 60, 10, 5000, 75, 300

Joseph Mock, 140, 26, 10000, 200, 830

Miram Smith, 162, 50, 15000, 300, 1325

Fenton M. Love, 200, 10, 16500, 400, 1750

Levi James, 200, 25, 10000, 200, 1150

Jonathan Brown, 135, 45, 1800, 125, 600

Burr Brown, 90, 40, 7800, 100, 625

Bushrod Brown, 108, 41, 9000, 400, 825

Archibald McDaniel, 135, 40, 16000, 400, 1000

Jas. M. Kilgour, 310, 50, 22000, 700, 2200

John T. Debutts, 210, 40, 1150, 250, 1400

Wm. Orrison, 100, 11, 6600, 50, 420

Isaiah Beanes, 141, 48, 8500, 150, 540

Eli C. H. House, 102, 60, 8500, 100, 670

Eli A. Love, 125, 75, 9000, 200, 1100

Bernard Hough, 210, 60, 18000, 225, 1100

Craven A. Copeland, 100, 36, 8000, 150, 730

John R. White, 230, 70, 18000, 300, 1250

Elizabeth R. White, 220, 60, 16000, 250, 1800

Mary Howell, 110, 30, 5000, 50, 300

Malen Beans, 135, 40, 7000, 150, 600

Thomas J. Nichols, 180, 32, 10000, 200, 1000

Eans Purcell, 158, 60, 9800, 250, 950

Jas. H. Purcell, 90, 83, 9000, 100, 575

C.M. Vandervenre, 180, 100, 11500, 30, 150

Maley Thomas, 275, 115, 11800, 200, 900

Owen Thomas, 225, 120, 14000, 200, 1400

Jerred Chamblin, 178, 60, 12500, 150, 1275

G. G. Gregg, 80, 20, 5000, 150, 500

John W. Best, 80, 15, 5500, 200, 2500

Amos Beanes, 100, 25, 6500, 150, 725

Richard Adams, 269, 50, 17000, 200, 900

Edward Preston, 80, 50, 6500, 150, 800

Jesse Pickett, 120, 40, 9000, 100, 800

Jesse J. Dillen, 147, 149, 14800, 350, 960

Stephen Gregg, 147, 70, 11500, 150, 800

Lewis Taylor, 80, 69, 7300, 80, 600

Mary S. Taylor, 100, 43, 7500, 60, 450

R. C. Littleton, 300, 60, 14400, 800, 4325

Hannah Littleton, 150, 20, 8500, 200, 915

Joseph Gouchman, 185, 25, 11550, 200, 1142

John Wernel, 165, 20, 7500, 350, 920

John P. Newlen, 95, 20, 6750, 100, 600

George W. Hoge, 100, 40, 10000, 200, 900

Wm. H. Benton, 700, 250, 52000, 1500, 5000

Craven James, 300, 207 20825, 300, 1700

Washington Beavers, 145, 40, 9500, 400, 925

George W. Beavers, 80, 20, 500, 25, 400

Wm. H. Krantz, 100, 10, 4400, 200, 600

Isaac Pickett, 363, 125, 21500, 500, 1300

Joseph Pancoast, 105, 15, 6000, 100, 600

Jonah Nichols, 250, 120, 20000, 500, 2000

Jonathan Goodding, 140, 40, 9900, 200, 700

John W. Loveless, 143, 45, 10000, 150, 1400

Gibson Gregg, 314, 90, 20000, 200, 550

Fenton Updike, 101, 30, 4550, 100, 500

Samuel Thomson, 60, 20, 3000, 50, 225

Leonidas Peugh, 42, 12, 3350, 75, 280

Malen Tavenner, 120, 45, 8300, 200, 860

John Tavenner, 215, 115, 19800, 200, 1375

Bazzle Shoemaker, 60, 25, 5500, 50, 450

Heston Hirst, 117, 40, 9420, 200, 700

James H. Jones, 60, 30, 5400, 150, 325

John Pancoast, 120, 80, 8000, 200, 800

Hannah J. Pickett, 271, 70, 17000, 200, 1000

Wm. Otley, 156, 20, 8800, 200, 1000

M. P. Watson, 50, 9, 2500, 40, 500

Benjamin Taylor, 90, 23, 6780, 150, 650

Samuel Nichols, 150, 50, 10000, 150, 950

Joseph Nichols, 250, 80, 20000, 500, 3600

Amos Hughs, 185, 65, 16240, 200, 1740

Joshua Pancoast, 130, 20, 9500, 25, 350

John Mead, 83, 12, 6500, 60, 500

Wm. Gregg, 168, 32, 11000, 200, 1600

Isaac Willson, 136, 40, 7000, 150, 2580

Isaac Willson, 110, 10, 6000, -, -

Thomas Nicholas, 240, 120, 22000, 150, 1500

John Pitzer, 211, 60, 15000, 200, 1200

George Tracy, 80, 50, 5000, 100, 625

Richard Wyukorp, 148, 20, 6500, 150, 800

Elizabeth Hill, 60, 50, 4000, 50, 400

Mason James, 141, 48, 7500, 150, 820

Burr P. Chamblin, 75, 32, 4500, 150, 800

Enos T. Best, 110, 40, 6750, 150, 800

Mason Chamblin, 62, 40, 7500, 40, 500

Isaac Nichols, 125, 35, 9000, 500, 1300

Isaac Nichols, 55, 35, 9000, -, 50

Tazwell Lovett, 200, 50, 10000, 200, 250

Wm. Willson, 200, 45, 1200, 100, 800

Poley Birdsall, 110, 40, 7500, 200, 1200

James McDaniel, 200, 50, 15000, 300, 1520

Benj. Walker, 30, 20, 2500, 150, 500

James Welch, 50, 10, 3000, 100, 450

Samuel Mageath, 110, 12, 6100, 150, 800

Philo Crane, 135, 25, 8000, 200, 1275

Ludwell Luckett, 500, 215, 21500, 200, 2500

Edgar Russle, 100, 30, 5000, 100, 800

John E. Mount, 350, 100, 18000, 400, 3000

Arminda Davis, 127, 73, 4000, 175, 1300

Henderson Coe, 200, 40, 9600, 250, 900

John H. Brown, 110, 55, 9100, 250, 1200

Chas. Powell, 100, 30, 600, 100, 300

David Hughs, 175, 75, 10000, 200, 600

Chas. W. Henderson, 150, 150, 3000, 100, 700

Jacob F. Cost, 179, 50, 4600, 200, 1200

Samuel Carr, 180, 50, 10000, 300, 1500

David A. Wilbourn, 141, 40, 7000, 100, 500

Joseph Davis, 130, 38, 7000, 200, 900

Chas. H. Hidgen, 113, 25, 7000, 50, 400

Elijah Hanes, 175, 80, 16000, 200, 950

James Laycock, 86, 30, 6000, 125, 400

Jesse Hoge, 102, 30, 8000, 100, 600

Eli J. Hoge, 80, 30, 7700, 100, 600

Thomas Gore, 95, l400, 9500, 100, 800

Wm. Nichols, 72, 14, 6000, 125, 800

Henry Taylor, 215, 100, 19000, 300, 2100

Richard Taylor, 180, 20, 12000, 200, 800

Edmond M. Janney, 90, 10, 700, 40, 250

Elijah Janney, 43, 14, 5000, 25, 200

Mary A. Bolen, 33, 20, 3000, 50, 200

William Homes, 78, 25, 7500, 200, 1000

Enoch Fenton, 230, 40, 11200, 200, 1200

Jowos Janney, 160, 40, 11000, 250, 1300

David Hesser, 116, 40, 6300, 125, 600

Sarah Moore, 130, 20, 6000, 75, 700

Eli H. Nichols, 131, 40, 8500, 150, 725

Thomas Hesser, 65, 15, 7000, 50, 180

Chas. N. Taylor, 42,-, 6000, 100 400

Timothy Taylor, 150, 50, 13000, 500, 900

Eli Tavenner, 179, 35, 12800, 400, 1850

Aquilla Janney, 137, 35, 11000, 250, 800

Jonah Hatcher, 200, 100, 18300, 300, 900

Benj. Birdsall, 240, 63, 18000, 200, 1200

Amos Whitecer, 55, 5, 4000, 100, 900

Bernard Taylor, 85, 30, 7500, 300, 1000

Cecelia Heaton, 162, 71, 11500, 200, 1100

Lydia Taylor, 90, 47, 8300, 250, 750

Geo. W. Noland, 90, 25, 7000, 300, 1400

Thomas E. Hatcher, 114, 45, 9000, 150, 700

James Allder, 110, 40, 6000, 150, 500

Joshua Reid, 113, 20, 5500, 150, 500

Levin Ogden, 52, 18, 2800, 80, 482

Saml. Beans, 90, 10, 5000, 150, 600

Malinda Dowell, 160, 20, 11000, 200, 1000

Wm. N. Everhart, 175, 25, 12000, 200, 1000

David Brown, 180, 90, 15500, 200, 1900

Edwd. Brown, 100, 13, 7400, 250, 1400

James M. Frame, 20, 5, 1900, 75, 350

Eli T. Reese, 200, 45, 17200, 150, 700

Wm. F. Mercer, 185, 80, 17200, 250, 850

Richd. Brown, 246, 50, 19000, 350, 1100

Soloman Lucas, 70, 30, 6500, 100, 200

Washington Vandervanter, 275, 105, 20000, 700, 1200

Gabriel Vandervanter, 175, 75, 15000, 400, 1600

Elenia Rogers, 235, 50, 18800, 150, 1100

Sarah Sands, 75, 28, 6200, 25, 300

Samuel Brown, 160, 100, 14000, 200, 900

Wm. McCray, 165, 70, 14000, 250, 750

Nancy Hatcher, 212, 100, 18700, 400, 1300

William Tavenner, 54, 17, 5000, 100, 500

Jane Rogers, 130, 60, 9500, 100,800

Joseph Heaton, 172, 75, 13500, 200, 1200

William Caruthers, 125, 25, 10000, 200, 1500

Philip H. Wyncoop, 100, 100, 5000, 100, 400

Thomas Hogue, 160, 49, 9000, 250, 700

Wm. Nicholls, 100, 23, 6700, 200, 800

Nancy R. Shepherd, 220, 20, 12000, 25, 400

John Pancost, 100, 80, 16800, 150, 200

Nailor Shumaker, 100, 30, 7800, 100, 700

Joseph Gibson, 85, 10, 4300, 75, 500

John Nickolls, 350, 150, 24000, 250, 3000

Wm. Young, 80, 13, 5000, 100, 400

Thomas Young, 115, 45, 9600, 100, 1000

John Young, 35, 5, 2500, 50, 175

Thomas Nickolls, 165, 70, 1550, 400, 1600

Tho. R. Smith, 150, 50, 12000, 200, 500

Isaac Nicholls, 120, 30, 9000, 250, 1400

Saml. N. Brown, 115, 25, 9100, 250, 600

Robert A. Ish, 332, 32, 8000, 200, 1200

John F. Allen, 350, 250, 12000, 200, 1130

Lewis Birch, 275, 65, 14500, 200, 800

Wm. Fulton, 100, 33, 4500, 200, 500

David Carr, 230, 70, 21000, 200, 2000

Josephus Carr, 400, 200, 22800, 200, 2000

Jonah Tavener, 175, 100, 12000, 125, 500

Wm. Hall, 343, 57, 18000, 100, 1200

John Aldridge, 485, 130, 34000, 300, 2800

Wm. H. Brown, 130, 70, 10000, 200, 1050

Joshua Hatcher, 234, 100, 18000, 150, 2400

Abraham Skellman, 120, 40, 8500, 150, 920

John Jones, 100, 34, 5000, 25, 250

Wm,. Lickey, 110, 12, 6000, 100, 460

John Smith, 230, 30, 14300, 250, 250

Wm. Hogue, 105, 65, 11200, 200, 600

Nancy R. Donahoe, 95, 6, 6000, 150, 400

Benj. Davis, 150, 70, 8800, 150, 1100

John White, 125, 100, 9000, 200, 700

Isaac Nicholls, 261, 125, 15000, 150, 1300

Levi White, 200, 100, 13000, 200, 1500

Elisha Holmes, 138, 40, 9000, 200, 830

Eden White, 165, 40, 9500, 200, 900

Richd. White, 146, 45, 8600, 200, 1025

Benj. Brown, 60, 15, 3700, 100, 400

Eli. Nixon, 115, 32, 8000, 150, 700

Garret Wyncoop, 35, 5, 1500, 100, 500

Jonah Nixon, 116, 75, 9500, 150, 800

Jas. White, 272, 40, 12500, 200, 1100

Thos. Powell, 150, 50, 6000, 150, 850

Armanda Poindexter, 195, 5, 3500, 100, 550

Benj. Saunders, 80, 10, 3500, 50, 200

Asbury Nixon, 100, 75, 7000, 100, 500

Fenton Hampton, 75, 50, 5000, 100, 800

Chornelus Vandervanter, 226, 100, 14000, 150, 900

George Rhodes, 380, 120, 17500, 20, 2000

Isaac Vandervanter, 120, -, 10, 9500, 150, 600

Francis E. Shrewd, 270, 80, 17500, 300, 2000

Gustavus Elgin, 140, 60, 5000, 75, 400

Nathaniel J. Skinner, 600, 230, 18500, 200, 1350

Thos. L. Ellzey, 250, 125, 13000, 200, 1350

Wm. Polton, 95, 5, 1000, 100, 500

Alex. Elgin, 200, 75, 6800, 100, 1000

Isaac Carr, 146, 80, 4000, 100, 850

Lewis Donahoe, 190, 60, 6250, 150, 700

Reed Poulton, 103, 40, 4300, 75, 624

Burr Harrison, 600, 100, 17500, 300, 2500

John T. Lynn, 237, 45, 8500, 125, 820

Peter Etcher, 68, 30, 2500, 75, 500

Wm. McPherson, 150, 50, 5000, 100, 800

Mathew E. McPherson, 100, 25, 2500, 100, 600

Wesley S. McPherson, 100, 350, 13000, 200, 1100

Peter Myers, 325, 75, 10000, 175, 1200

Thomas N. Claggett, 300, 120, 25000, 400, 2300

Thomas H. Claggett, 500, 160, 20000, 500, 3500

Isaac Hawling, 170, 110, 8000, 200, 1000

George A. Dodd, 151, 25, 7000, 300, 1500

Robert Elgin, 200, 100, 9000, 200, 1000

Dennis McCarty, 250, 50, 9000, 125, 1000

Wm. D. Havener, 250, 55, 10000, 175, 900

Ignatius Elgin, 154, 50, 3000, 75, 500

James Higdon, 50, 8, 1800, 60, 300

Samuel Laycock, 80, 20, 3500, 100, 550

Sanderson Thrift, 210, 100, 12000, 150, 2000

Francis M. Carter, 370, 88, 25000, 800, 3900

Thornton Saffer, 270, 80, 7000, 200, 1200

Stacy Gulick, 200, 200, 5000, 100, 300

James W. Taylor, 170, 30, 5000, 30, 400

Francis Daniel, 54, 27, 3600, 125, 375

Lemuel Daniel, 272, 70, 8000, 200, 500

John W. Fairfax, 650, 85, 40000, 200, 5000

Robert Bently, 600, 120, 22000, 500, 5000

Jno. J. Hogueland, 40, 10, 4000, 50, 200

Jesse McIntosh, 85, 85, 1700, 50, 200

Francis Elgin, 360, 110, 14000, 200, 1500

M. C. Shumate, 215, 70, 8000, 95, 690

Wm. Donahoe, 120, 40, 5000, 100, 500

Sanford Gulick, 330, 143, 15000, 200, 1600

Benj. F. Taylor, 200, 166, 7500, 200, 1500

Westley Brooks, 130, 30, 3000, 50, 400

John Morgan, 220, 130, 10500, 150, 800

John Daniel, 143, 60, 6000, 100, 600

John Lee, 100, 50, 3000, 175, 400

Almond Birch, 150, 50, 5000, 100, 500

A. G. Smith, 225, 95, 4000, 100, 700

C. F. Hempston, 300, 30, 20000, 300, 675

B. F. Carter, 618, 100, 35000, 460, 3600

Elizabeth O. Carter, 1200, 2400, 75000, 1500, 6000

Elizabeth O. Carter, 700, 300, 40000, 500, 3500

L. W. Swart, 200, 67, 5500, 150, 400

Louisa County, Virginia
1860 Agricultural Census

The Agricultural Census for Virginia 1860 was microfilmed by the University of North Carolina Library under a grant from the National Science Foundation from original records at the Virginia Department of Archives and History in 1963.

There are forty-eight columns of information on each individual. Only the head of household is addressed. I have chosen to use only six columns of information because I feel that this information best illustrates the wealth of individuals. The columns are:

1. Name
2. Improved Acres of Land
3. Unimproved Acres of Land
4. Cash Value of Farm
5. Value of Farming Implements and Machinery
13. Value of Livestock

The first half of this county is very faint and difficult to read.

S. G. Trice, 200, 239, 4265, 65, 891
Silas Bagley, 277, 298, 6660, 65, 819
Gudard T. McCaddy, 1100, 463, 13504, 720, 3013
Chas. E. Jones, 50, 210, 2600, 70, 133
Jno. M. Johnson, -, -, -, -, 135
S. W. Clarke, -, -, -, 110, 72
George A. Louis, -, -, -, 4, 35
Wm. E. Mambleton, 50, 247, 3712, 200, 390
Orrison Gibson, 80, 10, 450, -, -
France M. Gibson, -, -, -, 40, -
Jos. C. Bagley, 240, 260, 7500, 170, 2170
Jas. Bagley, 210, 275, 4750, 200, 2285
William Bagley, 175, 149, 2592, -, -
Drucilla Bagley, 400, 600, 10000, 125, 740
Benjamin T. Trice, 200, 102, 4000, 200, 1348

Wm. T. Crupton(Cruxton), 200, 185, 2305, 75, 527
Mrs. E. Hart, 250, 281, 10620, 200, 1329
Anderson Hart, 350, 150, 7500, 150, 713
Alexander V. Trundixler, 125, 119, 2440, 100, 522
Jno. L. Collins, 613, 150, 15260, 275, 2096
Herman Harris, 250, 187, 8000, 400, 1040
Mrs. A. M. Johnson, 100, 108, 2180, 75, 646
Andrew J. Welson, 100, 440, 9340, 200, 948
J. A. Vaeden, -, -, -, 48, 213
Wm. A. Kupee, 200, 100, 9000, 445, 1248
Robert Goodwin, 550, 600, 13800, 445, 1425
Caius M. Carpenter, 300, 1221, 6135, 200, 1015

Harman Barnage Sr., 538, 330, 12920, 550, 1560

Jno. B. Carpenter, 150, 50, 3000, 150, 278

Jno. E. Bumpass, 75, 25, 1320, 10, 120

J. W. Bagley, 100, 100, 4000, 75, 387

F. N. Trice, 183, 200, 5743, 175, 1039

Semple Goodwin, 470, 470, 400, 250, 876

M. Buckner Trustee for Mrs. S. Mills, 420, 510, 13350, 400, 1904

Chas. E. Cornasean, -, -, -, 5, 3

Chas. T. Cosby, 132, 131, 2630, 110, 1172

Francis E. Moss, 300, 217, 5170, 125, 570

Lewis M. Harris, 120, 130, 3012, 100, 675

Humphrey J. Parrish, -, -, -, 30, 100

Jno. G. B. Coleman, 39, 1, 860, 50, 346

Elizabeth J. Hannah, 450, 450, 900, 300, 1433

Wm. Cooke, 574, 240, 10582, 225, 478

Jas. Quisenberry, 23, 27, 3500, 75, 397

John A. Harris, 120, 80, 2000, 50, 370

Bently Fleming, 74, 2, 456, 25, 184

John D. Jones, 330, 100, 5160, 150, 741

Thomas A. Jones, 175, 151, 2608, 75, 615

Frances M. Moss, 360, 180, 5400, 75, 437

Mrs. B. Tellius, 425, 428, 10200, 200, 412

John M. Tulla Sr., 80, 105, 1850, 65, 488

Henry A. Lewis, 35, 12, 282, 4, 93

Jno. G. Bumpass, 110, 100, 4000, 50, 390

Larkin Luck, 24, 14, 570, 10, 69

Priscilla Tulla, 150, 150, 2100, 50, 306

Jno B. Cook, 119, 100, 2628, 40, 482

Drucilla F. Smithy, 114, 100, 3200, 75, 583

J. S. Cooke, -, -, -, -, 250

Anderson Puller, 18, 1, 380, 10, 94

Wm. McAlister, 60, 13, 285, 30, 182

Wm. A. Freeman, 200, 65, 3168, 50, 382

Garland Duke, 400, 400, 8000, 80, 447

James Hall, 16, 3, 216, 6, 34

Jno. W. Darvin, 25, 18, 344, 60, 83

Jno. Campbell, 100, 120, 2200, 50, 295

Geo. W. Duke, 70, 72, 1702, 100, 222

W. H. Kinney, 120, 78, 1584, 50, 259

Chas. Kinney, 60, 33, 755, 40, 190

Jas. A. Claybrooke, 800, 830, 24750, 425, 1939

Charlotte Dickinson, 300, 287, 10566, 500, 2035

Jas. G. Whiller Sr., -, -, -, 25, 695

Elizabeth Bird, 163, 40, 2030, -, 23

Adison T. Coales, 300, 136, 6540, 100, 669

Sarah D. Cooke, 106, 100, 3090, 55, 515

Jas. T. Dickinson, 200, 300, 12500, 500, 3492

Jas. M. Trice, 400, 600, 15000, 600, 2478

B.M. Buckner, 300, 200, 5000, 204, 1222

Mary Tyler, 25, 48, 1314, 15, 140

M. A. Bumpass, 100, 550, 1922, 60, 150

E. M. Sims, 65, 20, 1105, 65, 194

Jno. M. Bullock, 280, 139, 7123, 240, 983

A. J. Goodwin, 240, 90, 5085, 315, 782

Wm. E. Longdon, 400, 481, 9691, 100, 1003
Wm. T. Harris, -, -, -, 200, 1055
Laben R. Swift, 260, 200, 7000, 60, 610
Carter Hancock, 250, 222, 7080, 150, 1261
Goin Gibson, 5,-, 200,-, 73
Wm. C. Saunders, 2,-, 250, -, -
John Richardson, 100, 75, 1320, 10, 209
Wm. B. Cocke, 200, 187, 7740, 220, 968
Wm. J. Mills, 100, 118, 3270, 150, 1003
Elias F. Harris, 80, 197, 3324, 50, 423
W. M. Hall, 40, 170, 252, -, 30
Jno. T. Bibb, 500, 476, 19520, 400, 1824
D. M. Kridlston, 600, 400, 12000, 800, 1805
Jno. B. Kridlston, 116, 125, 2210, 75, 592
Ro. S. Cosby, 200, 70, 2350, 40, 544
Wm. Waddy, 280, 286, 10188, 203, 1524
Miss Clara L. Garland, 700, 300, 24000, 250, 1392
Jno. Garland, 125, 100, 2636,-, -
Jno. L. Nuckolls, -, -, -, -, -
L___ton Nuckolls, 300, 287, 5860, 178, 1117
A. F. Gentry, 250, 150, 1800, 30, 830
_____ Lockheart, -, -, -, -, -
Sarah Harris, -, -, -, -, -
Dosia Duckman, -, -, -, -, -
Matthew C. Lacy, 226, 100, 3912, 130, 831
Wm. Terry, 200, 70, 3240, 40, 130
Thomas Terry, -, -, -, -, -
Alpheus Parsons, 75, 60, 1300, 20, -
Mary K. Henshaw, 125, 100, 2250, 20, -

L. B. Harrison, 104, 50, 1540, 150, 200
Wingfield Cosby, 600, 400, 10000, 120, 1008
Cordeane E. Brimly, 138, 40, 1780, 50, 246
Martha L. Loellin, 40, -, 300, -, 30
Wm. T. Snarp, 200, 163, 2451, 25, 93
Luch Tate, 60, 10, 490, 26, 163
Lansford Duke, 350, 168, 4144, 100, 617
Wm. J. Parsons, -, -, -, 25, 140
Henry W. Lasseter, 69, 69, 728, 40, 259
Henry A. Lasseter, -, -, -, -, -
Louisa Tate, 100, 70, 1700, 65, 267
Albert Henshaw, 54, 15, 640, 60, 203
_____ K. Tate, 60, 60, 960, 30, 154
Carter Tate, -, -, -, -, 24
Wm. Cauker, -, -, -, -, 24
Sallie Kersey, 30, 30, 300, 20, 30
Lindsey Richardson, 75, 30, 840, -, -
Thomas Richardson, 100, 15, 920, 23, 75
Elizabeth Sharp, 170, 125, 3540, 130, 604
Garland Sims, 120, 60, 1440, 30, 267
M. D. Sharp, -, -, -, -, -
Jno. G. Gardner, 44, 6, 900, 15, 290
Thos. P. Goodsy, 252, 30, 2538, 50, 415
Eli M. Hall, 130, 43, 1211, 50, 265
Carter M. Greely, 50, 30, 800, 80, 586
Abner Greeley, -, -, -, -, 18
Ann Deals, 75, 25, 600, 15, 216
Ro. J. Sharp, -, -, -, -, 35
Nancy M. Hambleton, 125, 75, 3000, 15, 100
George Tyler, 307, 500, 8000, 340, 1400
Jas. R. Goodwin, 240, 256, 11720, 187 970

Sallie R. Holladay, 500 200, 14000, 100, 907
Ann C. Carter, -, -, -, -, 20
John Carter, 315, 130, 22265, 40, 336
Prudence Talley, 275, 125, 2400, 65, 307
Richd. F. Talley, -, -, -, -, 75
Geo. W. Harris, 125, 125, 6800, 75, 511
John Foster, 5, 12, 136, -, 75
Alexander M. Barret, 40, 10, 500, 50, 137
Lewis Harrison, -, -, -, -, 37
Edwin J. Baker, 620, 580, 20000, 350, 1120
Roy___ S. Morrison, 100, 120, 3000, 100, 429
Thomas B. Harris, 700, 407, 12000, 250, 1299
P. B. Knudlston MD, 640, 420, 10700, 550, 2415
Unity Y. Knudlston, -, -, -, -, 60
Lewis S. Knudlston, -, -, -, -, 150
John Thacker, -, -, -, -, 33
Sally A. Jones, -, -, -, -, 20
Thomas Barish, 106, 50, 1560, 60, 332
John W. Kinnsy, -, -, -, 95, 128
Mordica Thompson, 50, 10, 360, 5, 66
Wm. F. Seilst, 123, 60, 1098, 50, 180
Geo. J. Waldrop, -, -, -, -, 338
R. B. Davis, 110, 360, 10000, 325, 1270
Wm. Arensking, 300, 320, 7440, 100, 453
Jas. A. Wharton, -, -, -, -, 35
Wm. Camden, 40, 60, 600, 15, 182
Henry Kenady, 45, 55, 600, 15, 259
Cundif _. Soiles, 25, 75, 500, -, 97
Harwood T. Estas, 60, 181, 1246, 15, 180
Ro. Hart Jr., 90, 39, 800, 40, 228

Lewis H. Pruller, 136, 30, 1162, 40, 436
Andrew J. Wheeler Sr., -, -, -, 10, -
John Hancock, 400, 370, 9240, 400, 1990
Thomas Nelson, 360, 225, 3850, 172, 936
Garland Haynes, 100, 90, 1000, 150, 350
John Tally, 100, 90, 1000, 50, 288
Richd. Talley, 200, 49, 1494, 40, 326
Saml. C. Harris, 200, 40, 2000, 50, 495
Nathan Ware, 900, 500, 8000, 125, 797
Ro. G. Mansfield, 113, 30, 1300, 55, 350
James Heler, 120, 60, 1800, 25, -
Jas. N. Patterson, -, -, -, -, 235
Samuel Talley, -, -, -, 50, 518
Jno. _. Marshall, 90, 44, 1200, 45, 393
Thomas L. Gesselor, -, -, -, -, 132
Madison Grady, 60, 10, 420, 16, 74
Ro. Hart, 300, 413, 5700, 50, 704
Edward S. Wien, 30, 66, 672, 10, 144
Edmond N. Ealer, 50, 169, 2190, 75, 303
Lucy Bird, 150, 50, 1000, 10, 78
John Harper, 150, 90, 1200, 65, 459
Lewis Harper, -, -, -, -, 76
Mary A. Harper, 20, 3, 161, -, -
Alexander C. Hambleton, -, -, -, -, 115
Henry Williams, -, -, -, -, 40
Stephen Darrington, 150, 410, 3360, 300, 1110
James H. Hesler, -, -, -, 10, 28
Jno. S. Spicer, -, -, -, -, 96
G. E. M. Wallass, -, -, -, 150, 619
Geo. W. Greene, 62, 16, 390, 15, 75
Martha Shiffler, 10, 153, -, -, -
Jos. T. Cosby, 100, 166, 2660, 40, 297
Ro. M. Baker, 70, 30, 3000, -, -

Wm. Baker, 800, 1100, 28500, 250, 6091

E. F. Gunter, 225, 75, 4500, 106, 133

Chas. M. Dickinson, 163, 20, 2196, 25, 346

Jos. Timberlake, 121, 55, 1760, 40, 266

Lewis Smith, 46, 15, 488, 85, 271

____ Whitlock, 30, 10, 400, 15, 258

Bartholomew Whitlock, 50, 100, 1600, 20, 252

Abraham E. Estes, 20, 60, 500, 25, 92

Andrew L. Mills, 75, 55, 1320, 100, 630

Martha A. Nuckolls, 300, 260, 4480, 117, 789

Wm. H. Goodwin, 251, 250, 5010, 350, 1494

Jno. T. Goodwin, 273, 225, 5000, 170, 1333

Thomas Hall, -, -, -, -, 410

Martha Acros, 2, 1, 30, -, 88

Nathan H. Crawford, 425, 425, 9500, 300, 1202

Jno. W. Walker, 200, 215, 4980, 75, 1192

Wm. A. Robinson, 100, 20, 1440, 55, 733

Caleb Gray, -, -, -, 35, 216

Anderson M. Trice, -, -, -, 15, 81

Saml. S. Dickinson, 5, 5, 100, -, -

Safronia W. Maupin, 123, 100, 3345, 60, 420

Aderlia H. Trice, 200, 108, 3080, 20, 219

Ro. N. Trice, 331, 100, 10775, 80, 1180

Jas. L. Gunter, 275, 125, 10000, 100, 877

Bushrod W. Baker, 100, 49, 2780, 100, 1060

Jacklin D. Gillum, 118, 20, 3380, 50, 398

Ira S. Gillum, 170, 30, 2400, 50, 433

Archibald D. Arnett, 231, 231, 4620, 30, 327

Alfred M. Baker, 120, 43, 2471, 125, 600

Dicy H. Boxley, 575, 200, 11625, 200, 1164

Patey C. Edwards, 90, 50, 1400, -, -

Elliott E. Darrell, -, -, -, -, 156

Albert G. McGehee, 3, -, 330, -, -

Garrett E. McGehee, 200, 220, 3360, 75, 420

Wm. D. Mansfield, 780, 332, 13344, 383, 2605

Mary B. Mansfield, 100, 50, 3000, -, -

Edward S. Mansfield, 200, 100, 6000, 350, 850

Lewis Ballard, 39, 10, 414, 15, 94

John Talley, -, -, -, 20, 147

Littleton Talley, 110, 87, 1960, 25, 388

John Q. Graves, 40, 60, 700, 10, 76

Robert Long, 52, 35, 435, 7, 131

WM. Bellamy, -, -, -, 110, 175

James Smelsy, -, -, -, -, -

Henry Harris, -, -, -, -, -

N. W. Harris, 900, 1100, 32800, 1000, 3500

Alexander Thomson for Wm. Thompson, 400, 400, 12000, 100, 386

Jos. E. Young, 6, 13, 400, 15, 100

Charles Harris 169, 72, 1014, 15, 170

George Kennen, 13, 11, 250, 20, 155

Joseph Tribble, 25, 48, 730, 25, 67

A. W. Thompson, 200, 95, 5291, 75, 308

Phillip Timberlake, 108, 108, 3888, 70, 240

Frances Arnett, 73, -, 1460, 50, 185

Hugh L. G. Hiler, 235, 120, 7000, 150, 1205

Wm. Y. Hiler, 30, -, 600, 250, 1220

Wm. Goodwin MD, 375, 350, 18125, 350, 1692

P. M. Daniel, 125, 61, 5000, 75, 484

Jabez Massie, 240, 120, 7200, 75, 518

Jos. B. Goodwin, -, -, -, -, -

Jas. J. Trice, 65, 96, 2415, 20, 195

Alfred W. Trice, -, -, -, 75, 290

Timden Smelsy, 60, 67, 2500, -, -

Jno. L. Burruss, 270, 132, 12060, 600, 1533

Edmond P. Goodwin, 517, 115, 18900, 200, 2060

J. M. Baker, 300, 131, 10775, 300, 1458

Julia A. Holladay, 677, 170, 25000, 500, 2000

R. N. Dickinson, 262, 125, 6966, 200, 770

N. Smelsy, 110, 40, 3000, 100, 340

Ro. Hesler, 200, 38, 1190, 40, 485

John Hesler, 140, 130, 3200, 50, 730

Est. T. Lipscomb, 330, 170, 15300, 225, 1584

Henry Harris, 666, 150, 22900, 300, 1346

John Biggers, 271, 100, 7420, 150, 944

Sarah T. Shisler, 100, 100, 4000, 100, 177

James L. Kean, 300, 220, 10000, 340, 1075

Wm. A. Gillespie, 370, 750, 12400, 1050, 1771

John C. Harris, -, -, -, 100, 476

Thomas Norris, 246, 300, 5520, 75, 524

W. W. Downer, -, -, -, 125, 138

W. T. Eawter, 150, 55, 1640, -, 145

C. T. Timberlake, -, -, -, 25, 150

Joel Estes, 292, 100, 7800, 100, 248

M. W. Quarles, 249, 127, 10000, 400, 2360

Clayton Matthews, 400, 50, 5400, 125, 491

Richd. G. Bibb, 500, 206, 7000, 250, 982

Geo. DE. McGehee, 450, 288, 11000, 400, 1836

Wm. E. Woolfolk, 444, 60, 15120, 250, 2270

Anderson Talley, 146, 200, 3460, 125, 580

Jas. R. Robertson, 250, 250, 3000, 150, 529

Dabney Lockie, 100, 50, 1500, 100, 530

John Bibb, 60, 270, 1980, 50, 422

Hardin Chambers, 150, 50, 1200, 60, 141

Garret Atkins, 200, 100, 1800, 40, 160

Ann E. Robertson, 100, 30, 1300, 50, 542

Wm. G. Duke, 100, 500, 6000, 250, 715

Wm. H. Harris, 70, 30, 1000, 75, 210

Nancy Butler, 68, 25, 930, 50, 296

Abner Coy, 75, 48, 1200, 250, 245

Jno. Trainheour, -, 75, 370, -, -

Richd. T. Ogg, 75, 120, 1940, 50, 96

Jerrelin Dernes, 100, 71, 1710, 100, 410

Ed. T. Hawkins, 125, 225, 4500, 150, 1415

Archibald Shiflet, 30, 207, 1185, 50, 175

Ro. T. Gentry, 200, 60, 2600, 75, 169

Lewis D. Robertson, 1131, 150, 1578, 100, 553

Jas. M. Norford, 8, 22, 250, -, -

Wm. L. Hall, 10, 22, 192, -, -

James Hall, 120, 230, 2000, 75, 249

Thos. Sutherland, 60, 240, 5500, 80, 319

John B. Strange, 200, 365, 11300, 225, 1032

Albert Spicer, 25, 35, 960, 75, 210

Chas. G. Goodman, 500, 500, 20000, 660, 1639

Wm. S. Carter, 400, 650, 10500, 1000, 2310

W. G. Turner, 125, 96, 1890, 50, 435

Wm. G. T. Nelson, 1300, 1130, 32710, 300, 2415

Catharine Mills, 300, 140, 4800, 50, 470

Alex. Netherland, -, -, -, 50, 612

C. G. Trevelian, 350, 222, 5720, 75, 350

Thomas Ogg, 342, 684, 5130, 190, 723

Nancy Lewis, 22, 125, 1460, 50, 209

Reuben T. Chewning, 120, 170, 7250, -, 729

Christopher Valentine, 123, 100, 4460, 160, 810

James Vest, 700, 500, 48000, 485, 4048

W. K. T. Michie, 400, 200, 18000, 450, 1952

John R. Quarles, 700, 1231, 22250, 300, 2336

C. H. Boulware, 115, 180, 8850, 320, 1934

Henry W. Jones, 100, 176, 5520, 200, 760

James Fielding, 10, 27, 1345, 40, 140

Robert M. Sims, 5, 20, 300, -, -

W. T. Reynolds, 300, 822, 21220, 350, 2784

F. M. Leake, 2, 3, 700, -, -

James Mahames, 55, 30, 510, 70, 329

Est. J. Lindsay decd, A. F. Taylor trustee, 2000, 1440, 86000, 1500, 10394

Matthew Butler, 5, 45, 150, -, -

Annie Brancharm, 200, 308, 10160, 30, 940

Wm. M. Baker, 500, 400, 27000, 225, 1734

Thos. N. Gregory, 5, 5, 100, -, -

Sarah Brainha___, 550, 200, 22500, 225, 1874

Charles Quarles, 700, 690, 26300, 350, 2926

Matilds Wash, 100, 175, 2750, 40, 386

Richard Kenison, 250, 150, 2400, 120, 667

Edwards S. Wash, 100, 150, 1150, 30, 323

George Gibson, 150, 100, 2500, 45, 470

Robert P. Bibb, 75, 90, 990, 40, 374

Charles Denne, 10, 15, 1800, -, -

A. F. Butler, 201, 135, 5000, 75, 814

Wm. W. Beadles, 82, 164, 3680, 20, 636

Robert T. Gooch, 75, -, 6200, 5, 250

J. T. Cowherd, 488, 300, 17410, 225, 1280

Sarah K. Jones, 225, 125, 7000, 150, 756

Benjamin Ogg, 75, 25, 500, 25, 88

James Harris, 8, -, 80, -, -

Elisha Harris, 11, 11, 220, 10, -

James D. Pooler, 250, 175, 8500, 150, 818

Joseph Powel, 160, 165, 9750, 50, 419

Elijah Bibb, 220, 220, 6600, 100, 641

C. T. Poindexter, -, -, -, -, 332

Jesse Bibb, 67, 67, 670, 20, 252

G. W. L. Harper, 350, 685, 7460, 150, 919

John S. May, 500, 700, 18000, 300, 1245

John Graves, 325, 325, 9735, 200, 978

Jas. R. Poindexter, 85, 100, 1104, 75, 555

D. Poindexter, 200, 400, 3600, 50, 162

Robert Bibb, 30, 20, 400, 20, 155

John N. Grovin, -, -, -, 10, 76

David T. Butler, -, -, -, 76, 394

Thomas J. Barret, 400, 580, 9800, 300, 1762

Judy Knapper, 10, -, 100, 4, 6

Ro. L. Jones, 110, 100, 1470, 112, 328

Samuel Ogg, 100, 34, 3350, 150, 421

G. T. B. Hesler, 200, 204, 2900, 60, 949

John T. Hesler, -, -, -, -, 70

George Grady, 50, 50, 1000, 20, 310

Richd. Poindexter, 53, 40, 558, 30, 89

Saml. Poindexter, 60, 30, 720, 40, 188

James Terry, 50, 15, 640, 12, 160

Mary A. Melton, 250, 230, 5760, 160, 568

Thomas R. Dunn, 150, 50, 2000, 100, 518

Richd. A. Perkins, -, -, -, -, 41

Henry W. Quarles, -, -, -, 150, 537

Wm. C. Scales, 600, 360, 24000, 525, 2026

Gen. C. G. Coleman, 1600, 2565, 83300, 2020, 7524

M. A. Hope, 250, 188, 8760, 450, 1519

R. C. Bowles, 300, 112, 4180, 250, 896

Chas. Q. Goodwin, 300, 100, 12600, 285, 1594

Dr. Wm. J. Pendleton, 600, 250, 12750, 600, 2904

John Hunter, 450, 2250, 32000, 300, 1300

Jas. Walton, 175, 237, 5000, 150, 922

John Nunn, 280, 280, 10000, 200, 945

Jos. A. Nunn, 200, 125, 4875, 100, 684

Wm. Q. Thomson, 300, 200, 12500, 500, 2648

Henry S_iles, 350, 175, 10480, 325, 1310

Edwin J. Boyd, 300, 400, 14000, 200, 1270

Jas. Timberlake, 300, 90, 5880, 150, 638

Dr. E. L. Smith, 66, 63, 7750, 100, 495

Wm. Walton, 400, 200, 7200, 100, 231

Edward Walton, 100, 30, 2000, 50, 592

Wm. W. Bradley, 70, 65, 1300, 10, 225

Geo. W. Trice, 411, 2415, 22040, 200, 1238

Patrick H. Jones, 250, 305, 10275, 200, 861

S. D. Gooch, 400, 231, 12620, 400, 1979

Wm. Rennolds, 240, 241, 7215, 225, 1292

William Bagby, 35, 108, 2250, 20, 385

Joseph T. Smith, 100, 58, 1880, 100, 350

The heirs of Anur F. Dukerson, 45, 55, 700, 25, 180

Tandy Baughan, 40, 40, 1200, 25, 75

Pettus W. Sims, 60, 40, 1280, 57, 500

Louisa A. C. Sims, 120, 40, 1280, 57, 500

John R. Chick, 100, 62, 1620, 125, 500

John Swift, 400, 100, 6000, 130, 700

John T. Smith, 200, 130, 3000, 150, 900

Robert Duggins, 100, 69, 1690, 40, 370

James M. Duggins, 200, 28, 2280, 200, 650

Agnes F. Foster, 10, 24, 350, 20, 200

James T. Hall, -, -, -, 40, 100

Marshall V. Henderson, 175, 120, 2950, 200, 600

Francis W. Hope, 120, 210, 2640, 50, 350

Thos. W. Hope, 150, 106, 2560, 120, 350

Thos. B. Johnson, -, -, -, -, 800

William D. Johnson, 1000, 638, 16380, 200, 1060

William Gammon, 100, 100, 1400, 10, 150

George Adams, 60, 140, 200, 60, 107

Richard K. Bowles, 50, 60, 5000, 150, 350

Richard J. Atkison, 130, 2, 1510, 100, 300

William Smelson, 20, 95, 1150, 30, -

George Thomas, 350, 370, 7200, 300, 1200

Edwin Perkins, 30, 50, 750, 20, 170

William Turner, 90, 105, 1950, 30, 220

Augustus K. Bowles, 200, 165, 6000, 100, 500

James T. Duval, 127, 67, 2000, 20, 400

Joseph K. Bowles, 250, 200, 2000, 100, 500

Susan D. Foster, 30, 90, 720, 10, 181

Elizabeth Perkins & others, 250, 100, 4200, 140, 780

Mathew G. Anderson, 300, 240, 7000, 150, 1000

Joseph S. Meredith, 275, 70, 4142, 150, 750

Mary E. Duke, 300, 150, 5300, 100, 700

Spotswood L. Attkison, 250, 150, 4000, 150, 900

George R. Lindsay & others, 700, 300, 2000, 1000, 2000

Albert G. Bowles, 400, 150, 5000, 200, 1460

Martha M. Armstrong, 10, 17, 150, 2, 20

George Gammon, 20, 50, 350, 3, 75

Elisa T. Anderson, 130, 150, 1800, 40, 375

Elanah B. Aukeson, -, -, -, -, 60

Mathew Loyd, 300, 359, 4931, 100, 250

Robert Harris, 5, 25, 240, 10, 60

Robert Loyd, -, -, -, 5, 200

Thos. Loyd & others, -, -, -, 10, 400

John Loyd, -, -, -, 75, 290

Elkanah Brooks, 340, 250, 4000, 200, 640

Peter Johnson, 85, 44, 1400, 10, 130

John B. Anderson, 230, 30, 5000, 250, 1000

Archibald H. Anderson, 175, 207, 6500, 225, 1000

Archibald Anderson, 300, 400, 6000, 200, 600

Archibald Anderson agent for James Johnson's estate, 250 416, 6000, 50, 300

Mary A. Anderson, 200, 141, 11080, 40, 595

Nancy Turner, 100, 120, 2000, 25, 350

Fredrick S. Duke, 300 100, 6000, 140, 1000

Frances Loyall, 200, 132, 3200, 15, 360

Andrew Jackson, -, -, -, 125, 300

Claibourn W. Gentry, 100, 95, 1950, 100, 365

Charles D. Meredith, -, -, -, 75, 400

Wm. G. Corker, 50, 18, 680, 10, 50

David W. Isbell, 100, 112, 3000, 50, 500

Robt. T. White, 150, 150, 3000, 100, 340

George Turner, 300, 150, 8000, 400, 2000

Nathaniel H. Turner, 300, 170, 5000, 150, 1022

Caroline G. Bowles, 150, 50, 1600, 15, 210

John C. B. Goodwin, 120, 70, 3000, 2, 220

Charles Z. Smith, 180, 120, 3000, 100, 800

Joseph Sims, 85, 85, 1360, 30, 270

Polly Alestork, 8, 29, 300, -, 75

William H. B. Goodwin, 600, 400, 12000, 100, 1000

John Thacker, 20, 12, 248, 15, 75
Oscar Herring agt. For Frances E. Brooke, 900, 700, 32000, 975, 5300
Warner M. Noel, 125, 100, 1800, 75, 400
Meriwether S. Sims, 100, 82, 1820, 20, 165
Archibald T. Goodwin, 400, 120, 5200, 200, 920
William S. Walton, 375, 416, 7210, 75, 1470
Wm. P. Tate, 40, 50, 900, 30, 200
Robert Thacker, 40, 58, 980, 40, 175
Martha H. Isbell, 300, 400, 5600, 30, 210
Daniel C. Foster, -, -, -, -, 170
Nelson Gibson, 25, 25, 500, 10, 200
Sarah W. Foster, 35, 38, 720, 40, 170
David M. Foster, 40, 90, 1950, 150, 300
Theodosia Armstrong, 90, 41, 1300, 50, 310
Sarah Talley, 80, 15, 950, 15, 120
Francis Anthony, 75, 25, 800, 20, 130
Lancelot W. Hill, 60, 50, 1220, 30, 350
Samuel Meredith, 500, 220, 10000, 400, 1500
Elizabeth Dabney, 250, 150, 8000, 200, 1030
Oswold T. Parsons, 150, 50, 2500, 80, 310
William S. Johnson, 250, 160, 8200, 150, 540
John Walton, 325, 75, 8000, 150, 800
Elizabeth Meredith, 250, 206, 9120, 100, 700
Samuel O. Clough, 400, 130, 6360, 250, 970
William _. Walton, 130, 241, 7200, 200, 700
George W. Mallory, -, -, -, 40, 380
James N. Sharp, 75, 225, 3000, 150, 200

William O. Harris, 1000, 354, 10000, 500, 1500
Rebecca Winston, 325, 75, 12000, 200, 1020
Richard C. Carpenter, 300, 83, 6000, 200, 840
Samuel Harris, 200, 100, 4500, -, 245
William F. Harris, 250, 140, 5850, 750, 1180
Wilson Herring, 200, 166, 7320, 280, 925
Clora A. Sims, 300, 200, 6000, 200, 800
Addison L. Tate, 29, 29, 444, 15, 135
George J. Gardner, 400, 300, 14000, 600, 1470
Benjamin J. Walton, 100, 77, 2670, 150, 465
Alexander S. Jackson, 500, 400, 15000, 120, 240
Betty Jackson, -, -, -, 15, 230
Nancy Cristmas, 175, 25, 3000, 20, 585
Atwoold Wash, 200, 160, 7200, 200, 705
William G. Walton, 146, 100, 3690, 180, 990
James Burnley, 350, 130, 9600, 175, 1100
David Richardson, 400, 400, 8000, 100, 570
Martia H. Swift, 275, 125, 4800, 75, 580
Charles Brooks, 20, 13, 265, 6, 100
William N. Gentry, 130, 70, 1500, 75, 200
Stephen Farrar, 300, 186, 6720, 150, 1125
Robert A. Duncan, 300, 155, 10000, 200, 1020
William Meredith, 400, 280, 8160, 200, 725
Robert Foster, 50, 56, 636, 2, 160
Charles F. Gentry, -, -, -, 15, 200

James Brooks, 200, 234, 4340, 150, 270

Fredrick A. Perkins, 600, 500, 25000, 500, 2150

Joseph K. Pendleton, 750, 380, 15000, 800, 2000

Coleman S. Goodwin, 350, 250, 7200, 50, 624

John R. Robertson, 100, 100, 1600, 100, 300

John M. Thomas, 13, -, 2500, 40, 630

Littlebery J. Haley, 45, 50, 2500, 25, 250

Thomas Whitlock, 40, 100, 2100, 50, 210

Charles B. Cosby, 80, 60, 1680, 30, 350

Edward Cosby, 60, 36, 1120, 10, 110

John Yearman, 40, 141, 1810, 50, 280

John M. Timberlake, 175, 100, 3000, 40, 400

John Waldrope, 60, 60, 1200, 5, 150

Benajah Butler, 80, 100, 910, 15, 230

James F. Fleshman, 160, 100, 2600, 60, 360

James M. Butler, 50, 104, 770, 10, 50

Roda Poindexter, 20, 20, 220, 40, 280

Thomas Waldrope, 75, 175, 1500, 30, 480

Chapman White, 150, 76, 1816, 60, 260

Elisha Butler, 40, 60, 1210, 5, 175

James R. Boyd, 100, 150, 2000, 20, 200

John S. Burk, 600, 657, 12510, 300, 1400

William W. Pettus, 100, 150, 3300, 50, 490

Samuel P. Hacket, 400, 600, 25000, 1000, 2030

Henry L. Francisco, 300, 250, 16000, 700, 1480

Benjamin M. Francisco, 600, 500, 20000, 800, 2500

Beverly R. Fox, 400, 80, 10000, 300, 1320

Henry W. Johnson, 207, 140, 4000, 75, 350

Jermiah A. Roberts, 50, 10, 300, 30, 250

Mary B. Lipscomb, 60, 40, 600, 20, 150

Umphrey Bukley, 160, 140, 3600, 25, 400

Agnes Bukley, 100, 50, 1500, 20, 200

William C. Harlow, 75, 79, 1272, 15, 290

Richard T. Roberts, 45, 5, 400, 6, 100

James A. Davis, 50, 300, 7000, 150, 945

Wilson M. Carter, 65, 135, 1300, 75, 300

Drury Wood, 300, 150, 2500, 50, 550

Nicholas M. Ware, 400, 270, 4690, 50, 550

William W. McGehee, 320, 176, 5720, 200, 1060

George P. Hacket, 300, 85, 8000, 150, 1620

Mary C. Hickman, 200, 100, 9000, 200, 550

William J. Winston, 300, 152, 12000, 210, 1170

Elijah Butler, 100, 83, 2190, 40, 320

Thomas Nelson, 300, 150, 13000, 350, 1200

Chapman Sergeant, 280, 103, 11430, 45, 820

Thadeus Dickinson, 155, 50, 7000, 100, 1000

Robert. F. Fox, 220, 80, 9000, 220, 800

Ralph Dickinson, 350, 220, 15000, 150, 1800

William Beadles, 140, 59, 5970, 150, 400

John H. Wood, 500, 300, 6000, 50, 500

Hardin Perkins, 150, 100, 3750, 25, 500

Elizabeth Hunter, 400, 200, 6000, 250, 650

Joseph H. Perkins, 350, 250, 12000, 200, 795

Albert Chewning, 200, 50, 5000, 30, 600

John Carroll, 225, 68, 5000, 200, 800

Samuel Groom, 125, 73, 1980, 120, 400

Elizabeth M. Gooch, 100, 200, 3200, -, -

Miletus T. Gooch, -, -, -, 100, 600

Horrace C. Gooch, 30, 20, 700, 100, 185

John B. Poindexter, -, -, -, 70, 100

Maria Tisdal, 110, 10, 840, 10, 90

James Turner, 100, 250, 2120, 130, 370

William Crawford, 250, 324, 5000, 150, 810

James E. Smith, 18, 31, 1300, 50, 200

James M. Hart, 250, 260, 10000, 580, 990

Ann Watson, 1000, 1266, 75960, 500, 4160

David Eastham, 137, 80, 6340, 125, 685

Maria C. Morris, 650, 350, 60000, 300, 3341

David Watson, 450, 150, 36000, 550, 1920

Thomas S. Watson, 412, 440, 30640, 500, 3000

D. Watson agt for Henry Talor, 440, 260, 36000, 500, 1550

Francis T. West & brother, 6750, 350, 60000, 600, 2700

Robert B. Watkins, 300, 200, 20000, 300, 1500

Joseph W. Morris, 600, 320, 32200, 1000, 3040

Ambros F. Belamy, 40, 160, 1200, 20, 120

John D. Fielding, 70, 80, 750, 7, 110

Joseph A. Ballman, 42, 14, 4000, 130, 550

George L. Gordon, 200, 430, 9000, 200, 820

William Mahanes, 4, 3, 500, 105, 350

Samuel Thaddock, 100, 217, 3000, 100, 450

Albert G. Watkins, 150, 127, 4080, 100, 337

Robert W. Williams, 200, 153, 3530, 350, 890

Sarah E. Williams, 50, 62, 770, -, -

Samuel O. Bunck, 100, 106, 1530, 130, 350

James B. Harris, 30, 40, 420, 20, 70

William B. Belamy, -, -, 25, 1, 130

Thomas Corr, 70, 60, 1230, 10, 180

Manoah L. Lanford, 140, 225, 3650, 50, 460

Brightberry B. Bruice, 150, 250, 2400, 100, 670

Lewis W. Landram, 25, 75, 1000, 25, 190

John N. Hughson, 130, 230, 11000, 100, 800

Oliver P. Winston, -, -, -, 50, 250

Samuel Parrish, 65, 115, 2000, 100, 500

John M. Crew, -, -, -, 15, 120

John A. Robards, -, -, -, 50, 375

Ellis G. Hughson, 200, 150, 3570, 100, 550

Jessee Hawkins, 60, 65, 750, 70, 140

Bartlet A. Hanson, -, -, -, 100, 500

Lucy Henson, 70, 80, 1200, 40, 300

George Vest, 1000, 500, 37000, 1300, 2850

Bickerton Winston, 200, 78, 18000, 250, 1200

Burwell B. Bunck, 170, 264, 20000, 300, 1000

Benjamin Henson, 70, 500, 36000, 500, 2500

Richard Gilbert, -, -, -, 5, 50

Daniel A. Saunders, 150, 150, 8000, 500, 900

Wellington Gordon, 960, 350, 35000, 1000, 4000

Charles R. Dickinson, 600, 200, 16500, 600, 1900

John T. Sergeant, 75, 75, 1200, 50, 280

Elizabeth Sergeant, 130, 10, 840, 30, 240

Samuel L. Diggs, 160, 280, 3360, 30, 400

Garland Rennolds, 50, 152, 3600, 5, 70

William M. Diggs, 60, 116, 1750, 20, 250

Joseph Payne, 30, 70, 500, 20, 150

Samuel F. Desper, 100, 100, 1200, 20, 190

Thomas M. Diggs, 30, 250, 1680, 20, 150

John J. Bragg, -, -, -, 15, 140

Jno. C. W. Roberts, 40, 80, 1800, 60, 200

Hiram B. Sanders, -, -, -, 170, 450

William Overton, 800, 400, 33000, 600, 4200

William Morris, 850, 767, 48280, 700, 2280

Richard _. Morris, 900, 1057, 60000, 1130, 5270

Jane Saunders, 150, -, 1800, 60, 250

James M. Morris, 700, 1000, 55000, 1000, 4580

Thomas Harlour, 60, 40, 600, 75, 300

Garland Walker, 200, 90, 1850, 50, 400

John Anderson, 200, 250, 3600, 100, 620

Lucy Darnel, 45, 45, 720, 10, 70

William S. Stephens, 50, 30, 800, 10, 150

John B. Lastly, 150, 50, 2400, 100, 740

Acheles B. Sanford, 25, 127, 1520, 30, 280

James D. Wheeler, 125, 152, 3120, 100, 460

John H. Wheeler, 400, 260, 7700, 70, 1000

John Haden, 25, 55, 560, 5, 140

John W. Davis, 100, 280, 3800, 150, 220

William Walker, 180, 129, 2392, 100, 280

David Jones, 150, 450, 4800, 75, 260

Richard C. Cole, -, -, -, 60, 220

James B. Armstrong, 25, 7, 640, 60, 150

John Richardson, 250, 45, 3560, 40, 580

James E. Farrar, 200, 135, 4750, 100, 550

Joseph D. Cambell, 200, 100, 4500, 40, 725

Lucy A. Harris, 600, 400, 15000, 400, 2800

William Waddy, 500, 208, 8496, 150, 1210

Julian Kean, 450, 233, 8196, 125, 1210

Charles J. Thompson, 550, 200, 10000, 70, 960

Martha S. Dickinson, 300, 200, 5000, 130, 680

Robt. D. Wash, -, -, -, 25, 120

James M. Wash, 275, 99, 4468, 30, 500

William Johnson, 800, 750, 18900, 200, 900

John P. Smith, 300, 300, 7000, 50, 740

Nathaniel Long, -, -, -, 20, 60

William J. Cocke, -, -, -, 10, 100
Layfaytte R. Riordon, -, -, -, 15, 180
James Webb, 130, 107, 1422, 65, 480
James F. Lay, 30, 73, 824, 15, 200
William C. Thompson, 50, 57, 700, 15, 210
Rosa C. Hughson, 75, 73, 1060, 15, 220
Edmond Sachra, 30, 134, 1640, 20, 220
Jonathan Payne, 240, 588, 8280, 50, 890
Thomas M. Jones, 200, 157, 23000, 30, 250
Powhatan L. Jones, 150, 90, 2280, 15, 400
David M. Hope, 200, 75, 3500, 60, 500
Juluan P. Jones, 200, 50, 1900, 30, 525
Julian C. Martin, -, -, -, 20, 200
George A. Payne, -, -, -, 50, 510
William Edwards, 85,188, 1884, 75, 380
William N. Jackson, 120, 170, 2900, 30, 450
Daniel H. Grubbs, 50, 88, 789, 15, 280
Samuel Bunch, 150, 158, 3080, 20, 370
William Perkins, 400, 400, 4856, 50, 795
William Grinstead, 250, 250, 9000, 90, 490
Mary W. Winston, 600, 500, 16500, 300, 2000
James D. Bragg, 25, 30, 500, 10, 150
Lurna Newman, 100, 200, 3000, 100, 250
Henry Wilkerson, 30, 124, 1078, 10, 250
James M. Carpenter, 35, 175, 4200, 25, 530
Thomas C. Talley, 275, 75, 2820, 100, 545

David Bullock, 40, 35, 800, 30, 135
James D. Nuckolds, 200, 400, 6000, 175, 750
Granvil Bullock, 450, 250, 13000, 300, 2000
Sarah & Mary Thompson, 150, 83, 4660, 125, 700
William M. Ambler, 800, 600, 28340, 850, 4000
Mary P. Kimbrough, 600, 270, 17400, 100, 1200
Josephus Wash, 300, 367, 8000, 200, 1380
Jacob Payne, -, -, -, 35, 500
John Mitchel, 400, 110, 4080, 50, 610
Mercer Corley, 60, 13, 571, 15, 100
Joseph A. Bowles, 150, 148, 2880, 80, 310
William Armstrong, -, -, -, 20, 40
Peter King, 50, 10, 480, 25, 135
Joseph Thompson, 3, 7, 250, 15, 44
David Armstrong, 100, 70, 2450, 25, 485
David T. Armstrong, 200, 50, 3200, 80, 475
Thomas Turner, 175, 65, 3600, 40, 425
Fredrick H. Sims, -, -, -, 100, 700
Benjamin G. Higgason, 175, 49, 2140, 50, 1110
George B. Talley, 100, 60, 1900, 110, 390
James M. Adams, 60, 60, 1200, 45, 230
Benjamin Z. Richardson, 250, 196, 7000, 85, 940
William J. Walton, 350, 224, 11440, 50, 910
Charles U. Nuckolls, 193, 41, 3042, 130, 560
William L. Nuckolls, 230, 70, 3600, 50, 650
Addison L. Nuckolls, 175, 91, 3460, 50, 520
Levy Parrish, 85, 15, 1000, 20, 100

James Parrish, 50, 50, 1000, 15, 150
Burras Diggs, 8, 8, 240, 25, 105
James Sims, 150, 150, 2700, 75, 665
Micajah Parris, 150, 65, 2150, 50, 240
Nelson Anderson, 200, 130, 3300, 50, 550
Joseph Shelton, -, -, -, 15, 450
George W. Payne, 180, 70, 3000, 65, 550
James M. Wright, 200, 200, 4800, 75, 870
William S. Fowler, 400, 150, 10000, 300, 1010
John A. Parrish, 100, 30, 1300, 20, 150
John M. Hollins, 130, 131, 2610, 100, 590
Mary C. Anderson, 340, 48, 3880, 75, 620
John B. Shelton, 280, 320, 12000, 600, 1900
Overton Talley, -, -, -, 75, 310
David R. Shelton, 510, 170, 10000, 150, 1350
William S. Atkison, -, -, -, 5, 60
Samuel F. Fresher, 100, 88, 1880, 30, 100
Nancy S. Crosby, 67, 293, 3500, 100, 220
Joseph W. Sanders, -, -, -, 50, 135
David Saunders, -, -, -, 35, 140
Cornelius S. Dabney, 72, 72, 1440, 30, 100
Cornelius Diggs, -, -, -, 25, 175
Edward _. Shelton, 5, 1, 3, 350, 570
James _. Shelton, -, -, -, 190 880
Thomas B. Brooks, 195, 100, 4000, 60, 570
Meredith Goodwin, 700, 550, 15000, 250, 1070
E__ Bagby, -, -, -, -, 420
John B. Sigps, 95, 47, 365, 70, 100
David Anderson, 500, 1034, 2840, 250, 2260
John C. Houchins, -, -, -, 15, 140

Hamilton Johnson, 100, 253, 19575, 200, 1040
Chisdon __. White, 90, 91, 1500, 20, 220
James M. Wright, 300, 140, 5040, 250, 970
James M. Morris, 275, 122, 3175, 40, 530
Nathaniel Heolland, 100, 40, 1600, 75, 140
Nancy T. Perkins, 300, 100, 7200, 75, 580
Andrew J. Perkins, -, -, -, 100, 700
George Parrish, 92, 92, 1000, 15, 180
Samuel T. Watkins, 100, 59, 1272, 75, 300
James A. Gunter, 50, 109, 2030, 300, 400
William P. Watkins, 75, 75, 1500, 75, 410
William T. Wright, 91, 45, 2380, 75, 595
Albert G. Wright, 47, -, 470, 6, 130
Joseph M. Wright, 80, 50, 2500, 20, 250
Samuel Hasher, 1565, 40, 3420, 35, 260
George W. Christmas, 75, 160, 2350, 35, 2500
Granvil S. Wright, 60, 7, 1540, 35, 290
Horatio Johnson, 11, 2, 130, 20, 140
Meriwether S. Hoggan, 200, 154, 2564, 60, 350
James Bishop, 150, 75, 1500, 50, 250
David W. Perkins, -, -, -, 60, 450
Wm. S. Harlow, 25, 77, 582, 5, 250
Rubin Gilbert, 200, 300, 5000, 7, 300
John R. Crank, 100, 75, 1500, 35, 450
Cornelius O. Gooch, -, -, -, 50, 350
William Shepherd, 75, 46, 1240, 25, 250

Thomas Desper, -, -, -, 25, 250
Thomas H. Mallory, 60, 140, 3000, 100, 450
Willis Parrish, 80, 25, 2100, 100, 500
William Reynolds, 100, 200, 3000, 25, 500
Elizabeth Weldy, 50, 25, 900, 40, 300
Lewis W. Leake, -, -, -, 26, 90
Ned. Wilkerson, 50, 32, 820, 40, 250
Richard A. Loving, 50, 33, 1130, 35, 350
James Williams, 80, 120, 2000, 60, 400
Richard Johnson, 25, -, 200, 15, 75

Cyrus Wilkerson, 175, 469, 3205, 50, 400
Andre G. Walton, 400, 557, 10445, 250, 990
Wm. A. Winston, -, -, -, 200, 900
Robert Meredith, 250, 250, 6120, 200, 1080
Woodson Parsons, 400, 340, 7400, 120, 670
Benjamin M. Hall, -, -, -, 20, 100
Jefferson Diggins, 125, 75, 3000, 60, 410
William C. Thomas, 200, 200, 3000, 50, 450

Lunenburg County, Virginia
1860 Agricultural Census

The Agricultural Census for Virginia 1860 was microfilmed by the University of North Carolina Library under a grant from the National Science Foundation from original records at the Virginia Department of Archives and History in 1963.

There are forty-eight columns of information on each individual. Only the head of household is addressed. I have chosen to use only six columns of information because I feel that this information best illustrates the wealth of individuals. The columns are:

1. Name
2. Improved Acres of Land
3. Unimproved Acres of Land
4. Cash Value of Farm
5. Value of Farming Implements and Machinery
13. Value of Livestock

Paul Wilson, 300, 250, 700, 150, 1300
Elisha J. Handy, 200, 166, 2000, 20, 550
Francis Vorter, 100, 97, 750, 25, 125
Cephas Pennington, 60, 52, 336, 15, 287
N. Penninton, 40, 85, 375, -, -
Daniel E. Gunn, 180, 180, 2450, 100, 350
Jon Hazlewood, 50, 50, 1000, 70, 295
Robt. D. Philips, -, -, -, -, 40
Samuel E. Loe, 150, 150, 300, 50, 356
Asely Robertson, 100, 97, 300, 30, 200
Nancy Pulliam, 60, 35, 400, -, 65
James McAlister, 200, 325, 3000, 75, 555
Mary B. Smithson, 360, 340, 4200, 200, 548
E. B. Smithson, 20, 63, 500, 20, 250
T. S. N. Smithson, 50, 102, 1500, 120, 456
John Williams, 20, 130, 700, -, -

Wm. J. Smithson, 150, 350, 3500, 130, 475
Wm. B. Lester, 15, 4, 1500, -, 140
Isaiah Slaughter, 25, 30, 250, 6, 107
Cephas White, 1, 9, 100, -, 80
Eliz. Skaton, -, -, -, -, 63
Wm. L. Willson, 305, 305, 3050, 170, 900
H. H. Bohanan, 40, 77, 700, 30, 255
Thomas L. Russell, 50, 75, 500, 10, 200
Mary A. Townsend, 28, 80, 540, 10, 195
E. F. Ellis, 100, 250, 3000, 30, 350
Wm. Philips, 140, 146, 1200, 30, 441
Wm. Ashworth, -, -, -, -, 20
Newman Ashworth, -, -, -, -, 17
Thomas E. Lacy, 12, 13, 20, -, 51
Jno. Williams, 70, 71, 800, 20, 170
Richd. Vashnoth, -, -, -, -, -
Alex. J. Watson, 230, 200, 3000, 210, 783
Wm. Parson, 130, 70, 1000, 30, 200
Hartwell Marable, 100, 100, 2080, 90, 500

George C. Ellis, 150, 162, 2000, 50, 436

Joseph P. Townsend, 100, 223, 1000, 30, 122

Wm. S. Couch, 400, 580, 10000, 310, 1225

Edwin R. Smithson, 100, 200, 3000, 50, 515

Joseph B. Gregory, 300, 456, 10000, 350, 1361

Saml. A. Bruce, 200, 383, 5000, 100, 500

Robt. H. Williams, 60, 240, 3000, 15, 365

Jesse Watson, 250, 250, 2000, 78, 480

G. W. Knight, 30, 128, 1200, 25, 125

Robt. E. Knight, 100, 150, 2000, 75, 390

Mary Thompson, 90, 100, 1400, 25, 295

Edna S. Pollard, 125, 203, 1640, 50, 390

Wm. Thompson, 58, 30, 500, 50, 130

Jas. A. Rutledge, 93, 93, 800, 15, 108

Mary J. Nucomb, Jno Roland's farm, 8, 50

Charles Hudson, 50, 52, 408, 5, 136

Edwd. W. Nucomb, Hudson farm, 43

Hilary Hudson, -, -, -, -, 48

Allen Duffe, 32, 18, 500, 20, 110

John B. Barns, 183, 367, 4000, 75, 345

Ann Lipscomb, 200, 287, 4860, 60, 444

Mary Barns, 80, 80, 900, 25, 144

R. E. Burnett,-, -, -, -, 15

Ann A. Burwell, 250, 230, 3960, 135, 818

Clomus J. Thompson, 200, 300, 5000, 120, 710

Joseph F. Ellis, 163, 55, 2180, 60, 800

Wm. E. Dodson, 250, 564, 14000, 200, 1320

E. M. Hepleburne, 600, 1500, 16800, 300, 1100

Martha Dance, 160, 320, 3000, 100, 471

Isauc E. Holmes, 80, 90, 850, 12, 455

Wm. S. Dance, 300, 400, 4200, 105, 693

Nancy Philips, -, -, -, -, 50

Jno. R. B. Tisdale, 225, 220, 4240, 125, 970

Fred Lester, 500, 260, 6080, 130, 474

E. A. J. Winn, 100, 100, 1200, 130, 333

Isaac Winfree, 160, 197, 1500, 55, 282

James A. Jeter, 50, 55, 600, 22, 103

Wm. B. Winn, 50 69, 600, 10, 166

J. Q. A. McKinny, 240, 250, 2000, 80, 380

Susan Wrenn, 75, 23, 400, 15, 80

Thos. J. Hardin, 100, 108, 1000, 40, 285

Edward G. Baily, 112, 13, 150, -, 42

Caroline E. Wood, 70, 20, 1500, 18, 148

David Williams, -, -, -, 5, 150

David Williams Jr., 70, 155, 1200, 25, 400

Joseph E. Davis, 200, 200, 3200, 135, 473

Mary E. Davis, 80, 50, 1200, 15, 248

Wm. Townsend, 250, 200, 4000, 175, 723

John C. Smithson, 200, 205, 3600, 217, 674

Philip T. Jeter, 40 70, 1000, 47, 640

Wm. S. Marshall, -, -, -, 8, 321

Jno. G. Green, -, -, -, -, 4

Wm. White, -, -, -, 5, 65

Thos. J. Marshall, 150, 180 3000, 40, 478

John Pamplin, Inge's farm, -, 45, 112

Gill Watts, 350, 250, 3100, 165, 540

Wm. P. Jerdan, 150, 160, 2400, 150, 378

Edmund C. Winn, 250, 440, 10000, 300, 1074

James Lambert, 30, 70, 1000, 20, 195

Creed T. Scruggs, 200, 296, 2500, 50, 514

Joseph Ryland, -, -, -, -, -

George St. John, -, -, -, -, 66

Lewis Davis, 100, 384, 3500, 75, 985

Dr. E. A. Willson, 300, 630, 6000, 150, 1028

Elyat Lester, -, -, -, 10, 25

Robert W. Land, 90, 67, 1000, 65, 335

Ann Smith, 300, 220, 4000, 225, 855

John L. Coleman, 160, 251, 4100, 225 855

Wm. E. Willson, 300, 325, 10000, 125, 1326

G. B. Wagstaff, 150, 550, 3500, 20, 322

Ellen J. Berwell, 175, 325, 4000, 20, 738

Mary Jefferson, 400, 400, 7000, 173, 963

Jno. D Petty, 60, 126, 1900, 40, 425

David B. Bragg, 30, 85, 1000, 50, 250

Joel G. Wall, 150, 165, 3500, 195, 722

George W. Terry, 300, 300, 4000, 60, 940

Thomas R. Tisdale, 200, 300, 2500, 60, 940

Tabitha Key, 40, 116, 800, 12, 43

Washington Davis, Key farm, -, 27

Mary Walker, 20, 60, 600, 12, 79

Mark Mize, 55, 60, 600, 37, 264

Saml. Thomas, 230, 249, 4000, 140, 744

Elisabeth J. Hix, 110, 340, 4000, 50, 320

Hanisser J. Elder, 190, 190, 3000, 40, 442

John D. Bell, 150, 220, 3500, 155, 468

John D. Tisdale, 200, 342, 5000, 230, 622

Wm. B. Smithson, 100, 147, 1500, 30, 171

John Campbell, 75, 162, 1600, 75, 427

Archibald Campbell, -, -, -, -, 60

Thomas Pearcy, 50, 130, 640, 35, 110

Mack Procise, -, -, -, -, 25

James F. Crowder, 90, 180, 5700, 60, 698

Henry Freeman, 50, 100, 800, 22, 192

Lucretia Freeman, 400, 50, 3000, 40, 381

Washington Brown, -, -, -, -, 18

G. W. Crowder, 75, 390, 2500, 185, 575

Robt. B. Smith, Blount's farm, 5, 180

Richard Crowder, 50, 50, 500, 10, 383

Lucy Williams, 10, 16, 120, 3, 46

Caroline Smith, 50, 50, 500, -, 15

Peter Smith,-, -, -, 25, 89

Wm. Pearcy, 56, 70, 600, 20, 363

Jno. R. Garland, 400, 425, 20000, 540, 2000

Est. of W. Blount, 200, 384, 7000, 100, 355

Merritt Lester, 20 7, 75, 10, 143

Sally G. Robertson, 30, 130, 1500, 5, 74

John G. Almond, 90, 95, 2000, 50, 225

Charles A. Sharp, -, -, -, -, 14

John H. Smith, 100, 288, 2000, 50, 525

Beverly R. White, 2, 20, 250, -, 75

Wm. Inge, 125, 375, 3500, 20, 340

Wm. J. Jeter, 35, 20, 350, 35, 122

J. J. A. Moore, 100, 86, 1000, 20, 384

Jeana B. Moore, -, -, -, -, 125

Wm. T. Davis, 125, 275, 3000, 100, 536

Silas Shifleman, 750, 410, 3500, 100, 762

Thomas G. Crowley, 200, 200, 2500, 30, 171

Daniel Smith, 40, 988, 1000, 14, 118

James Cole, 300, 110, 2000, 100 683

M. S. Hurt, 500, 750, 2900, 90, 610

S. A. Staples, 200, 250, 3500, 90, 515

C. Anderson, 50, 53, 500, 10, 245

David White, -, -, -, -, 35

P. H. Hurt, 500, 800, 10400, 432, 960

Wm. J. Townsend, 140, 288, 1500, 38, 130

Drewry Townsend, 45, 15, 250, -, 27

Wm. Williams, -, -, -, 20, 104

D. T. McCormick, 150, 500, 12000, 15, 645

Tyra Cooksy, 100, 287, 2500, 55, 462

Rob Bragg, 1000, 1846, 25000, 365, 2713

Jennings Love, 100, 135, 1750, 70, 232

Josiah B Wilson Sr., 150, 278, 3500, 45, 360

Jonah B. Wilson Jr., -, -, -, -, -

R. J. Hatchett, 300, 824, 10000, 200, 1179

Wm. H. Love, -, -, -, 50, 280

Joshua Smith, 250, 162, 3500, 55, 885

Henry Gary, 3, 95, 1000, 5, 50

James Gary, -, -, -, 100, 147

Charles May, 80, 25, 2500, 85, 441

James Keater (Kater), 30, 211, 1500, 30, 155

Wm. A. Nash, 1, 19, 150, 30, 162

Ellison W. Ellis, 130, 26, 1200, 20, 230

Joel M. Ragsdale, 200, 579, 6000, 150, 1145

John H. Ragsdale,-, -, -, 100, 145

C. M. Smith, 400, 400, 8000, 165, 976

Griffin Bayne, -, -, -, -, 144

Francis Raney, 75, 71, 500, 20, 180

Lewis M. Gee, 175, 325, 4000, 172, 767

Jesse H. Gee, 150, 350, 4000, 100, 450

Wm. Rainey, 200, 404, 4000, 85, 600

Thomas J. Love, 200, 180, 3000, 165, 400

Lewis Andrews, -, -, -, -, 20

Wm. Banks, -, -, -, -, 34

Charles Lewis, 125, 135, 1000, 15, 65

Sally Russell, 5, 25, 150, -, -

James Hawthorne, 100, 207, 1200, 15, 218

Cal__ Tucker, 11, 24, 1200, 5, 218

Wm. M. Bagley, 75, 45, 5000, 300, 600

Mrs. P. Bagley, 300, 200, 6000, 200, 850

Thomas L. Pearcy, 50, 150, 800, 35, 245

Mrs. T. Rudd, 160, 205, 3000, 125, 600

Saml. W. Reece, -, -, -, 10, 30

Jno. B. Philips, 300, 323, 6000, 206, 1066

Henry C. Hawthorne, 200, 396, 3600, 120, 633

Parks Tucker, 100, 60, 1500, 10, 165

James E. Hazlewood, 15, 380, 3000, 80, 408

James H. Andrews, -, -, -, 5, 77

L. T. Andrews, 110, 23, 1200, 18, 300

Nancy Gill, 100, 140, 2000, 38, 273

George W. Andrews, 10, 90, 400, 5, 164

Charles B. Cole, -, -, -, 12, 19

Peter Raney, 32, 18, 400, 25, 65

Charles B. Raney, -, -, -, 5, 15

Mary A. Tucker, 40, 35, 400, 15, 128

Dennis Gee, 70, 75, 1500, 12, 155

Wm. A. Harriss, 280, 283, 4500, 255, 775

Lewis Plantation, 300, 435, 9500, 45, 730

James R. Philips, 150, 72, 1500, 30, 276

James Smith, 80, 100, 1800, 10, 140

Anderson Harriss, 425, 425, 6800, 200, 958

Wm. Claton, -, -, -, -, 2

Richard Jones, 100, 130, 2000, 58, 215

Wm. T. Andrews, 83, 167, 7250, 85, 475

Richard H. Baley, 100, 150, 1500, 40, 485

David R. Stokes, 500, 236, 9000, 285, 1498

Richd. Epps farm, 550, 561, 9000, 287, 956

Eliza C. Gregory, 40, 63, 1200, 20, 246

Benj. Harriss, -, -, -, -, 165

Robt. Sanders, 383, 160, 9000, 200, 915

Miss Cabranss, 75, 28, 1000, 5, 150

R. H. Davis, 40, 80, 1200, 20, 226

T. L Crow, 30, 78, 1400, 40, 200

Wm. J. Niblett, 400, 800, 15000, 350, 1662

A. E. Ragsdale, -, -, -, 10, 161

Charlotte Gier, 8, 52, 500, -, -

Eliz. Stewart, 13, 17, 100, 6, 38

Wm. A. Ingraham, 400, 400, 5000, 135, 364

Edward Ingraham, 160, 140, 2000, -, 300

A. T. Smith, 100, 410, 3500, 77, 342

T. H. Ingraham, -, -, -, 20, 270

D. T. Davis, 8, 2, 50, -, 43

J. H. Marable, 150, 195, 3000, 80, 351

Wm. F. Moore, 140, 140, 2500, 105, 609

James Rutledge, -, -, -, -, -

Joel M. Parish, 100, 295, 1800, 100, 359

Wm. B. Parish, 15, 198, 1200, 30, 225

Mrs. M. Ralls, 50, 210, 1000, 5, 116

Paschal Rufner, 50, 277, 2000, 25, 390

James W. Hardy, 200, 412, 4000, -, -

Misse A. E. Saunders, 500, 450, 9500, 350, 645

Sophia Pettus, 250, 250, 3000, 15, 126

John Thompson, 150, 550, 5000, 50, 531

P. T. Thomas, -, -, -, -, 125

Z. D. Burnett, 60, 40, 600, 5, 15

H. H. Love, 500, 900, 14700, 150, 1297

M. E. A. Walker, -, -, -, -, 65

D. W. Tisdale, 85, 200, 1500, 35, 329

C. W. Keaton, 35, 165, 1800, 45, 265

P. T. Robertson, 66, 127, 1000, 50, 350

M. B. Hyden, -, -, -, -, -

E. H. Bonan, 193, 190, 200, 30, 90

James Slaughter, -, -, -, 5, 37

Wm. L. Bragg, 50, 250, 1500, 20, 178

Edwd. Overby, -, -, -, 5, 41

Wm. T. Flippin, 1, 4, 1000, -, 565

Geo. W. Gee, 75, 220, 3000, 40, 400

Josiah Foster, 40, 100, 900, 55, 150

Nathan Gee, 67, 65, 900, 35, 475

Miss T. Gee, 200, 150, 1800, 45, 500

Clemt. Jackson, -, -, -, -, 90

R. B. Moore, -, 32, 300, 5, 48

A. G. Gee, 300, 400, 5000, 170, 556

D. A. Gee, 125, 125, 1500, 137, 628

T. B. Tomlinson, 150, 208, 3000, 30, 338

John Bishop, 250, 402, 3160, 200, 510

Philip Bowers, 60 77, 100, 16, 338
Nelly Overby, 4, 2, 80, 5, 30
James Snead, -, -, -, -, 37
James T. Robinson, 13, 50, 500, 7, 95
Alfred Bishop, -, -, -, 10, 224
Tho. W. Ragsdale, 5, 5, 100, 15, 70
Jno. W. Overby, -, -, -, 40, 230
Polly Overby, 15, 5, 200, 10, 80
Saml. K. Snead, -, -, -, 10, 180
Edmund Bishop, 52, 20, 1000, 50, 320
David T. Garland, 500, 700, 10000, 275, 1736
R. B. Atkinson, 350, 500, 9000, 375, 2915
Wm. Wilkinson, 275, 387, 6000, 185, 990
Henry Powers, 100, 150, 2000, 80, 543
Colin Niblett, 800, 763, 18740, 435, 2175
B. B. Jones, 275, 355, 5000, 85, 433
Washington Davis, 50, 22, 300, 5, 57
Wm. W. Holmes, 40, 40, 350, 5, 20
R. A. Allen, 280, 600, 8000, 310, 1787
Benj. E. Smith, 350, 380, 7000, 275, 1140
Jno. F. Rudd, -, -, -, -, 150
Wm. S. Floyd in trust, 61, 176, 1000, 24, 323
Anderson Steavrs, 20, 25, 280, 50, 318
Joe Edmunds, -, -, -, 5, 10
Randolph Thompson, 45, 47, 740, 25, 1347
George W. Cox, 200 177, 4250, 75, 590
Eliza Goodwin, 25, 219, 1250, -, 46
Fanny Stone, 5, 5, 50, -, 50
Jesse D. Abernethy, 870, 400, 7500, 75, 980
Wm. P. Tisdale, 258, 250, 3050, 55, 500
James E. Jeter, 65, 57, 600, 32, 173

Henry Bonan, -, -, -, -, 20
Mary L. Bagley, 220, 200, 2700, 60, 426
N. E. Davis, 400, 492, 7000, 187, 880
M. K. Jordan, 100, 113, 1000, 10, 20
R. M. Williams, 200, 130, 4000, 185, 1152
Col. H. Hatchet, 600, 415, 5000, 100, 1058
Hugh Walace, 50, 70, 700, 45, 270
Johnston Ashnoth, -, -, -, 5, 48
A. J. Ashnoth, -, -, -, 5, 48
N. P. Mangrum, -, -, -, 10, 35
A. R. Neal, -, -, -, 50, 726
Jno. A. Johns, 74, 149, 2000, 75, 355
C. E. Averett, 75, 150, 1600, -, 150
Wm. J. Foulkes, 200, 207, 5000, 205, 448
Wm. J. Foulkes upper farm with lower farm, 1000, 800, 10000, 200, 1318
Jnad. Cheathan, 460, 440, 8500, 225, 1064
Tho. Cheathan, 600, 1350, 15000, 225, 1540
Clarky Foulkes, -, -, -, 105, 425
Mrs. E. A. Epps, 813, 500, 10500, 280, 2424
W. W. Webb, 200, 320, 7000, 70, 566
L. R. Stokes, 200, 252, 3756, 100, 1000
John J. Parish, 20, 80, 500, 52, 120
James Cox, 100, 280, 2000, 165, 900
Claiborn Henderson, -, -, -, 25, 20
James Stewart, -, -, -, -, 60
Edwd. Goodwin, -, -, -, -, 105
Wm. Chatman, -, -, -, 5, 5
Edwd. Crade, 240, 244, 8000, 295, 1412
Jas. Bohanan,-, -, -, 5, 70
Henry Ware, 40, 156, 1900, 32, 295
Edwd. Bagley, 1000, 1000, 12000, 700, 3807

G. & B. Bridgeforth, 375, 260, 6500, 384, 1482

James Walker, 600, 300, 3000, 300, 1405

S. W. Winn, 300, 360, 8000, 215, 610

Col. J. Marshall, 400, 400, 9000, 360, 1080

Abram Marshall, 200, 400, 6000, 175, 1045

Mrs. A. R. Marshall, 300, 360, 8000, 425, 935

Jas. L. Hite, 400, 450, 9000, 300, 1084

Dabney Hardy, 300, 200, 5000, 75, 1305

Mrs. M. Blackwell, 250, 175, 7000, 100, 1056

Wm. & R. Blackwell, 350, 339, 7000, 75, 1280

Tho. Robertson, -, -, -, -, 20

Samett Oslin, 70, 139, 2000, 103, 300

Brooken Elder, 500, 226, 9000, 270, 1835

T. C. Elder, 200, 115, 1700, 100, 338

Wm. A. Hines, 100, 109, 1000, 10, 146

Thos. T. Hines, 70, 130, 1200, 20, 425

E. L. Turner, 30, 110, 700, 10, 137

R. N. Keller, 30, 70, 1500, 23, 187

Wm. L. Hines, 40, 30, 500, 20, 183

_. C. Lambert, 4, 6, 200, 2, 8

Joe Winn, 1, 12, 250, -, 8

Wm. Dixon, 75, 85, 2000, 65, 600

Jno. L. Lambert, 2, 11, 200, 10, 32

W. E. Walker, 55, 100, 1600, 45, 320

A. B. Hines, 150, 450, 3600, 190, 845

Millington Hines, -, -, -, -, 150

Mary E. Hawthorne, -, -, -, 5, 25

Turner T. Hines, -, -, -, -, -

Warner P. Hines, -, -, -, 10, 90

J. Bonan, 40, 4, 800, 40, 215

Grey Thompson, 80, 170, 2500, 87, 384

Jas. R. Hines, 10, 40, 400, 10, 256

Saml. Davis, 75, 110, 4000, 65, 386

Wm. Thomas Blackwell, 200, 460, 7000, 75, 1435

Wm. L. Hite, 300 700, 10000, 125, 2570

Wm. Laffoon, -, -, -, -, 21

Lew. Hite, 400, 500, 7500, 65, 838

Howell Taylor, 165, 140, 1200, 50, 283

M. Sturdivant, 100, 60, 1500, 40, 692

C. C. Hoskins, 400, 250, 7000, 285, 1080

Mrs. A. Featherston, 200, 200, 3200, 75, 717

Jas. Johnson, 300, 340, 5500, 145, 735

Wm. E. Jackson, -, -, -, 15, 61

Wm. B. Featherston, 75, 64, 1000, 25, 481

P. Taylor, H. Taylor's farm, 10, 148

A. F. Edmunds, -, -, -, 7, 75

James Taylor, 20, 40, 305, 17, 100

Beery Hammons, 75, 55, 900, 15, 180

Anderson Overby, 47, 50, 600, 15, 70

Jas. L. Scoggins, 240, 10, 2500, 230, 880

R. W. Turner, 95, 157, 2000, 37, 567

Peter Thompson, 300, 350, 4200, 200, 1673

Dolly Bolton, 75, 75, 1500, 15, 173

A. V. Rash, -, -, -, 15, 180

Jacob Rash, 100, 114, 2100, 30, 321

James T. Turner, 80, 40, 1200, 75, 385

Tho. Johnson, 40, 56, 800, 12, 261

Afred Johnson, 75, 137, 1700, 40, 385

Spencer Mills, -, -, -, 5, 28

Nancy Lambert, 40, 104, 1000, 4, 97

John Heton, 100, 30, 600, 5, -

T. Nate J. Dix farm, 5, 2, 70, 4, 29
E. W. Jackson, 45, 12, 400, 15, 125
Miles Taylor, 120, 113, 3000, 80, 602
John & Jones Matthew, 300, 326, 3000, 150, 960
P. W. Hawthorne, 466, 233, 7000, 180, 750
Wilkins Edmonds, 25, 25, 250, 8, 73
B. W. Jackson, -, -, -, -, 14
Jno. R. Featherston, -, -, -, -, -
Jonah Roberts, 36, 14, 500, 12, 123
R. G. Garland, 50, 146, 1500, 33, 268
Mortimer Laffoon, 30, 70, 800, 35, 271
Wm. Calaham, -, -, -, -, 20
Wm. Moore, 17, 8, 200, 17, 70
Wm. Epperson, 36, 50, 600, 40, 140
J. W. Trumell, 15, 50, 500, -, 62
Ganst Waller, 100, 25, 800, 47, 137
Saml. A. Peace, 100, 172, 3000, 133, 739
Thomas Parish, 175, 179, 3500, 150, 605
Jas. Skinner, -, -, -, -, 60
Geo. W. Potts, 15, 25, 400, 15, 152
Mary Dixon, -, -, -, -, 45
Wm. Edmunds, 80, 20, 400, 5, 69
Patty Dixon, 70, 30, 1000, 15, 72
Peter Matthews, 75, 20, 1000, 18, 116
Lew Matthews, 50, 25, 600, 27 98
Wm. Laffoon, 280, 50, 2000, 130, 346
Parks Laffoon, 208, 100, 2200, 75, 248
Vincent Inge, 146, 50, 1500, 50, 342
James W. Mare, 30, 20, 400, 35, 90
Susan J. Skinner, -, -, -, -, 21
Thomas Laffoon, -, -, -, 80, 202
G. W. Moore, -, -, -, 65, 57
Asa Moore, 65, 72, 1400, 30, 133
B. J. Wilkinson, 125, 125, 2500, 115, 528

Lazarus Burnett, 140, 130, 2000, 75, 488
M. J. Perkins, 25, 15, 320, 27, 109
Tho. Stone, 20, 10, 350, 8, 160
Tho. Vaden, -, -, -, 7, 93
M. B. Anderson, 10, 13, 300, 2, 45
Sterling Cardle, 60 40, 600, 85, 216
Wm. M. Cardle, 20, 5, 200, 10, 75
Jesse Morgan, 100, 200, 2400, 140, 354
B. Skinner, -, -, -, 5, 12
George L. Bagley, 250, 450, 7000, 385, 1890
T. H. Manson, 400, 375, 7000, 160, 822
H. T. Gilliam, 166, 343, 4000, 50, 540
Claiborne Jones, 400, 500, 9000, 160, 1480
J. G. Laffoon, -, -, -, 7, 122
Wm. J. Blackwell, 400, 400, 4800, 145, 692
Washington Maddox, 405, 545, 7600, 374, 1000
J. L. A. Smithson, 300, 650, 9500, 125, 829
Virginia Goodwin, -, -, -, 5, 25
S. S. Hawkins of T. S. Hawkins Farm, 50, 150, 1500, 40, 324
T. S. Woodson, 600, 700, 12000, 450, 1930
Thos. S. Shackleton, 1250, 1264, 28600, 535, 2190
George R. Robertson, -, -, -, 5, 28
German Bailey, 104, 100, 1600, 144, 350
A. R. Atwell, 56,118, 700, 25, 25
Wm. T. Ellis, 6, 132, 1100, 45, 152
Wm. L. Jude(Inde), 70, 130, 1100, 30, 103
Jno. T. Wood, 440, 400, 5440, 55, 438
Jno. T. Shelton, 80, 100, 900, 45, 190
J. W. Dalton, 75, 95, 1500, 35, 132

Jas. D. Barns, 125, 312, 4370, 25, 585

Masten Barnes, 225, 481, 9000, 205, 626

Sarah S. Arms, 90, 43, 1295, 28, 142

Jeny Russell, Cheatham's farm, 22

John Harding, 300, 320, 5000, 100, 832

R. Harding Sr., 80, 232, 2000, 36, 283

Jordan R. Hardy, 80, 250, 2300, 80, 445

Geo. S. Smith, 100, 240, 2000, 150, 295

Martha Stokes, -, -, -, -, 32

Thomas H. Gee, 250, 1107, 8142, 110, 1176

J. W. Bagby, 14, 46, 2000, 50, 419

H. O. A. Atwell, 25, 200, 1200, 35, 316

Jno. Destin(Dostin), -, -, -, -, 60

Wm. S. Clark, -, -, -, -, 75

Cass T. Clark, 100, 220, 2100, 35, 385

David Flowers, -, -, -, -, 18

George Harvey, -, -, -, -, 61

Wm. D. Nowell, 100, 600, 5000, 85, 882

J. L. Waddle, 10, 29, 600, 43, 134

Stephen Justice, -, -, -, -, 58

Mrs. Nancy Cox, 250, 250, 2500, 150, 875

Tho. Moherne(Mohane), 140 160, 1000, 15, 38

Tho. Williams, 48, 50, 900, 27, 235

Robt. Bruce, 150, 470, 5000, 120, 504

H. W. Wall, 150, 210, 3000, 132, 626

J. Penne, 2, 48, 140, 45, 382

F. Higgins, -, -, -, 60, 35

David Pully, 200, 375, 4000, 240, 892

B. T. Figg, 100, 145, 1500, 25, 211

Ed. Dilon, 35, 100, 900, 15, 221

J. D. Davis, -, -, -, 160, 1008

J. Carwiles, -, -, -, -, 92

Tho. Rutledge, -, -, -, 30, 262

Nancy Croftion, 15, 77, 900, 23, 105

Saludd Stokes, 200, 400, 5000, 68, 710

Lucinda Rowlett, 300, 200, 4000, 110, 850

W. H. Stokes, 130, 260, 3000, 25, 423

S. Stokes, 250, 250, 3000, 30, 333

S. Bates, 180, 180, 14000, 30, 135

B. Harding, -, -, -, -, 43

E. A. Staples, 180, 190, 3000, 35, 576

Mrs. J. C. Staples, 260, 540, 4800, 215, 585

Jas. Staples, -, -, -, -, 230

W. W. Webb, -, -, -, 15, 122

Richd. Crofton, 200, 160, 3500, 30, 700

Jos. Holmes, -, -, -, 30, 86

Wm. Flowers, 35, 15, 500, 27, 61

John T. Robertson, 20, 20, 300, 25, 90

Martha Hood, 30, 5, 200, -, 30

Jno. E. Overton, 150, 200, 2500, 60, 565

E. E. Foulkes, 80, 336, 3300, 100, 384

T. Regals, 30, 49, 400, 5, 96

Morning Crofton, 60, 167, 1100, 25, 268

Theodorick Cole, 380, 160, 3900, 35, 591

B. Frost, 100, 50, 600, 50, 117

J. L. Watkins, 100, 198, 2000, 22, 110

W. H. H. Williams manager of Dupus farm, 40, 83, 400, 12, 54

Capt. Jno. Eubank, 500, 500, 10000, 275, 2560

Patrick Egleston, -, -, -, -, 2

J. Y. Hardy, 300, 312, 7000, 330, 1275

Jno. Crymes, 200, 400, 12000, 130, 800

Paul A. Farley, 80, 83, 1200, 10, 170

Drewry A. Smith, 360, 418, 9600, 600, 1340

Elizabeth Rux, 300, 350, 4300, 35, 655

F. A. Dixson, 170, 163, 4000, 110, 798

Wm. Y. Neal, 250, 390, 11000, 110, 888

Jas. Neal, 700, 416, 9400, 145, 1429

John A. Stokes, 200, 200, 6000, 180, 1295

G. W. Hardy, 300, 453, 10000, 180, 1144

R. H. DeJarnett, 800, 500, 18000, 235, 1460

Thomas J. Ellis, 70, 70, 1600, 70, 350

W. Hill's farm, 150, 204, 2500, 30, 374

Laban Hawkins, T. A. Hawkins Farm, 40, 700, 2800, 7, 148

Richd. J. Jeffriess, 350, 300, 6000, 130, 1000

Wm. Arvine Jr., 400, 455, 8000, 258, 1216

Liubbery Harding, 140, 50, 960, 35, 335

Wm. Harding Sr., 125, 125, 1700, 20, 154

Wm. Passmore, 100, 50, 1500, 5, 366

Martha Bragg, -, -, -, -, 36

John Smith Sr., 40, 63, 600, 10, 148

Wyatt H. Pettus, 200, 400, 5000, 125, 993

G. A. Egleston, -, -, -, -, 25

Jas. Smith, -, -, -, 20, 75

R. L. B. Williams, 80, 150, 1400, 60, 378

George T. Knight, 100, 100, 1500, 75, 310

Wm. G. Bailey, 75, 133, 1400, 20, 197

Jno. P. Wootton, Ashworth's farm, 50, 83, 600, 10, 132

Susan Ashworth, 43, 23, 200, 18, 120

Ebenezer Crofton, -, -, -, 40, 340

R. B. Willson, 200, 272, 8000, 110, 418

Zebulon Clark, -, -, -, -, 37

Mrs. M. E. Blackwell, 250, 450, 9000, 150, 1270

Jno. H. Lean, 500, 500, 10000, 350, 1328

Jno. Orgain, 400, 500, 12000, 200, 1479

B. Stewart, -, -, -, 5, 28

A. Jackson, -, -, -, 15, 149

Wm. Jones, -, -, -, -, 31

Jas. P. Street, 400, 430, 8000, 200, 1479

V. R. S. Bagley, 1300, 700, 10500, 210, 1690

Daniel Laffoon, 100, 100, 1200, 28, 168

Richard Edmund, H. Gillians farm, 28, 168

Wm. A. Reese, 100, 145, 1900, 35, 317

Saml. Skinner, -, -, -, 5, 31

Kitty Parish, 30, 13, 400, 18, 80

John Skinner, 18, 7, 250, 10, 110

George Inge, 300, 100, 4000, 110, 627

Wm. B. Moore, 50, 110, 1200, 25, 195

Jno. Ragsdale, 15, 23, 400, 10, 117

Robt. Laffoon, 60, 90, 900, 12, 128

Wesley Inge, -, -, -, 10, 110

Gholson Skinner, 36, 17, 300, 8, 19

Morrison J. Peace, 138, 100, 1600, 40, 266

Saml. Hammock, 150, 170, 2000, 42, 230

C. W. Kirk, 95, 190, 1500, 34, 300

T. M. Kirk, 150, 183, 2300, 50, 480

Mary Wilkes, 3, 3, 150, 3, 57

Lewis Hammock, 200, 240, 3000, 30, 450

R. J. Morgan, 50, 50, 1000, 27, 265

Lucy Hammock, 150, 100, 1764, 215, 630
N. H. Stewart, 40, 22, 500, 20, 172
Mrs. Jackson, -, -, -, -, 6
R. B. Brydie, 200, 250, 3000, 115, 504
M. E. Morgan, -, -, -, -, 40
Amos Johnson, -, -, -, 5, 44
Joseph Parish, 100, 118, 1500, 33, 338
E. B. Gee, 400, 350, 4000, 85, 533
Jno. L. Matthews, -, -, -, 5, 25
Jos. Peace, 50, 122, 1000, 28, 174
Jno. L. Peace, 130, 82, 1200, 40, 560
Martha Inge, 180, 190, 1900, 40, 500
Anderson Moore, 95, 6, 900, 32, 164
Wm. T. Justice, 133, 260, 2750, 87, 708
R. Singleton, 35, 169, 700, 42, 119
Louise Marshall, 200, 130, 1640, 70, 424
Ann E. Kirk, 100, 252, 1800, 30, 471
George E. Brander, 60, 440, 3500, 65, 482
Jno. R. Jones, 60, 170 1840, 30, 471
Wm. C. Hamlin, -, -, - 80, 517
A. B. Skinner, -, -, -, 17, 280
James A. Bell, -, -, -, -, 25
Mrs. C. Lambert, 200, 670, 7000, 32, 200
Aaron Brown, 200, 207, 3250, 100, 525
Miss M. Flinn, 100, 122, 1800, 55, 100
Garner Webb, 300, 300, 6000, 304, 846
Malichi Dupsan, Tho. Jefferson, 15, 120
Thomas Justice, 100 75, 1200, 50, 380
Wm. Irby, 800, 700, 24000, 375, 2500
Robt. Rash, 400, 400, 12000, 210, 1459
R. C. Gregory, 300, 304, 7500, 170, 650

G. A. Lidy, 325, 325, 5200, 75,580
E. K. Barns, 200, 452, 5260, 130, 458
Anderson Bradshaw, 107, 100, 1650, 55, 450
M. S. Rux, 100, 60, 1100, 20, 348
U. Edmundson, 180, 184, 2500, 30, 390
Wm. Gills, 180, 195, 2050, 40, 241
Wm. Crowder, 20, 121, 821, 15, 168
James Inge, 150, 297, 2200, 60, 454
Wm. L. Ruse, 20, 25, 300, 15, 146
J. Sneed Jr., 25, 75, 300, 8, 77
Miss C. Wallace, 25, 5, 400, 7, 75
Mrs. T. W. Snead, 20, 75 700, -, 70
J. Sneed Sr., 35, 79, 410, 5, 35
Wm. Elder, 60, 40, 700, 20, 200
A. G. Barns, 300, 241, 4375, 195, 758
P. H. Roney, 60, 60, 1200, 45, 312
Ponhoun Farm, 30, 170, 800, -, 60
B. A. Hatchells upper farm, 75, 325, 4300, 105 825
Wm. J. Bragg, 300, 490, 8000, 200, 915
B. Hatchell lower farm, 200, 225, 7000, 95, 42
Wm. Brown, 80, 109, 1200, 25, 200
Jesse Brown, 400, 400, 6000, 288, 1100
C. A. Dupriest, 80, 100, 1000, 45, 326
Dr. Jones Farm, 130, 181, 2500, 90, 440
Jno. C. Read, 300, 700, 10500, 245, 660
Jno. R. Hatchell, 300, 700, 12500, 140, 830
N. Matthews, 400, 350, 12800, 645, 2632
P. & P. Hatchell, 300, 660, 11250, 275, 1080
D. C. Fitzgerald, 200, 260, 3900, 140, 655
Jno. R. Foster, 200, 360, 5000, 65, 492

Wm. F. Blackwell, 600, 1200, 16200, 200, 1715

N. Blackwell farm, land included above, 115, 940

John Rash, 200, 400, 2500, 115, 500

Jos. A. Pugh, 150, 295, 3000, 130, 650

Jno. R. Bayne, Gee Hardy farm, 20, 120

Mrs. P. Hardy, 300, 499, 8000, 420, 1323

R. R. Calle, 275, 135, 4000, 200, 1295

Est. A. B. Cralle, 175, 180, 4260, 75, 545

Alex. Cralle farm, 392, 196, 4200, 90, 808

Wm. E. Estice, 80, 185, 2500, 125, 312

Robt. Bolling, 200, 955, 5775, 130 620

L. H. Hill, 40, 110, 1200, 55, 250

David Hamilton, 8, 92, 700, 23, 345

Thomas Crafton, 45, 17, 250, 5, 38

Henry Estes, -, -, -, 5, -

James H. Pully, 10, 40, 200, 7, 102

Charles C. Cox, 4, 60, 320 8, 20

Mrs. Mary Tucker, 50, 65, 720, 20, 257

Wm. T. Gary, 30, 290, 3100, 60, 460

Lucy Edmunds, 20, 30, 600, 5, 28

Tho. Osburn, 150, 210, 2700, 25, 391

Wm. M. White, 400, 200, 6000, 165, 942

C. B. Foulks, 35, 163, 1150, 65, 282

E. J. Estis, 20 63, 498, 10, 173

F. E. Winn, 150, 350, 3000, 25, 3897

Colin Stokes, 1000, 1106, 21000, 720, 3390

Wm. H. Walton, 150, 350, 3000, 115, 225

Edwd. C. Scott, 450, 900, 13300, 150, 1719

J. J. Jordan, 250, 252, 4000, 275, 1128

James Winn, 20, 63, 320, 25, 150

Jno. B. Barlow, 20, 38, 500, 20, 155

P. F. Foust's farm, 187, 186, 4000, 100, 758

John Arvine, 500, 700, 8400, 110, 1045

J. C. Jackson, 200, 450, 4000, 130, 794

Jincy Tisdale, 60, 65, 625, 43, 100

H. W. Tisdale, 65, 285, 6000, 65, 660

T. K. Lear, 1, -, 550, -, 20

E. B, Jackson, 275, 289, 5000, 135, 950

B. M. Adkinson, 150, 150, 3000, 82, 565

Branch Cheatham, 200, 498, 7000, 152, 1484

Isaac East, F. Watson's farm, -, 20

Wm. Weatherford, 40, 40 600, 48, 264

Jno. S. Bayne, 150, 150, 1500, 50, 540

Celia Smith, 50, 100, 900, 5, 8

A. W. Weatherford, 120, 52, 1400, 50, 292

A. Townsend, 200, 325, 3300, 20, 310

G. A. Wood, 250, 412, 6000, 270, 914

Ro. H. Crawley, 250, 250, 5000, 238, 1565

Thomas Arvine, 500, 850, 13500, 280, 1800

Nathan Gee, 250, 300, 3600, 65, 917

George Crymes, 300, 440, 4500, 145, 700

Jas. Satterfield, 33, 67, 1500, 20, 286

Joshua Coleman, 20, 3, 150, 30, 9

Eliza. Coleman, 67, 100, 900, 30, 193

Martha Crymes, 150, 260, 4000, 70, 242

Wm. H. Hatchell, 355, 800, 6000, 360, 1458

Barbary Hudson, -, -, -, 4, 52

Mary Street, 300, 420, 5000, 475, 1211

Mary White's farm, 133, 67, 1200, 28, 301

M. Buckwell, -, -, -, 3, 60

Joseph Jennings, 300, 720, 10200, 195, 1104

Mrs. M. Chimney, 200, 140, 3400, 175, 900

Wm. H. Peny, 300, 465, 10000, 240, 1330

L. Tunstale manager for R & others, 400, 719, 10200, 215, 1830

E. A. Pool, 100, 270, 3330, 233, 315

L. Tunstale's farm, 12, 15, 200, 15, 126

Wm. M. Woodson, 150, 122, 1000, 28, 126

Joel Johns, 500, 328, 10000, 220, 1320

Daniel Bentley, Johns farm, -, 22, 222

John B. Gaulding, 200, 240, 4000, 50, 744

Drewry E. Gaulding, 200, 191, 2000, 30, 629

M. L. Spencer, 200, 390, 6000, 218, 1050

John A. Johns, 500, 420, 7400, 217, 893

A. N. Johns, 200, 200, 3200, 250, 795

John T. Merriman, 200, 557, 7000, 341, 1152

Leonard Crymes, 40, 185, 2000, 23, 21

Euc___ Sterne, as virs farm, 10, 141

Jno T. Dowdy, 100, 245, 2000, 65, 619

John Foulkes, 500, 700, 11400, 180, 1345

Buck Holmes land given in B. D. Stoks, 15, 84

Tho. Gee Sr., 240, 400, 3600, 45, 622

Mrs. M. Bruce, 100, 180, 2500, 12, 135

J. Fuqua M. Crymes farm, -, -, 5, 54

Hatcher Clark, 200, 250, 2700, 100, 628

James Clark, 110, 100, 1300, 25, 260

James Neal, 300, 300, 8000, 230, 675

Benj. E. Ward, 120, 127, 1600, 103, 335

Frank Watson, 250, 365, 6000, 195, 720

G.L. Bayne, 700, 389, 10000, 250, 2319

Jno. T. Eubank, 350, 230, 5220, 100, 1120

Mildred Wood, 250, 330, 5000, 145, 584

Hatcher Clark Roach's farm, 8, 95

Jas. C. Lovn, 180, 200, 3000, 59, 607

Wm. H. Eubank, 500, 1063, 20319, 390, 1522

Dr. C. M. Knight, 400, 400, 9600, 300, 1450

J. G. Hardy, 500, 500, 7000, 190, 890

E. P. & V. R. Williams, 1000, 1600, 39000, 800, 4129

G. O. Hardy, 200, 615, 6000, 160, 810

Henry A. Vaughan, 300, 440, 8880, 135, 942

Richd. Philips, 25, 25, 4000, 110, 340

L. B. Crafton, 100, 300, 1500, 70, 270

J. J. Deshazer, 75, 90, 1600, 15, 383

S. Niblett Sr., 400, 200, 600, 100, 950

S. Niblett Jr., 600, 733, 9500, 520, 1500

J. & E. White, 12, 80, 400, 8, 223

Carter White, 5, 25, 120, 4, 33

Sady(Lady) Jones, 250, 250, 6500, 150, 738

Adam Bell, 300, 400, 7000, 175, 1020

A. Wallace, 71, 30, 600, 10, 170

Saml. Crawley, 200, 100, 3000, 75, 814

Wm. Fram, 18, 2, 100, 5, 30

Wm. Bates Reese's farm, -, -, 25, 163

Wm. Arvine Sr., 700, 330, 6900, 285, 1292

Tho. Jefferson, 400, 500, 9000, 350, 2135

L. H. Knight, 500, 300, 14000, 365, 1613

Jno. R. Gaulding, 133, 267, 2000, 45, 560

Jno. J. Robertson, 60, 46, 1200, 90, 486

Wm. Shelton, 30, 20, 400, 7, 55

Madison County, Virginia
1860 Agricultural Census

The Agricultural Census for Virginia 1860 was microfilmed by the University of North Carolina Library under a grant from the National Science Foundation from original records at the Virginia Department of Archives and History in 1963.

There are forty-eight columns of information on each individual. Only the head of household is addressed. I have chosen to use only six columns of information because I feel that this information best illustrates the wealth of individuals. The columns are:

1. Name
2. Improved Acres of Land
3. Unimproved Acres of Land
4. Cash Value of Farm
5. Value of Farming Implements and Machinery
13. Value of Livestock

Wesley Long, 330, 79, 8000, 450, 1000
William Walker, 300, 166, 8000, 450, 1200
William Walker Jr., 300, 441, 7000, 300, 700
Eliza Burton, 960, 600, 38000, 500, 1160
Benjamin Burton, -, -, -, 100, 500
William Simms, 105, 95, 7000, 200, 700
Wm. M. Simms, -, -, -, -, 275
Burwell Melone, 307 160, 7000, 100, 555
Edward T. Herndon, 75, 50, 1500, 100, 200
T. R. Wallace, 300, 191, 4000, 50, 1500
Cudden Simon, 100, 52, 1500, 50, 350
James Rinnell, -, -, -, -, 10
E. G. Wayland, 300, 200, 11000, 200, 650
M. M. Sprinkle, 4, -, 800, -, 20
A. S. Milton, 3, -, 500, -, 25
J. S. Scott, -, -, -, 10, 150

L. Kennedy, -, -, -, -, 20
A. Kennedy, -, -, -, -, 15
Jas. Collins, 400, 50, 9000, 300, 1000
B. C. Wayland, 200, 90, 4350, 200, 1000
Moses S. Weaver, 400, 240, 5000, 150, 1000
Thos. R. Collier, 125, 50, 3500, 100, 340
Willis Gaar, 400, 341, 11100, 330, 660
Michel Eheart, 90, 40, 1010, 100, 330
B. S. H. Stockdell, 150, 143, 2400, 50, 340
Henry Fry, 200, 380, 5220, 150, 640
Adam C. Eheart, 100, 84, 1300, 50, 100
Wm. E. Jackson, 130, 131, 2610, 150, 390
Thos. M. Jackson, 200, 114, 3100, 150, 350
Mary D. Jackson, 150, 80, 1840 150, 400

Lucretia Jackson, 100, 125, 4500, 100, 350

John H. Jackson, -, -, -, -, 150

Eliza Seal, 100, 51, 5000, -, 20

Ro. C. Garnett, 300, 260, 18000, 300, 1000

E. G. Shiss, 300, 150, 13500, 300, 1200

R. D. Tuyman, 500, 337, 23000, 500, 1800

R. R. Hicks, -, -, -, -, 20

E. C. & B. R. Davis, 350, 194, 16320, 800, 1300

Wm. O. Smith, 252, 100, 9000, 250, 1100

John T. Martin, -, -, -, -, 20

E. B. Ambler, 400, 398, 23000, 500, 1500

Jas. A. Reid, 350, 225, 18000, 700, 2000

Samuel Arrington, 1, -, 100, -, 20

Jos. W. Waller Sr., 1300, 500, 25000, 1000, 2000

Jos. W. Waller Jr., 450, 19200, 25700, 500, 2000

Fitzhugh Taliaferro, 181, 50 9240, 500, 500

R. C. Booton, 20, 580, 700, -, -

R. G. Booton, 400, 305, 11500, 500, 2215

Thos. T. Slaughter, 500, 230, 18250, 500, 1800

Samuel Spicer, 7, 7, 150, -, 20

Harlow & Jones, 8, -, 7000, -, 100

Julius B. Harlow, -, -, -, -, 165

B. F. Jones, -, -, -, -, 60

Jas. W. Raddish, 225, 53, 3140, 250, 360

J. W. Marshall, -, -, -, -, 170

Thompson Shepherd, 300, 100, 13000, 200, 1000

J. S. Walker, 620, 321, 28230, 600, 2000

E. W. McIntire, 177, 100, 2500, 50, 200

Henry Carpenter, 400, 300, 5300, 150, 800

Noah Henkel, 377, 70, 8940, 300, 2000

J. H. Tucker, -, -, -, -, 20

L. M. Zerkel, 350, 226, 11520, 500, 450

Martin Lohr, -, -, -, -, 450

W. H. Clarke, 100, 58, 370, 50, 300

J. W. Scott, 130, 126, 3840, -, 220

Benj. Scott, 500, 197, 9560, 130, 985

John Skinner, 50, 50, 1200, 50, 150

Catharine Earley(Easley), 200, 70, 2160, 300, 700

J. A. Earley(Easley), 150, 47, 2000, -, -

Joseph Good, 200, 260, 5900, 500, 570

Wm. Clore Sr., 200, 101, 9030, 300, 660

Nancy Clore, 100, 30, 1300, -, 80

Catharine Sparks, 50, 30, 960, 15, 50

Wm. J. Sparks, -, -, -, -, 40

Michael H. Gaar, 330, 330, 6600, 100, 600

Oliver J. Utz, 75, 73, 2500, 100, 200

Thos. Collins, 160, 40, 2000, 150, 250

Jas. Duff, -, -, -, -, 20

Buford Bursougles, 20, 60, 300, -, 5

Sam. W. Nichol, -, -, -, -, 50

Jordon Floyd, -, -, -, -, 30

Philip Lohr, 100, 96, 3525, 350, 480

Olly Ravig, 30, 5, 500, 10, 10

Wm. K. Kean, 268, 100, 2954, 100, 520

Alexr. Lowry, -, -, -, 5, 35

R. N. Dobbs, -, -, -, -, 50

Robert Davis, -, -, -, -, 35

Jacob Weast, -, -, -, -, 125

Robert Warren, -, -, -, -, 15

Eliza Herndon, 40, 11, 700, -, -

George Herndon, 400, 12, 700, -, -

Nath. G. Herndon, 40, 12, 700, -, -

Easther Herndon, 110, 40, 2700, 60, 260

Leonard Herndon, 20, 33, 700, -, -
D. B. Herndon, 28, 20, 700, -, -
Mary H. Bradford, 45, 7, 1500, 25, 200
Thos. T. Yager, 180, 60, 3500, 100, 640
Jas. S. Carpenter, 60, 20, 1200, 100, 675
Pleasant Tinsley, 150, 52, 3030, 100, 700
John C. Utz, 270, 230, 8000, 300, 920
Edmond Rowzee, 100, 94, 1900, 50, 300
Thos. Jones, 350, 150, 5000, 200, 1000
Jesse Jones, 75, 65, 1800, 150, 500
Thos. W. McMullan, 100, 62, 2500, 100, 250
Thos. S. Gravis, -, -, -, -, 120
A. G. Grinnan, 187, 3400, 22000, 150, 1000
Isaac Walters, 180, 140, 16000, 200, 500
J. W. Twyman, 800, 500, 25000, 750, 4445
Jane B. Madison, 300, 146, 13380, 250, 575
Wm. H. Twyman, 640, 200, 29400, 600, 1000
J. S. Twyman, 550, 250, 19775, 1000, 900
Ellen Lovett, 600, 300, 25000, 350, 1965
R. E. Lightfoot, 300, 424, 21000, 1500, 1225
Geo. W. Clark, 840, 772, 31610, 1000, 2350
H. T. Sparks, 200, 160, 5400, 200, 1065
Lucy Frey, 230, 90, 8610, 500, 850
W. P. Yowell, 25, 15, 450, -, -
James Lloyd, 98, 35, 990, 60, 170
George Racer, 130, 71, 1500, 20, 85
Walker Cobler, 8, 2, 100, -, 5
Sally Clark, ¼, ¼, 50, -, -

Joshua Canntly, ¼, ¼, 50, -, -
Wm. A. Rose, 366, 100, 5600, 200, 350
Wm. A. Rose, -, -, -, -, 150
Simeon Marshall, 372, -, 1500, -, 150
Benj. G. Wilhoit, 70, 60, 4000, 100, 250
George Tinsley(Teasley), 4, 1, 600, -, 2
Geo. W. Sprinkel, 28, 17, 900, 60, 200
Alonzo Miller, 13, 1, 800, -, 50
Asa W. Gre___ 215, 50, 6000, 550, 600
Geo. Benton, 500, 381, 15858, 500, 1600
Jas. Benton, 120, 80, 2400, -, 100
Geo. N. Shufts Est., 1000, 1270, 34600, 350, 1720
Thos. N. Harrison, 300, 180, 14400, 250, 1250
Clarissa Buckner, 300, 150, 8000, 200, 400
Ann P. Booton, 540, 179, 17975, 500, 1075
Benj. F. Walker, 337, 150, 12000, 400, 1125
Richard Richmas Jr., 4, 1, 200, 20, 100
Richard Richmas Sr, 65, 67, 1370, 15, 100
John Risor, 80, 62, 1420, 40, 135
William Terrey, 80, 50, 1300, 10, 220
Wm. Gilmore, 40, 25, 650, 10, 150
Wm. F. Nichol, 400, 240, 12800, 500, 1390
Robert Long, -, -, -, 50, 150
Willis Clayton, 40, 35, 1500, 50, 250
J. S. Breeden, -, -, -, -, 25
Michael Bazil, 3, 38, 800, -, 50
Noal Price, -, -, -, -, 300
Thos. R. C. Graves, -, -, -, 25, 30
Samuel Lusing, 100, 36, 4488, 75, 230

John Dodson, 3, 3, 120, 10, 50
Elizabeth Dodson, -, -, -, -, 30
Benj. F. Eheart, -, -, -, 10, 200
Benj. F. Graves, 20, 36, 560, 100, 540
John S. Horton, -, -, -, -, 15
Fontaine Coats, -, -, -, -, 30
Reuben Tucker, 100, 100, 2400, 30, 240
Adam J. Utz, 49, 40, 534, 5, 75
R. P. Barr, -, -, -, -, 100
Robert Davis, 1, -, 100, -, 15
Lucinda Thompson, 80, 20, 600, 35, 125
R. B. Sullivan, -, -, -, -, 158
Alphonso Jones, 250, 180, 6000, -, 100
N. H. Carpenter, 80, 40, 3000, 200, 450
Betsy Delph, 1, -, 100, -, 40
David Seal, 100, 35, 1620, 50, 150
John Seal,-, -, -, -, 175
Hiram C. Reid, 245, 120, 4000, 100, 500
Thos. A. Sparks, 45, 9, 650, 10, 100
Eliza H. Sparks, 25, 25, 500, -, -
J. S. Sparks, 110, 90, 3000, 150, 500
Jese Yager, 300, 150, 6750, 250 900
H. F. Yager, 250, 100, 3000, 100, 720
Charles Hume dec., 350, 100, 6780, -, -
L. T. Carpenter, 500, 250, 7800, 200, 200
J. A. Johnson, -, -, -, -, 150
Mildred Carpenter, 200, 80, 2240, 100, 300
George Wise, 150, 100, 2000, 100, 500
A. Jacobs, -, -, -, -, 20
Jas. Tucker, -, -, -, -, 30
Wm. Tucker, 39, 6, 900, 12, 140
Lewis Layton, -, -, -, 70, 200
Henry Lohr, 20, 40, 300, 5, 50
Michael Lohr, 40, 50, 500, 5, 75
John Lohr, 30, 100, 700, 75, 300

Benj. Lowry, 40, 57, 1350, 100, 200
Delilah Bell, 170, 30, 1000, 50, 200
Nancy Smith, 510, 141, 35805, 400, 2000
B.G. Jones, 155, 25, 3600, 200, 560
A. W. Lacey, 22, -, 2860, 100, 260
Wm. Brown, 430, 100, 7950, 200, 800
Merry Aylor, 725, 562, 23975, 1000, 2384
F. T. Frey, 300, 160, 6980, 150, 780
James Banks, -, -, -, -, 30
James Bukers, 300, 120 5600, 100, 600
C. B. Ford, 23, 12, 200, 5, 30
Edward Dulin, 21, 12, 200, -, 40
Powhattan Massie, 700, 250, 9500, 150, 1000
Wm. W. Milton, -, -, -, -, 30
M. N. Strother, 140, 66, 2100, 70, 480
M. R. Newman, 175, 65, 2400, 100, 593
Zackry Favel, -, -, -, -, 50
W. J. Clatterbuck, -, -, -, 10, 200
John Bickers, 80, 45, 2500, 50, 300
M. E. Kemper, 500, 450, 7600, 200, 1685
Simeon Carpenter, 147, 100, 2800, 200, 800
Edwin Henshaw, 65, 10, 600, -, -
J. A. Cave, -, -, -, -, 20
A. T. Broyles, 148, 25, 1038, 50, 260
Eliza Simons, -, -, -, -, 33
Fielding Payton, -, -, -, -, 75
Wm. Taniel, -, -, -, -, 140
Thos. E. Aylor, 300, 160, 6900, 500, 1650
Elizabeth Sleet, 175, 57, 1380, 50, 300
Geo. W. Weaver, 25, 15, 480, 40, 157
Matthew Burke, 140, 60, 2000, 100, 450
A. G. Yager, 200, 70, 2700, 250, 405
Jas. W. Clore, 70, 70, 2100, 200, 350

Alfred Carpenter, 130, 130 2670, 100, 580

Mary Clatterbuck, -, -, -, -, 30

Nat. J. Wayland, 400, 400, 5000, 200, 1230

Christopher Rosson, -, -, -, -, 120

Mildred Cook, 100, 50, 1050, 50, 200

J. R. Tucker, 200, 150, 5395, 100, 550

G. A. Hume, 300, 150, 3600, 250, 785

Elizabeth Hume, 150, 45, 1560, 150, 450

A. H. Carpenter, 255, 200, 5370, 500, 990

Theopholus Smoot, 35, 13, 200, 60, 270

Jas. W. McMullan, 135, 15, 1500, 50, 275

Elizabeth Tanner, 20, 11,300, -, -

H. P. Tanner, -, -, -, -, 100

Cornelius Delph, 13, -, 100, -, 20

Wm. B. Raier, 50, 47, 970, 200, 150

Francis Brooking, 80, 61, 1692, 100, 200

Thos. B. Jackson, 15, 6, 336, 40, 80

Ella Carpenter, 10, -, 100, -, 25

Alfred Cloid(Close), 119, 125, 2005, 75, 250

John Richards, 35, 15, 400, 20, 150

Robert Carpenter, 300, 200, 4790, 200, 800

Alexr. Carpenter, 100, 100, 3250, 50 450

Lewis Miller, 70, 60, 3000, 100, 450

Capt. Wm. Thomas, 400, 144, 15232, 795, 1515

A. H. Carpenter, 680, 300, 14050, 500, 1050

Charles W. Swann, 50, 27, 1200, 150, 675

J. C. Aylor, 6540, 250, 24000, 680, 2205

John Fishback, 256, 80, 6048, 300, 500

Sturmton Fishback, -, -, -, -, 246

W. H. Harrison, 70, 407, 3000, 150, 587

G. K. Harrison, -, -, -, -, 55

J. F. Strickler, 300, 191, 10000, 670, 1550

Wm. Clatterbuck, -, -, -, 15, 200

Urial Carpenter, 400, 300, 20000, 500, 1350

S. E. Wayland, 200, 42, 9000, 400, 2000

J. O. Wayland, 89, 30, 4705, -, -

W. T. Utz, 145, 70, 5985, 100, 400

Thos. J. Aylor, -, -, -, -, 120

Catesby Aylor, 78, 75, 1500, 60, 250

Betsy Lloyd, -, -, -, -, 20

D. L. Crigler, 418, 200, 6000, 75, 410

Wm. M. Blankenbeker, 185, 125, 5000, 500, 1310

Albert Margriess, 300, 80, 1000, -, 12

Fielding Fleshman, 110, 110, 4000, 50, 300

Angus Garrett, -, -, -, -, 75

J. H. Lloyd, -, -, -, -, 85

Harriet Collins, 70, 30, 5000, 200, 350

George Bickers, 104, 200, 2432, 250, 400

Benj. H. Sparks, 37, 40, 3080, 65, 515

J. L. Grayson, 1, -, 500, -, 100

A. J. Bickers, -, -, -, -, 20

Thomas Aylor, 60, 70, 1560, 60, 200

Jas. M. Fray, 600, 250, 10000, 200, 1860

Pricilla S. Yager, 8, -, 1500,-, -

S. W. Yager, -, -, -, -, 50

F. M. Henshaw, 130, 20, 2500, -, 500

Robert Henshaw, 100, 109, 1500, 300, 420

Johnathan Roberts, 400, 327, 12800, 200, 900

Mrs. Bery Roberts, -, -, -, 10, 150

James Cave, -, -, -, -, 50

Wm. H. Weakley, -, -, -, -, 50
Wm. T. Garth, 500, 280, 6240, 400, 830
Thomas Pratt, 200, 144, 5175, 250, 744
Elizabeth Skinner, 146, 52, 1584, -, -
Mariah L. Yager, 200, 90, 5220, 100, 820
Delilah Thomas, 11, 1, 2000, -, 280
Wm. H. Thomas, -, -, -, -, 100
Michael Riffs, 100, 35, 3400, 100, 875
Isaac Burton, -, -, -, -, 20
J. F. Colvin, 35, 3, 850, -, -
H. C. M. Colvin, 35, 3, 850, -, 125
N. E. Colvin, -, -, -, -, 140
Triplett L. Estes, 150, 164, 7280, 150, 556
Michael O'Neal, -, -, -, -, 30
Alexr. Reid Est., 330, 100, 10750, 150, 1110
Isham Tatum, 420, 200, 7440, 200, 600
Nat. Saturn, 600, 326, 9300, 150, 800
Alexr. B. Hunton, 173, 8, 2715, 150, 835
Stetair Graves, 150, 54, 4000, 250, 630
D. J. Smoot, 400, 400, 8000, 250, 1130
W. T. Gower, 300, 122, 3376, 60, 466
E. F. Sprinkle, 106, 30, 2070, 200, 240
W. A. Hue, 400, 250, 15000, 950, 2500
Edwin Nichols, 400, 150, 11000, 150, 820
James Coatney, -, -, -, 100, 450
Alfred Utz, 140, 182, 4990, 100, 450
Hiram Yager, 250, 390, 14200, 250, 1130
Landford L. Carpenter, 200, 200, 10000, 100, 380

Elias Blankenbeker, 200, 232, 6020, 100, 600
Nick Blankenbeker, 150, 100, 1750, 100, 250
Ambrose Jones, 300, 100, 8000, 100, 1640
W. H. Dickson, -, -, -, -, 20
Jesse C. Tatum, 150, 44, 2320, 100, 657
Wm. Rouse, 80, 45, 1200, 50, 250
Fielding Carpenter, 140, 128, 9050, 150, 350
Wm. Thompson, 30, 120, 800, 100, 200
Fielding Dur, 89, 20, 1635, 125, 530
D. B. Steigle, 200, 175, 7480, 300, 1065
J. C. Bowman, 200, 50, 6250, 100, 350
Augustin Hawkins, 4, 5, 60, 2, 2
A. F. Aylor, 150, 50, 3000, 100, 530
Wm. H. Carpenter, 100, 85, 2000, 200, 450
J. C. Clore, 250, 115, 7000, 300, 985
R. A. Yowde(Zowde), 170, 60, 2300, 150, 800
Jere Aylor, -, -, -, 15, 100
Aaron Jenkins, -, -, -, 100, 400
Wm. Thomas, -, -, -, 10, 150
R. W. Blankenbeker, -, -, -, 25, 200
J. Y. Shotwell, 100, 100, 1000, 25, 200
Robert Lindsay, 200, 100, 2000, 150, 545
John Smith (B), 75, 200, 350, 10, 300
Thomas Sisk, -, -, -, -, 75
John Rush, 60, 70, 1346, 75, 200
Thos. P. Simms, 140, 60, 6000, 75, 580
Jas. B. Willis, 300, 287, 10000, 600, 1340
Newton Kirtley, -, -, -, 10, 60
Abram Kirtley, 20, 2, 150, 8, 187
John Harrison, 400, 600, 8000, 500, 1400

Henry Floyd, -, -, -, -, 50
Arthur W. Bowles, -, -, -, -, 100
L. F. Slaughter, -, -, -, -, 70
Banks Goodall, 150, 300, 1500, 50, 500
R. M. Milton, 3, -, 800, -, 20
Zackry Smith, 25, 75, 300, 5, 100
Moses Smith, 100, 200, 300, 5, 40
J. L. Lindsay, 4, -, 1200, -, 150
Jas. S. Bewey(Beney), -, -, -, -, 35
M. C. Strickler, 165, 35, 10000, 800, 1440
J. M. Weaver, 200, 100, 10000, 1500, 1610
R. S. Thomas, -, -, -, -, 238
J. K. Rosser, 50, 7, 1500, -, 115
Chadwell Bewey(Beney), 20, 27, 275, 10, 120
S. E. McAllister, -, - -, -, 200
Jas. R. Finks, 36, 13, 1000, -, 100
Emanuel Weakley, 75, 80, 600, 40, 225
S. M. McAllister, 60, 40, 600, 10, 190
Wm. Bates, -, -, -, -, 180
Elizabeth Bates, 150, 156, 3060, 200, 545
R. A. Bates, -, -, -, -, 125
Thornton Paul, -, -, -, -, 30
Jas. H. Thomas, 50, 50, 500, 20, 300
J. H. Thomas, -, -, -, -, 100
Simeon Hart(Hurt), -, -, -, -, 35
Benj. Broyles, -, -, -, -, 150
A. N. Blankenbeker, -, -, -, 150, 600
Fielding Jenkins, -, -, -, 20, 150
Benj. Jenkins, -, -, -, 5, 100
Wm. F. Hurt (Hart), 10, 18, 100, 5, 230
J. H. Graves, 50, 50, 300, 5, 50
Aesey Bewey(Beney), 300, 240, 12000, 280, 1080
Judith Clore, 160, 100, 4000, 200, 1010
Geo. M. Bohannon, 700, 220, 25000, 600, 2520

Thos. W. Chapman, 600, 792, 10000, 500, 1740
David Storey, 33, 12, 2000, 75, 215
Wm. R. Beney, 150, 290, 6000, 500, 944
Henry O'Neal, -, -, -, -, 20
Wm. H. Clore(Close), 200, 188, 5300, 450, 1035
Wm. H. Weatherall, -, -, -, 125, 350
Jere Rush, 3, -, 250, -, 25
Jas. Ranalas, 23, 3, 4000, -, 450
Wm. H. Weaver, 175, 150, 6000, 150, 732
A. F. Carpenter, 200, 250, 5000, 300, 920
J. N. Lindsay, 140, 45, 3000, 275, 1100
Benj. F. Gads, 200, 140, 8000, 250 810
George Lillard, 29, 100, 516, 20, 150
Andrew Carpenter, 300, 180, 6000, 250, 984
Lucy Carpenter, 35, 15, 2000, -, 105
Geo. B. Taylor, 4, 1, 100, -, 100
J. T. Clore(Close), 240, 360, 6000, 15, 270
E. F. Hill, 800, 400, 22000, 600, 2140
Franklin James, 350, 350, 10500, 600, 2100
Jas. A. Jeffries, -, -, -, -, 25
Miland Graves, 250, 110, 9200, 300, 1125
Fielding Crigler, 500, 200, 10000, 500, 1375
Elisha Bewey(Beney), 70, 119, 1135, 150, 216
S. F. Jeffries, 40, 10, 2000, 100, 125
Dabney Minor, 7, -, 1000, -, -
Abner Hudson, -, -, -, -, 100
Jas.M. Jarrell, -, -, -, 20, 146
Henry Lysinger, 15, 22, 2000, 20, 100
Peter C. Lauck, 350, 354, 7000, 150, 500

Augustin Lillard, 100, 127, 1135, 60, 227

Ann Gaines, 80, 138, 1000, 100, 415

Churchwell Weakley, -, -, -, 10, 210

Jas. Floyd, 80, 178, 1000, 15, 260

Lemuel Cross, 60, 273, 700, 55, 260

Wm. W. Lillard, -, -, -, 60, 380

Robert Kary, 4, 6, 260, -, 50

Chadwell Beney(Berrey, Bewey) Sr.100, 200, 2000, 15, 140

John Brown, -, -, -, -, 46

Milton Rite_ous, 2, -, 1000, -, 170

J. L. Carpenter, 300, 150, 15000, 600, 1145

H. W. Carpenter, 100, 150, 3000, -, -

Irene Carpenter, 150, 100, 3000, -, -

Josiah Weaver, -, -, -, -, 25

A. Y. Yowell, -, -, -, 60, 274

J. H. Weaver, -, -, -, -, 120

Howard Yowell, 130, 65, 4875, 100, 467

Abram J. Blankenbeker, 320, 318, 7750, 250, 1130

Mary Utz, 50, 40, 2000, 50, 140

Nelly Crisler, 250, 200, 5200, 500, 850

Martha F. Crisler, 100, 125, 1500,-, -

Abram Blankenbeker, 100, 100, 2000, 300, 520

George Delph , -, -, -, -, 30

Julius Utz, 100, 400, 2500, 10, 200

George Utz, 170, 30, 3500, 200, 500

Lewis A. May, 2, 23, 150, 10, 128

Edward Hood, -, -, -, 10, 150

Fielding Utz, 100, 40, 2000, 150, 430

James Gordon, -, -, -, -, 50

Gabriel Aylor, 155, 75, 6000, 500, 755

G. M. Tanner, 73, 20 1000, 75, 325

Judith Tanner, 1, -, 100, -, 20

E. G. Chapman, 350, 60, 15000, 700, 1404

Y. L. Huffman, 130, 115, 2450, 75, 350

S. H. Thomas, -, -, -, -, 35

J. B. Marguiss, 5/8, -, 40, -, 75

Simeon Blankenbeker, 360, 215, 26500, 600, 1165

George Blankenbeker, -, -, -, -, -

J. B. Hague, -, -, -, -, 180

J. A. Clore(Close), -, -, -, -, 300

Jere Carpenter, 48, 65, 3100, 125, 520

Wm. B. Twyman, 400, 214, 10000, 150 700

George Weatherall, -, -, -, -, 50

E. C. Stover, 90, 5, 3800, 350, 450

H. Dodson, -, -, -, -, 6

Asa W. Smith, -, -, -, -, -, 10

Crisley Weakley, 5, 25, 100, 10, 30

Wm. R. Jenkins, 35, 108, 800, 50, 100

Gabriel Smith, 100 75, 3100, 40 225

Catharine Frenchman, 50, 400, 2000, -, 70

J. W. Jenkins, 15, 155, 350, 40, 80

Johnson Jenkins, 2, 48, 75, -, 33

James Hart, 50, 79, 645, 10, 40

Wm. Bewey, 50, 50, 250, 30, 100

Silas Weakley, 15, 35, 125, 10, 100

W. H. Bradley, 20, 50, 400, 20, 30

Isreal Hurt (Hart) Est., 112, 100, 1000, -, -

Wm. Bewey, 100, 300, 1200, 40, 130

Wallace Corbin, -, -, -, -, 21

Jeff Jenkins, -, -, -, -, 75

Adam Jenkins, -, -, -, -, 50

Russel Jenkins, -, -, -, -, 30

John Seal, 55, 10, 500, 20, 120

Lewis Seal, -, -, -, -, 30

Wm. Coppage, -, -, -, 15, 75

Robert Seal, 40, 460, 3000, 100, 175

Elcaney Jenkins, 8, 92, 600, 20, 60

G. R. Tanner, -, -, -, 5, 65

Arnold Neathers, 100, 20 2500, 150, 355

Strother Corbin, -, -, -, 10, 50

Edmund Corbin, -, -, -, 10, 30

William Corbin, 50, 50, 350, 10, 140

Edmund Nicholson, 25, 25, 100, 5, 95

Harrison Nicholson, -, -, -, -, 30
Wm. A. Dodson, 54, 100, 600, 20, 100
Peter Nicholson, 75, 50, 500, 10, 210
Aaron Nicholson, 25, 50, 100, 10, 90
Garrett Nicholson, 40, 60, 200, 10, 65
Joseph Nicholson, 10, 40, 100, -, 60
Jacob Nicholson, 10, 40, 50, -, -
Benj. Nicholson Jr., 40, 15, 350, 10, 120
Moses Nicholson, 25, -, 250, 10, 40
Crisley Nicholson, -, -, -, 5, 95
Benj. Nicholson Sr., 30, 49, 600, 10, 390
S. B. Nicholson, -, -, -, -, 40
Henry Nicholson, -, -, -, 10, 90
John Corbin, 70, 105, 800, 10 290
John Nicholson, 200, 383, 1900, 15, 384
Lindsay Jenkins, 50, 50, 600, 5, 272
Lowell Jenkins, -, -, -, -, 100
Thornton Nicholson, 125, 177, 1500, 5, 245
Francis Jenkins, 40, 60, 700, 10, 150
Lucinda Jenkins, -, -, -, -, 50
Lloyd Dodson, -, -, -, -, 50
Jermima Nicholson, 60, 30, 600, 5, 130
Reuben Ryder, 30, 55, 500, 5, 200
Joseph A. Weakley, -, -, -, -, 45
James Menbrag, -, -, -, -, 25
Layton Sisk, -, -, -, -, 15
William Sisk, -, -, -, -, 30
J. R. Gooding, 175, 75, 4500, 50, 610
Allen Carpenter, 300, 430, 5025, 200, 850
Albert Aylor, 5, 7, 250, -, 20
Eph. Rouse, 45, 50, 700, 30, 270
Jas. S. Rosser (Roper), ½, 4, 200, 5, 50
Martha Harlin, -, -, -, -, 75
M. D. Call, 175, 175, 4500, 50, 610
R. W. Kinsey, 100, 249, 3000, -, -

B. F. Kinsey, 300, 801, 7707, 2000, 944
Fielding Weakley, -, -, -, 10, 140
Robert Yowell, 6, 110, 1160, 15, 175
Eph. Dulaney, -, -, - 20, 60
Edmund Jenkins, -, -, -, 10, 70
Joseph Jenkins, -, -, -, 5, 95
Gustavius Utz, -, -, -, 10, 200
T. F. Bewey, 50, 75, 1400, 50, 330
W. P. Finks, -, 17, 40, -, 100
Thomas Smith, 30, 40, 700, 50, 190
Thomas Cubbage, 10, 10, 200, -, -
George Shuletice, 50, 26, 1000, 100, 225
J. M. McAllister, 20, 10, 150, 10, 50
Stewart Yowell, 80, 50, 1600, 50, 200
J. M. McAllister, 20, 50, 210, 20, 100
Elizabeth Dulaney(Dulancy), 5, 10, 200, -, 50
Rebecca Storey, 400, 400, 16000, 50, 150
E. M. Storey, 52, 13, 1100,-, -
Lucretia Storey, 35, 10, 1100, -, -
Alpheus Fletcher, 300, 424, 7240, 200, 960
Joseph Storey, 220, 70, 8000, 100, 725
William Harlon, -, -, -, -, 150
J. M. Rust, -, -, -, -, 150
A. G. Dulancy, 200, 180, 9000, 120, 550
T. F. Fletcher, 105, 67, 4000, 75, 700
J. I. Fletcher, 103, 67, 4000, 75, 400
Roland Cubbage, -, -, -, -, 60
Roland Yowell, 146, 404, 3040, 50, 150
Fielding Jenkins, 75, 125, 800, 30, 130
Gabriel Smith, -, -, -, -, 120
Lillard Jenkins, -, -, -, -, 25
A. W. Smith, 40, 55, 380, 5, 75
W. I. Thornhill, -, -, -, 10, 152
A. H. Brown, -, -, -, -, 172
Austin Smith, 60, 60, 250, 5, 70

Wesley Weakley, 200, 135, 4000, 100, 290

Jonas Huffman, 50, 21, 790, -, 25

J. M. Lillard, -, -, -, -, 120

Geo. L. Ri__comb, 210, 16, 4500, 100, 580

Paschael Jenkins, -, -, -, -, 96

Elzey Weakley, 30, 18, 650, -, 50

Alexr. Ryder, 60, 40, 1000, 20, 150

John Jenkins, -, -, -, -, 10

D. G. Thornhill, -, -, -, 7, 302

B. F. Smith, 60, 31, 910, 200, 350

Henry Lillard, 103, 7, 2500, 75, 542

Fielding Smith, 130, 71, 1600, 75, 170

Mary Fletcher, 150, 90, 4000, 75, 650

R. W. Fletcher, 100, 20, 4000, -, -

Robert Petty, -, -, -, -, 15

A. E. Mitchell, -, -, -, -, 130

Elizabeth Mitchell, 100, 40, 300, 50, 160

Loveless Mitchell, -, -, -, 100, 120

A. H. Simms, 250, 424, 14300, 220 1360

Sarah McAllister, 70, 44, 570, 10, 120

Miland Jenkins, -, -, -, -, 20

John Thomas, 200, 70, 4050, 100, 500

Edmund Rosson, 100, 23, 990 50, 400

Zackry S. Smith, 500, 500, 10000, 500, 1166

William Rosson, -, -, -, -, 60

Robert Yowell, 30, 52, 300, 5, 100

Lewis Finks, 500, 90, 5000, 150, 1030

Mary Hutcherson, 85, 86, 1710, -, 100

William Shotwell, 100, 100, 2500, 75, 220

P.M. Hutcherson, 120, 77, 3000, 40, 377

Betsy Garret, 5, -, 400, -, -

Smith Wilhoit, 150, 86, 3500, 200, 250

A. S. Wilhoit, -, -, -, -, 200

Julia Utz, 38, 30, 408, -, -

Charity McAllister, 73, 100, 519, -m, 80

J. R. Lillard, 30, 40, 900, 10, 135

Henry Utz, 50, 135, 1104, 50, 180

Allen Yowell, 100, 173, 1500, 50, 240

Mildred Smith, -, -, -, -, 80

Appleton Jenkins, -, -, -, -, 75

Elizabeth Henderson, 70, 56, 750, 50, 200

Petty Batton, 1, -, 50, -, -

Eliza Henderson, 15, 11, 100, -, -

F. E. Jenkins, -, -, -, -, 25

Morgan Jenkins, -, -, -, 10, 70

J. A. Dodson, -, -, -, 15, 120

G. A. Dulaney, 100, 154, 1000, 25, 351

Jacob Rouse, 35, 65, 500, 50, 150

Ann Yowell, 30, 47, 435, -, -

Milton Arrington, -, -, -, -, 60

Thomas Arrington, 40, 39, 800, 60, 90

Fountaine Deal, 100, 58, 1580, 150, 270

Rowland Yowell, 70, 20, 490, -, 100

Elliott Finks, 120, 57, 1256, 75, 290

Larkin Deal, 55, 50, 735, 100, 200

Joshua Miller, 500, 322, 16440, 840, 2385

Francis Broyles, -, -, -, -, 20

A. T. Tanner, 175, 181, 3000, 200, 497

Robert Henshaw, 109, 100, 624, -, -

Francis M. Henshaw, 62, 100, 624, -, -

A. L. Henshaw, 50, 100, 720, -, 100

M. A. Batton, 30, 16, 370, -, 45

Elizabeth Rosson, 10, 28, 200, -, 60

James Rosson, 12, 21, 165, -, 70

Christianna Rosson, -, -, -, -, 170

William Campbell, 25, 30, 400, 10, 120

Benj. Jenkins, -, -, -, -, 40
T. V. H. Tanner, 30, 10, 800, 40, 150
Wm. Deal, 30, 85, 2000, 10, 165
G. W. Goodall, 100, 63, 1500, 25, 425
Emily Arrington, 10, 15, 75, -, 15
Malinda Henshaw, 85, 85, 1070, 30, 120
A. M. Yowell, 175, 95, 300, 1000, 700
Nancy Aylor, 40, 15, 550, 15, 80
R. H. Aylor, 100, 160, 2196, 60, 210
James Henshaw, 300, 162, 3241, 50, 650
Coleman Brown, 150, 50, 1840, 200, 380
Henry Smith, 90, 27, 1770, 10, 50
James Colom, -, -, -, -, 60
Alfred Burke, -, -, -, -, 5
J. R. Sims, -, -, -, -, 40
J. M. Yowell, 100, 24, 1860, 100, 557
A. C. Curtise, 58, 15, 730, 10, 120
E. F. Murry, 45, 10, 925, -, -
M. B. Henshaw, 100, 79, 1790, 60, 200
Emanuel Price, 4, -, 25, -, 8
Simeon Huffman, -, -, -, -, 5
Lafayette Henshaw , 100, 56, 1560, 15, 920
A. H. Taylor, 85, 15, 600, 15, 200
M. B. Hill, 200, 80, 6000, 30, 275
Joshua Carpenter, 200, 116, 2530, 30, 130
Mrs. M. Gaar, 500, 394, 7152, 150, 650
Nelson Huffman, 31, 20, 510, 20, 100
Bombrey Buck, -, -, -, -, 35
A. B. Huffman, 67, 30, 455, 20, 170
William Huffman, 100, 42, 1420, 125, 270
Elizabeth Huffman, 60, 30, 630, -, -
M. G. Graves, 130, 257, 3000, 100, 620

Milton Edains, 300, 444, 9000, 150, 824
Robert Daniel, 275, 440, 4200, 250, 789
Robert Southard, -, -, -, -, 10
W. J. Bewey, 50, 42, 900, 30, 165
Silas Utz (Rafs), 250, 230, 4000, 200, 745
James Estes, 50, 50, 1000, 10, 175
Wm. Jenkins, 150, 210, 2000, 50, 250
L. A. Nicholson, 35, 80, 460, 5, 140
Philander Goodall, 200, 180, 4000, 200, 1520
Elisas Marks, -, -, -, -, 30
C. M. Earley, 200, 165, 10950, 500, 1000
B. F. T. Conway, 1200, 555, 42690, 500, 3980
R. A. Burnett, 160, 127, 4000, 200, 786
R. N. Rose, 88, 100, 2000, 150, 450
J. W. Cole, 100, 251, 1755, 50, 420
H. N. Wallace, 609, 608, 36500, 700, 2160
G. H. Allen, 267, 150, 8340, 200, 1000
W. T. Simpson, 150, 148, 3576, 100, 375
Benj. Simpson, 60, 26, 860, 30, 150
Agness Bledsoe, 15, 18, 660, -, -
Lawson Bledsoe, -, -, -, -, 50
J. W. Gullihugh, 70, 40, 2200, 30, 107
Elizabeth Hume, 330, 45, 13000, 75, 230
Sarah Lewis, 200, 260, 5520, 50, 220
Willis Austin, -, -, -, -, 50
J.M. Huffman, 108, 60, 3300, 50, 230
Allen May Sr., -, -, -, 10, 100
Lemuel Rush, -, -, -, 10, 120
John May, -, -, -, 10, 100
Jerward Gaar, 150, 250, 4000, 150, 200

G. W. Gibbs, 200, 130, 4000, 200, 650

W. L. Gibbs, 100, 160, 6000, 200, 615

Margaret Carpenter, 90, 70, 2500, 25, 65

G. W. Utz, 8, -, 900, 5, 30

Jane Weatherall, -, -, -, -, 40

J. H. Reynolds, 100, 20, 3000, 250, 765

Y. M. Yowell, 100, 60, 1920, 100, 300

J. N. Graves, 300, 500, 6000, 250, 1360

J. N. Blankenbeker, 300, 50, 12260, 100, 615

Wm. Huffman, -, -, -, 10, 218

J. B. Weaver, 35, 22, 450, -, 50

J. O. Henshaw, -, -, -, -, 175

Elliott Blankenbeker, 800, 544, 40000, 1000, 1600

Ann Huffman, 30, 20, 300, -, 70

M. H. Huffman, -, -, -, -, 150

M. F. Conway, 400, 180, 15660, 540, 1520

G. S. Conway, 100, -, 550, -, 200

John Booton, 150, 100, 6750, 600, 1340

James Harris, -, -, -, -, 20

Howard Berny, 100, 62, 1950, 50, 200

Southy Simpson, -, -, -, -, 8

Noah Anderson, 30, 212, 2000, 50, 150

Benj. Anderson, 20, 30, 1500, 15, 100

Obid. Anderson, 50, 21, 2840, 50, 300

Abram Yowell, -, -, -, -, 65

Jeff Jarrell, 200, 123, 3736, 15, 400

James Jarrell, 130, 210, 3600, 5, 110

Henry Jasper, -, -, -, -, 70

R. F. Mauck, 75, 125, 1000, 15, 120

James May, 100, 200, 1500, 50, 200

Jere S. Jarnell, -, -, -, -, 30

Simeon Slaughter, -, -, -, -, 70

Noel May, 300, 300, 3000, 200, 735

Milton Kirtley, -, -, -, 5, 150

Bashaba Jarrell, -, -, -, -, 50

Wesley Simpson, 20, 60, 300, 5, 100

Wm. R. Taylor, 30, 45, 200, 10, 130

Alexr. Nicholson, 60, 234, 1000, 30, 100

Reuben Booton, 160, 2100, 4000, 400, 500

Robert Taylor, 100, 170, 1000, 5, 120

F. R. Taylor, -, -, -, -, 100

Jere Jorrel, 60, 126, 700, 20, 120

Chadwell Berny(Bewey), -, -, -, -, -

Carnett Weakley, -, -, -, -, 20

John Collins, 350, 400, 4000, 100, 560

Richard Lancaster, 250, 354, 3624, 400, 1023

Ann Kinsey, 200, 500, 7000, -, 100

Sarah Graves, 125, 147, 5440, -, 270

Francis E. Graves, 150, 304, 9080, 300, 985

M. H. Utz, -, 40, 250, -, 50

Aaron Clore , 250, 526, 12915, 300, 1800

A. F. Clore, 25, 5, 1500, -, -

Sarah Utz, 100, 60, 8000, 150, 500

J. L. Kemper, 75, 125, 600, 75, 240

Frans F. Twyman, 100, 50, 7000, 800, 744

C. W. Tatum, 100, 69, 2500, -, 270

T. J. Allen, 80, 40, 3000, 100, 400

D. E. H. Sauders, -, -, -, -, 200

M. A. Rudleton, -, -, -, -, 40

T. H. Lewis, 15, 18, 3000, 7, 185

Benj. Carpenter, 1 ¾, -, 3100, -, 185

Tabitha Carpenter, 1 ½, -, 750, -, 8

L. E. Jackson, 1 2/4, -, 3500, -, 50

M. A. E. Thrift, 20, 16, 6560, 300, 350

Belfield Cave, 157, 60, 14000, 400, 380

R. A. Banks, 3000, 5028, 107650, 1500, 5300

W. E. Banks, 200, 116, 9000, 200, 930

C. H. Banks, 30, 20, 4500, -, 80

R. T. Graves, -, -, -, -, 100

Walker Jones, ½, -, 800, -, 95

G. W. Harrison, 37, 30, 6000, 125, 220

F. H. Hill, 1, -, 3500, -, 146

J. W. Taylor, 330, 70, 12500, 600, 1310

T. I. Humphreys, 4, 54, 3400, -, 225

R. N. Utz, 100, 142, 3000, 150, 400

Nel___N. Hume, 6, 131, 3500, -, 80

R. T. Hume, -, -, -, -, 150

R. A. Jackson, 5, 145, 4000, 10, 200

N. W. Crisler, 20, 12, 6500, 250, 400

O. B. Jenks, -, -, -, -, 150

G. P. Evans, 2, -, 2000, -, 25

J. A. Wagman, 2, -, 4000, -, 25

J. W. Lemmon, 6 ½, -, 1400, -, 140

A. R. Blakey, 100, 1024, 18980, 525, 700

M. C. Gordon, -, -, -, -, 100

J. I. Payne, 125, 85, 5250, 200, 685

Sarah Foushee, 1, -, 1800, -, -

S. I. Tyler, 1, -, 300, -, -

Sallie Gray, 2, -, 1200, -, 25

J. M. Blankenbeker, 1 ½, -, 1200, -, 60

Lindsay Thomas, 1, -, 400, -, 20

Robert Gooding, 5 ½, -, 1200m 75, 250

G. H. Leitch, 5, -, 3100, -, 100

Luteti & Blakey, -, -, -, 155

John Leitch, -, -, -, -, 40

G. W. Smith, 1 ¼, -, 1500, -, 50

W. L. Earley(Easley), 730, 820, 40000, 1000, 20400

Sarah Easley(Earley), -, -, -, 150, 250

Martha Garth, 60, 40, 8000, 150, 225

P. L. Thomas, 33,-, 4000, 100, 447

Richard Earley (Easley), 850, 850, 25500, 650, 1362

M. A. Easley(Earley), 130, 50, 2160, 75, 425

Francis Frinks, 160, 40, 12000, 700, 1500

Elizabeth & C. Frinks, 60, -, 4500, -, -

Eliza Thomas, 200, 118, 15000, 500, 1400

Robert Thomas, 50, 60, 1000, -, -

Ann C. Thomas, 70, 30, 1000, -, -

Francis E. Thomas, 80, 40, 1000,-, -

Reubin Thomas, 60, 70, 1000, -, -

James Thomas, 50, 40, 1000, -, -

Stewart Thomas, 70, 25, 1000, -, -

Sarah Thomas, 30, 50, 1000, -, -

Samuel Huffman, 100, 90, 1500, 25, 140

Elizabeth Huffman, 50, 40, 500, 10, 40

Isaac Huffman, 70, 69, 1670, 50, 250

Nancy Crow, 45, 45, 225, -, -

E. D. Fray, 100, 50, 2500, 150, 670

Lutheran Congrott, 120, 200, 8000, -, -

Joel M. Clore(Close), 46, 30, 1500, 100, 415

John Hawkins,-, -, -, -, 25

Peggy Fleshman, 25, 5, 240, -, 60

Joshua Fleshman, 20, 12, 1500, -, -

Moses Weaver, 100, 60, 960, 150, 320

Levi Keller, -, -, -, 40, 260

Peter Clore(Close), 60, 30, 1080, 50, 270

Thos. L. Hogle, 1 ¾, -, 4500, -, 175

Nimrod Heming, 10, 10, 100, -, 30

J. H. Marshall, -, -, -, -, 295

W. M. Storey, 50, 20, 1200, 50, 350

R. A. Seal, 40, 3, 6000, 200, 500

Joseph Leathers, 50, 49, 500, 20, 120

Susan Yowell, 30, 12, 300, 10, 50

Thomas Sparks Est., 200, 138, 5075, 150, 1012

J. H. Rush, 40, 24, 670, 50, 40

Rhodd Smith, 250, 300, 5000, 300, 780

W. I. Smith, 150, 250, 4000, 100, 540

Isaac Southard, -, -, - 10, 110
W. A. Samuel, -, -, -, -, 150
B. F. Weaver, -, -, -, -, 300
T. P. Wallace, -, -, -, -, 150
T. I. Huffman, -, -, -, -, 20
Jonas Weaver, 200, 30, 1840, 50, 240
Edward Gray, 300, 201, 17500, 500, 995
Madison Colvin, -, -, -, -, 20
Thos. R. Huffman, -, -, -, -, 15
Nancy Utz Est., 90, 270, 7039, 50, 258

Nat. I. Welch, 500, 200, 23500, 1000, 1220
John Welch, 150, 50, 4000, -, 150
James Mitchell, 50, 30, 1000, 30, 250
Stony Mann Ming & Co., -, 5371, 1793, -, -
Dark Hollow Ming & Co., -, 8000, 3600, -, -
Henry Huffman, 70, 41, 1128, 50, 200

Mathews County, Virginia
1860 Agricultural Census

The Agricultural Census for Virginia 1860 was microfilmed by the University of North Carolina Library under a grant from the National Science Foundation from original records at the Virginia Department of Archives and History in 1963.

There are forty-eight columns of information on each individual. Only the head of household is addressed. I have chosen to use only six columns of information because I feel that this information best illustrates the wealth of individuals. The columns are:

1. Name
2. Improved Acres of Land
3. Unimproved Acres of Land
4. Cash Value of Farm
5. Value of Farming Implements and Machinery
13. Value of Livestock

James R. Brooks, 170, 110, 3000, 30, 250
Samuel D. Hudgins, 40, 280, 1000, 20, 350
David Ripley, 15, 4, 800, 5, 20
Robert J. Billups, 200, 165, 7000, 125, 600
Humphry B. Davis, 4, 2, 300, 10, 150
Thomas B. Hudgins, 36, 43, 1000, 20, 70
Joel M. Hudgins, 8, -, 500, 5, 50
William K. Burroughs, 10, 8, 600, 12, 50
John P. Jarvis, 19, 13, 2000, 50, 320
John H. Jarvis, 42, 15, 1000, 50, 150
George L. Borum, 80, 20, 2000, 100, 300
William Armistead, 12, 6, 400, 25, 150
John E. Hudgins, 10, 8, 1000, 20, 150
Elsey B. Hudgins, 8, 8, 500, 20, 80
John W. Thomas, 50, 15, 2375, 30, 150

William H. Miles, 25, 20, 1500, 20, 175
Thomas A. Singleton, 30, 30, 1500, 10, 200
Joel Thomas, 29, 10, 800, 30, 125
Joseph N. White, 20, 25, 1200, 8, 50
Thomas D. White, 15, 15, 1200, 25, 120
William K. White, 24, 23, 1800, 15, 100
John Borum, 20, 38, 2000, 15, 150
Thomas D. James, 50, 50, 4000, 25, 100
Francis M. Hudgins, 25, 27, 2000, 15, 60
Mathew Thomas, 50, 50, 5000, 20, 250
Benjamin Diggs, 20, 20, 1500, 12, 75
James Thomas, 55, 43, 4500, 50, 350
Ralph A. Davis, 60, 21, 1500, 10, 200
Barzilia K. Kervan, 30, 20, 5000, 50, 200
John E. Kervan, 40, 60, 4000, 50, 350

Jacob Ramel, 100, 172, 6000, 200, 500

William T. Armistead, 40, 160, 4000, 10, 250

John Thomas, 35, 5, 500, 15, 90

Miles B. Hudgins, 18, 9, 750, 20, 200

John H. Armistead, 100, 36, 3000, 100, 500

Thomas J. White, 200, 168, 10000, 150, 200

Edward B. Thomas, 60, 20, 1200, 25, 250

Armistead J. Burroughs, 4, 30, 1000, 15, 75

William B. Hudgins, 25, 15, 300, 5, 25

Robert J. Morgan, 5, 4, 300, 5, 60

John L. Hudgins, 25, 9, 1000, 30, 175

Jeanna A. Jarvis, 50, 50, 1000, 25, 300

Thomas J. Hurst, 6, -, 500, 15, 40

Ann K. Thomas, 12, 21, 1000, 10, 45

George E. Thomas, 70, 25, 5000, 300, 500

William C. Borum, 55, 20, 3500, 20, 300

George K. Brooks, 70, 30, 2000, 30, 275

Jarvis Thomas, 65, 35, 3500, 25, 100

Lucy Brownley, 57, 58, 1400, 10, 215

Robert C. Miller, 35, -, 435, 20, 100

John W. Borum, 75, 25, 3500, 50, 375

Thomas A. Hudgins, 8, 9, 500, 15, 20

William Turner, 35, 25, 2000, 30, 150

James M. Diggs, 18, 10, 1400, 15, 40

Thomas R. Ransom, 38, 22, 3000, 100, 325

Sarah Hudgins, 35, 3, 2000, 75, 500

Jessey Hudgins, 250, 150, 8000, 150, 350

Mary Williams, 21, 21, 800, 10, 75

Thomas Smith, 370, 206, 10000, 300, 1200

Washington Brownley, 40, 22, 1500, 2, 100

John Foster, 250, 150, 10000, 200, 1000

Leroy Owens, 21, 4, 1000, 5, 150

Edward I. Thomas, 30, 10, 1500, 30, 200

William Hudgins, 60, 60, 2100, 200, 350

William M. Brownley, 60, 90, 5000, 50, 400

Sands Smith, 350, 169, 15000, 150, 900

George A. Hudgins, 4, -, 200, 5, 40

Elizabeth Foster, 15, -, 500, 20, 250

George Brooks, 90, 38, 8000, 75, 500

Chaplain B. Diggs, 14, -, 2000, 50, 225

George Hudgins, 100, 100, 7000, 100, 150

Winston Foster, 100, 270, 2500, 200, 500

Othniel Hudgins, 9, 1, 300, 5, 80

Francis Hudgins, 19, 16, 650, 30, 85

Augustin Brooks, 40, 60, 1000, 50, 300

Mary A. Hudgins, 10, -, 500, 20, 80

James Brooks, 30, 6, 1200, 15, 200

George Brooks, 50, 50, 1200, 45, 400

Henry S. Hudgins, 30, 20, 1000, 30, 75

Elizabeth R. Brooks, 29, 111, 1000, 15, 60

Martha A. Minter, 25, 25, 400, 3, 50

Susan Brooks, 25, 25, 675, 12, 250

James F. Brooks, 15, 35, 500, 5, 75

John Banks, 15, 17, 1000, 25, 175

John L. Brooks, 20, 60, 600, 40, 100

Sands Hudgins, 10, 10, 300, 3, 100

Dorothy Williams, 36, 8, 1200, 50, 225

Thomas J. Banks, 27, 30, 1500, 45, 250

George H. Forrest, 18, 1, 500, 4, 60

Argyle Hudgins, 15, 5, 400, 30, 75

Josiah Diggs, 40, 10, 1500, 60, 200

John W. Diggs, 40, 23, 1500, 55, 200

Isaac M. Diggs, 26, 26, 700, 25, 75

Polly Diggs, 20, 6, 1000, 6, 75

Thomas W. Diggs, 11, 1, 500, 17, 75

Thomas G. Weston, 115, 35, 6000, 100, 500

Sterling Borum, 105, 45, 6000, 150, 240

John H. White, 14, 10, 1000, 18, 30

William M. Brownley, 6, -, 1000, 10, 75

Elijah Hudgins, 20, 30, 1200, 30, 100

William D. Hudgins, 20, 8, 800, 50, 250

Johnson Hudgins, 45, 20, 1600, 40, 100

Thomas R. Hudgins, 20, 3, 1000, 10, 150

George Diggs, 40, 20, 1500, 25, 200

Bailey Diggs, 35, 15, 1500, 35, 400

William Diggs, 30, 18, 1800, 30, 70

John H. Diggs, 40, 20, 3000, 70, 350

Isaac Fosters, 50 15, 2500, 100, 420

Ann March, 52, 26, 3500, 15, 100

William L. Hudgins, 7, -, 400, 5, 60

Thomas Brooks, 17, 4, 1000, 35, 300

Lucy Minter, 5, -, 200, 5, 35

Armistead Steward, 16, 5, 600, 5, 45

John J. Minter, 8, 4, 800, 45,100

Anthoney D. Hudgins, 30, 12, 1000, 20, 150

Richard Steward, 17 -, 450, 25, 80

William Diggs, 40, 40, 2500, 12, 325

James Owens, 20, 10, 800, 50, 100

John W. Forrest, 17, 6, 700, 15, 80

Josiah D. Forrest, 37, 14, 1500, 50, 90

Richard W. Marchant, 160, 58, 5100, 110, 700

Sans S. Forrest, 30, 6, 1200, 20, 95

Hunley Hudgins, 30, 20, 1200, 40, 104

Shepard G. Miller, 433, 217, 30000, 600, 2100

William Shuttice, 400, 400, 20000, 800, 1450

John W. Jarvis, 350, 150, 16500, 400, 1600

Lucina S. Hudgins, 50, 20, 2000, 40, 170

John P. Hudgins, 112, 56, 3500, 175, 485

William Diggs, 50, 52, 3000, 60, 310

Elizabeth Miller, 22, 3, 1000, 30, 55

Archabald B. Hudgins, 77, 26, 3500, 55, 510

James Brownley, 7, 1, 750, 5, 70

Mary A. Billups, 12, 2, 1200, 5, 30

John Smith, 15,-, 500, 5, 70

Thomas Hall, 20, 22, 1000, 15, 70

Christopher T. Browne, 700, 160, 16000, 1000, 1700

William Williams, 100, 114, 12000, 200, 500

George E. Tabb, 450, 222, 18000, 525, 1930

Thomas Williams, 42, 20, 4500, 200, 350

Americus R. Bohannon, 48, 8, 2000, 19, 250

Francis Armistead, 27, 25, 3500, 50, 350

Robert J. White, 8, -, 1200, 30, 200

Francis T. White, 22, 38, 3500, 65, 450

Vatter G. Lane, 127, 40, 15000, 300, 800

James H. Garnett, 60, 50, 5000, 100, 600

John G. Bohannon, 60, 1, 4000, 30, 350

John Weston, 36, 28, 1200, 100, 180

Susan E. Parrot, 23, 4, 3000, 90, 300

Elizabeth A. Gayle, 7, 18, 2000, 5, 65

James B. Dare, 8, 8, 1000, 5, 75

Warner Lewis, 5, 11, 1000, 10, 75
Booker M. Miller, 100, 51, 7000, 300, 700
John Machem, 110, 50, 4000, 200, 500
Dee Miller, 19, 4, 2200, 15, 95
John P. Hurst, 17, 5, 1500, 50, 200
George W. Green, 8, 9, 2000, 5, 45
William Green, 14, 9, 1000, 10, 30
Miles P. Davis, 20, 10, 600, 15, 60
Cristopher G. Davis, 20, 10, 600, 15, 60
John White, 8, 10, 70, 35, 140
Theodorick Hughes, 18, 3, 500, 8, 40
Elijah Barnum, 210, 200, 12000, 150, 615
Sarah W. Hurst, 55, 29, 5000, 55, 150
Alexander James, 25, 35, 3000, 45, 250
Spencer H. Forrest, 12, 13, 2000, 100, 200
Zelica P. Bohannon, 30, 30, 5500, 50, 250
Edmond W. Bohannon, 90, 50, 4000, 100, 400
Robert Forrest, 12, 22, 700, 5, 75
William Bohannon, 104, 40, 6000, 50, 550
Silas C. White, 55, 5, 3000, 200, 500
James W. Smith, 65, 3, 2500, 175, 300
Joice R. Smith, 120, 110, 7000, 20, 100
John J. Callis, 22, 17, 1500, 40, 60
John T. Forrest, 150, 250, 5000, 75, 450
William H. Forrest, 50, 10, 1800, 20, 130
Thomas F. Morgan, 25, 15, 1000, 10, 55
Samuel Hudgins, 14, 13, 800, 15, 100
Joseph Smith, 16, 4, 800, 25, 90
John Diggs, 35, 28, 2000, 60, 600

Auguston Diggs, 30, 30, 2000, 75, 300
Wescom Hudgins, 20, 5, 500, 3, 45
Harriet Knight, 160, 227, 4000, 50, 400
Thomas S. Forrest, 40, 31, 2000, 35, 240
Michiel Dixon, 12, 5, 700, 8, 30
John R. Billups, 175, 221, 10350, 60, 820
William Brooks, 48, 54, 1500, 25, 225
Richard P. Forrest, 17, 6, 800, 15, 90
Ann Ward, 31, 7, 900, 4, 65
William J. White, 26, 6, 400, 3, 30
John White, 27, 12, 500, 5, 25
William H. Callis, 35, 15, 1000, 70, 200
James Callis, 5, 3, 300, 3, 30
William Morgan, 14, 6, 300, 5, 40
John S. King, 9, 8, 800, 12, 80
Susan C. Forrest, 29, 6, 1500, 25, 40
Thomas A. Hudgins, 15, 15, 500, 15, 50
Isaac M. Diggs, 47, 15, 1000, 15, 230
William H. White, 50, 30, 3500, 300, 500
Elizabeth White, 40, 28, 1800, 30, 300
John Foster, 20, 5, 500, 7, 80
Samuel E. Diggs, 59, 4, 3000, 125, 375
Samuel Walker, 45, 15, 1200, 25, 200
William White, 17, 3, 800, 20, 72
Frances White, 30, 11, 1500, 12, 125
Samuel B. White, 7, -, 500, 10, 75
Robert B. White, 28, 3, 1000, 20, 110
James B. White, 27, 37, 1500, 25, 107
William K. White, 21, 3, 800, 15, 75
Cyrus C. White, 42, -, 2500, 35, 70
John Thomas, 170, 30, 10000, 200, 830

Martha A. Minter, 100, 106, 3000, 50, 450

John R. Green, 10, 9, 1000, 15, 200

Joseph Hobday, 50, 19, 3000, 70, 310

Thomas Green, 15, 8, 2000, 25, 250

Thomas E. White, 16, 8, 1000, 8, 70

Thomas Hurst, 6, 8, 400, 2, 60

William H. Perkins, 16, 8, 2000, 50, 160

Lewis Rains, 200, 100, 10000, 500, 500

John B. Davis, 25, -, 1200, 15, 160

Mary E. Williams, 91, 5, 3500, 20, 170

Esther Williams, 25, 8, 1200, 20, 100

Robert T. Diggs, 70, 52, 2500, 70, 255

William Shakelford, 8, 4, 300, 10, 50

Larkin Miller, 201, 151, 12000, 150, 1000

John E. Miller, 150, 87, 10000, 157, 1365

Leatitia R. Miller, 25, 15, 1000, 25, 250

George P. Eavins, 59, 40, 8000, 100, 650

Mary Brownley, 70, 46, 4000, 30, 225

Jesse Diggs, 29, 16, 2500, 30, 200

James D. Brooks, 20, 10, 1500, 50, 250

Phillip Forrest, 22, 1, 500, 25, 70

Archibald B. Hudgins, 110, 40, 5260, 100, 380

Lucy A. Diggs, 60, 50, 3600, 20, 340

William L. Diggs, 36, 18, 2500, 100, 140

William A. Billups, 96, 25, 7000, 500, 1000

Henry Forrest, 105, 210, 10100, 150, 525

Richard G. W. Lilly, 45, 30, 2500, 60, 260

Peter Eavins, 27, 3, 1200, 15, 95

Andrew Diggs, 53, 10, 3000, 50, 400

John H. Davis, 17, 8, 1000, 20, 60

John M. Billups, 23, 17, 1200, 10, 130

Thomas Diggs, 200, 85, 10000, 100, 1500

Robert L. Sibley, 200, 133, 10000, 300, 1985

Willis Cake, 8, -, 500, 5, 70

William Diggs, 214, 58, 12000, 200, 800

Nancy Davis, 53, 2, 1500, 25, 110

William A. Ranson, 83, 25, 4000, 25, 300

Cornelious R. Begley, 60, 40, 3000, 37, 700

William James, 36, 12, 2000, 40, 375

William E. Hicks, 143, -, 5000, 90, 500

Thomas J. Machem, 55, 21, 1500, 30, 275

Elizabeth L. Foster, 167, 39, 10000, 600, 625

Polina F. Pool, 125, 22, 5000, 20, 350

Albert Diggs, 200, 100, 8000, 200, 870

John W. Dixon, 8, -, 1600, 30, 175

Fountain Green, 50, 35, 3000, 50, 250

William H. Brown, 176, 30, 10500, 500, 1088

Mathew Gayle, 320, 114, 15000, 650, 1150

John J. Burke, 561, 539, 30000, 225, 2000

Isaac S. Armistead, 75, 53, 2500, 75, 225

George W. Bohannon, 11, 17, 1500, 20, 125

Joseph J. Freeman, 225, 75, 7000, 90, 500

Yancey Sleet, 148, 65, 6000, 100, 350

Gabriel F. Miller, 165, 90, 12000, 250, 1400

Henry W. Tabb, 700, 1000, 50000, 2000, 3000

Euphan W. Roy, 900, 486, 45000, 1000, 3000

William P. Hudgins, 25, 20, 3000, 30, 100

Stephen Adams, 120, 49, 4500, 350, 1000

Daniel H. Foster, 87, 55, 5680, 90, 325

Samuel Trader, 40, 8, 3000, 100, 240

George H. Moughon, 130, 70, 6500, 150, 415

Joseph F. Foster, 130, 29, 6000, 50, 500

Joseph Bohannon, 48, 14, 3000, 50, 320

Warner, T. Taliaferro, 300, 187, 14000, 400, 1050

Moses Mathews, 15, 3, 1000, 25, 70

Lewis Hudgins, 341, 344, 25000, 600, 1000

Thomas M. Hunley, 125, 60, 7000, 200, 800

Joshua Gayle, 300, 88, 14000, 400, 1300

Robert Billups, 320, 130, 10000, 250, 1150

Edmond Jones, 130, 207, 10000, 250, 900

William G. Foster, 180, 87, 9000, 250, 1000

Rubin D. Huel, 40, 20, 1200, 20, 110

Mariah Dar, 6, 1, 500, 12, 140

Edward S. Gayle, 30, -, 3000, 25, 150

John R. Foster, 12, 16, 1300, 35, 250

Solomon Moore, 80, 30, 2500, 50, 430

William W. Ingram, 25, 55, 1500, 30, 75

Andrew C. Browne, 550, 187, 20000, 300, 1200

Lewis M. Hudgins, 40, 30, 4000, 50, 400

Edward T. Mallory, 50, 125, 1700, 25, 100

James Brooks, 80, 80, 5000, 200, 400

James Callis, 50, 10, 1200, 20, 200

Washington B. Callis, 120, 46, 3000, 15, 165

Benjamin F. Blake, 150, 50, 4000, 20, 300

Richard W. Foster, 70, 30, 3500, 200, 500

William J. Winder, 9, 4, 300, 10, 75

Leroy Bohannon, 22, -, 1200, 17, 100

William S. Jones, 25, 65, 1600, 25, 130

Henry W. Pratt, 45, 40, 1200, 50, 450

Ebenezar Bohannon, 120, 57, 4300, 150, 666

John Forrest, 16, 8, 1000, 30, 75

William H. Foster, 28, 10, 1800, 25, 86

Richard H. Respess, 15, 5, 1800, 50, 100

James H. Adams, 15, 5, 1500, 5, 100

Humphry H. Keeble, 193, 80, 2500, 100, 618

William J. Callis, 18, 7, 600, 25, 80

John Carney, 18, 7, 600, 25, 50

Mary T. Edwards, 150, 40, 3720, 100, 600

Robert E. Hudgins, 65, 20, 1600, 100, 525

George O. Hillon, 400, 100, 13000, 75, 560

George A. Hill, 140, 20, 3000, 100, 400

James Hill, 250, 80, 5000, 800, 500

William H. Hudgins, 90, 40, 10000, 300, 565

Thomas Carney, 20, 10, 1200, 40, 75

James R. Guyn, 7, 4, 1000, 10, 50

Augustus M. Hicks, 25, 20, 1350, 20, 200

William Lane, 200, 100, 10000, 150, 800

Noah Foster, 39, 15, 2000, 25, 160

George W. Dixon, 57, 27, 2200, 50, 250

Wickham B. Dixon, 14, 12, 1000, 10, 75

Pamly Dixon, 40, -, 1500, 10, 50

Richard Callis, 50, 17, 3000, 15, 125

John R. Callis, 18, 6, 1000, 10, 50

Thomas James, 84, 21, 3500, 100, 200

William James, 60, 45, 3000, 100, 200

John R. Winder, 31, 15, 2000, 100, 175

Mary L. Winder, 30, 3, 500, 10, 80

John J. Davis, 15, 10, 800, 20, 75

John S. Clark, 10, 3, 1000, 5, 50

James D. Marchant, 20, 26, 1500, 10, 100

James M. Lewis, 20, 2, 1500, 5, 120

Levi D. Marchant, 16, 7, 1000, 25, 160

Edmond Marchant, 17, 2, 1000, 5, 50

Mary Lewis, 11, 2, 800, 15, 25

Forrest B. Owens, 14, 11, 1000, 25, 75

Edmond Winder, 20, 13, 1000, 20, 60

Carter B. Hudgins, 40, 15, 1200, 40, 200

William H. Winder, 10, 20, 1200, 15, 165

Lewis Powel, 65, 10, 2000, 30, 250

Thomas J. Hudgins, 250, 100, 10000, 200, 900

William R. Williams, 6, 4, 600, 5, 75

William Kenner, 90, 30, 3500, 20, 275

John E. Terrier, 35, 15, 1200, 20, 150

Philip T. Adams, 55, 30, 2500, 60, 250

Lewis W. Sadler, 7, 3, 1000, 20, 110

Robert T. Lewis, 15, 5, 1000, 20, 50

John J. Marchant, 20, 20, 2500, 20, 50

Thomas J. Hudgins, 14, 5, 800, 25, 60

Holder Hudgins, 400, 200, 20000, 300, 2150

Henry Bell, 400, 558, 28000, 250, 1000

Bartlett Davis, 34, 2, 700, 10, 60

John W. Foster, 20, 5, 150, 25, 100

Albert Foster, 12, 2, 400, 30, 75

Peter W. Bridges, 50, 50 1000, 10, 100

Benjamin T. R. Wiatt, 20, 15, 800, 25, 100

William Blake, 210, 430, 7000, 30, 400

William H. Callis, 12, 15, 500, 10, 85

Henry Fleet, 40, 216, 4840, 20, 365

John A. Bassett, 15, 5, 600, 10, 335

Parker B. Richardson, 92, 16, 800, 10, 130

Amous Rainear, 12, 60, 750, 15, 175

William L. Edwards, 80, 130, 3000, 100, 320

William J. Minter, 80, 102, 4000, 75, 400

George F. Moss, 12, 5, 150, 4, 70

Carter B. Morgan, 25, -, 500, 10, 80

James W. Green, 9, 3, 600, 5, 50

Addison T. Lewis, 25, 40, 1100, 10, 25

Jacob H. Bell, 250, 268, 12000, 500, 825

Jonathan J. Collins, 73, 15, 2000, 20, 200

Joseph M. Haynes, 65, 43, 2000, 10, 150

James H. Marchant, 140, 100, 8000, 175, 500

Francis R. Haynes, 56, 6, 1500, 10, 80

John Shipley, 19, 5, 400, 15, 80

Peter Foster, 21, 8, 290, 25, 175

Richard Haynes, 37, 15, 600, 25, 100

George W. Simmons, 80, 120, 1800, 8, 250

James T. Carter, 25, 35, 1200, 20, 100

Albert Williams, 14, 6, 700, 10, 100

John M. Sadler, 22, 21, 860, 20, 100

Robert P. Simmons, 20, 5, 1500, 25, 150

John Hudgins, 25, 23, 800, 13, 100

James B. Wyatt, 42, 8, 800, 20, 100

Elizabeth Hern, 11,-, 500, 10, 65

John Spencer, 58, 45, 2000, 35, 328

William R. Smart, 500, 867, 35000, 1000, 1200

Henry L. Mathews, 77, 85, 3000, 60, 300

Hezekiah Blaylock, 35, 5, 700, 30, 61

Henry B. Dutton, 150, 100, 3500, 25, 200

Benjamin N. Bramhall, 150, 135, 5000, 20, 150

Ann J. Howlett, 225, 75, 10000, 100, 1000

Robert J. Venable, 15, 68, 9000, 30, 75

Milton S. Hodges, 30, 10, 1000, 40, 225

Benjamin B. Dutton, 150, 58, 5000, 250, 1500

John H. Blake, 60, 30, 2000, 40, 450

William D. Soles, 80, 28, 2000, 30, 450

Edward H. Sadler, 100, 100, 2500, 30, 450

Charles C. Duval, 127, 184, 5000, 100, 450

George M. Patterson, 10, -, 275, 3, 30

William H. Oliver, 80, 95, 3000, 30, 250

John H. Dunlavy, 220, 80, 14000, 500, 1000

Mecklenburg County, Virginia
1860 Agricultural Census

The Agricultural Census for Virginia 1860 was microfilmed by the University of North Carolina Library under a grant from the National Science Foundation from original records at the Virginia Department of Archives and History in 1963.

There are forty-eight columns of information on each individual. Only the head of household is addressed. I have chosen to use only six columns of information because I feel that this information best illustrates the wealth of individuals. The columns are:

1. Name
2. Improved Acres of Land
3. Unimproved Acres of Land
4. Cash Value of Farm
5. Value of Farming Implements and Machinery
13. Value of Livestock

Nathl. Talley Sr., 180, 60, 600, 150, 700
Ro. Bryson, 14, 29, 2500, -, 400
W. G. Garey, 200, 400, 7000, 50, 500
W. H. Fornhill, 34, 3, 600, -, 70
W. H. Gafford, 26, 25, -, -, 150
W. F. Beasley, 3, 36, 2400, -, 430
Booker Loyd, 3, -, 1000, -, 40
Alex. Lydner, 300, 225, 12000, 175, 1360
R. B. Bapteet(Basteet, Baptest), 350, 250, 10000, 500, 1280
J. J. R. Spencer, 100, 1200, 10500, 50, 1509
J. B. McPhail, 350, 250, 11000, 500, 1380
Richd. Boyd, 600, 500, 20000, 250, 1905
Dr. G. C. Venable, 475, 300, 16500, 190, 1091
T. T. Boswell, 200, 200, 7000, 200, 1112
B. W. Leigh, 400, 230, 13000, 200, 1310

A. B Crowder, 350, 219, 8535, 300, 1400
R. A. Puryear, 675, 675, 16500, 350, 2068
Dr. P. C. Venable, 800, 700, 35000, 600, 3714
E. D. Doggett, 100, 43, 3000, 147, 590
Ro. Gillespie, 95, 181, 2430, 20, 240
W. D. Gillespie, 92, 185, 2732, 15, 347
Wm. M. Moody, 77, 153, 4000, 55, 613
A. B. Lyle, -, -, -, -, 50
Jno. K. Jeter, -, -, -, -, 30
W. Hotterton Jr., 3, -, -, -, 75
E. A. Williams, 33, 100, 10000, 100, 215
J. J. Daniel, 300, 100, 10000, 275, 730
L. B. Daniel, 2, 78, 1500, -, 55
J. R. Tinch, -, -, -, -, 30
R. M. Scott, 500, 540, 13520, 300, 1350
B. D. Morton, 2, 10, 3530, -, 530

H. Skipwith, 2466, 2550, 11120, 2000, 3200

Wm. Russell, 510, 300, 30000, 1000, 3458

D. Shelton, 330, 90, 25000, 60, 1005

Richd. Russell, 800, 420, 45000, 1200, 3285

Royall Lockett, 433, 250, 10000, 450, 1114

E. S. Smitheon, 61, 61, 1200, 25, 83

John Newton, 250, 250, 3500, 25, 526

R. H. Moss, 5, 30, 6000, -, 610

T. Carrington, 1000, 105, 3000, 800, 2758

R. P. Yancey, 35, 35, 1200, 25, 300

Geo. R. Averett, 400, 474, 11120, 200, 1865

E. H. Howerton, 70, 60, 1500, 50, 550

H. Wood, 160, 71, 5000, 50, 675

F. Pollard, 420, 420, 10000, 150, 860

A. Oderbey, 500, 446, 20000, 100, 700

Dr. Jno. R. Leigh, 36, 20, 2000, 20, 250

J. A. Tarndlee, No Farm, -, -, 730

H. Overbay, 372, 68, 5000, 250, 810

C. Hardy, 400, 600, 10000, 250, 1545

P. Jones, 125, 125, 15000, 25, 283

Johnson, 650, 650, 25000, 500, 3060

G. Vaughan, -, -, -, 15, 143

D. Sizemore, 150, 150, 2000, 50, 888

D. Elam, 65, 560, 3600, 100, 700

L. Jones, 300, 325, 6000, 25, 902

C. Yancey, 300, 250, 5000, 300, 766

J. Walkins, 71, 62, 1000, 15, 60

R. H. Walker, 66, 134, 2000, 75, 520

Nancy Walkins, 600, 900, 75000, 200, 1580

Dr. B. G. Walkins, Included with his mother's, 100, 568

W. L. Harris, 130, 270, 6000, 100, 246

G. Roister, 100, 100, 2000, 150, 538

J. Vaughan, 115, 110, 750, 15, 225

J. C. Vaughan, -, -, -, 10, 40

E. Griffen, 200, 50, 2000, 20, 304

E. Owen, 50, 50, 1500, 20, 52

J. Ewoltz, 200, 207, 5000, 75, 483

J. Blanks, 350, 350, 4500, 35, 580

A. Puryear, 25, 26, 3000, 20, 51

M. W. Yancey, 8, 86, 500, 35, 160

R. C. Newton, 100, 50, 600, 20, 164

W. Smith, 75, 125, 800, 30, 182

G. I. Yancey, 60, 30, 300, 60, 203

H. W. Averett, 113, 113, 675, 20, 225

H. J. Newton, 66, 135, 600, 25, 205

W. Newton, -, -, -, 5, 20

W. Ramsay, -, -, -, 5, 20

J. J. Newton, 100, 200, 1000, 25, 278

J. Somerhill, 64, 64, 380, 15, 57

Jno. Lewis, 600, 3400, 40000, 600, 4930

L. Chandler, -, -, -, 5, 75

Jno. Nelson, 200, 200, 4000, 100, 843

Jas. Phillips, 50, 50, 500, 15, 188

Kemp. Mathews, 14, 28, 200, 15, 90

Lin. Wilson, 14, 28, 200, 15, 106

J. Culbreath, 60, 60, 600, 25, 285

C. Wilson, 37, 37, 175, 10, 110

A. Matheds, 23, 24, 161, 10, 50

Jno. Puryear, -, -, -, 10, 179

Frad R. Wilson, -, -, -, 10, 160

H. Griffin, 40, 139, 800, 15, 130

J. Yancey, 95, 193, 1160, 20, 159

J. R. Williamson, 140, 160, 1000, 100, 320

W. Lewis, 14, 42, 318, 15, 45

H. Phillips, 15, 18, 450, 20, 161

R. A. Phillips, -, -, -, 15, 490

E. Noblin, 32, 33, 200, 10, 34

J. Wiles, 50, 97, 560, 15, -

B. B. Belcher, -, -, -, 10, 25

R. Willbourne, 15, 15, 100, 10, 36

J. P. Welkins, 100, 100, 900, 20, 130

J. Yancey, 295, 591, 5316, 150, 905

Tho. Yancey, 200, 200, 2500, 50, 306

M. T. Hall, -, -, -, 20, 155
J. D. Davis, 375, 375, 3750, 20, 330
J. H. Yancey, 72, 144, 1080, 20, 236
D. H. Chandler, 30, 30, 600, 120, 492
R. H. Moody, 85, 173, 1560, 50, 511
Jno. Smith, 67, 135, 1616, 50, 525
D. Hayes, 34, 66, 1000, 100, 190
Wm. Williamson, 42, 83, 624, 10, 130
J. R. Yancey, 25, 50, 525, 20, 76
H. R. Ligon, 47, 93, 1260, 20, 100
Wm. Tucker, 37, 73, 300, 15, 97
W. H. Carter, 10, -, 70, 10, 34
J. P. Pattillo, 105, 105, 1680, 50, 24

T. B. Wall, 150, 300, 4500, 350, 2028
W. Mitchell, 125, 125, 2500, 50, 510
L. Hendrick, -, -, -, 10, 86
H. Puryear, 150, 150, 2000, 40, 550
J. R. Riggins, 75, 100, 900, 10, 45
W. Naimy, 10, -, -, 10, -
Jas. Bowers, 492, 493, 9850, 220, 1310
A. Yancey, -, -, -, 20, 175
H. Davis, 54, 54, 540, 15, 166
Rich. Noblin, -, -, -, 10, 86
R. A. Carter, -, -, -, 10, 98
W. Wilkenson, 266, 267, 4264, 50, 533
J. T. Carter, -, -, -, 10, 20
Archie Clark, 210, 210, 3360, 10, 147
Edmd. Wilkinson, -, -, -, 10, 91
J. A. Gregory, 200, 400, 4200, 200, 695
G. R. Puryear, 20, 40, 600, 10, 5
Wm. Jones, 33, 65, 784, 10, 185
T. W. Owen, 37, 113, 1700, 50, 575
J. H. Hayes, 191, 192, 1522, 76, 194
J. B. Gregory, 20, 128, 1025, 85, 203
A. A. Smith, 354, 705, 5300, 200, 1155
T. C. Keeks(Reeks), 222, 444, 3966, 200, 1140

J. L. Winn, 181, 182, 1815, 130, 440
B. T. Winn, 83, 83, 1660, 50, 530
A. Smith, 267, 533, 6400, 250, 1388
J. S. Couch, 235, 466, 7300, 200, 1460
J. I. Beame, 184, 366, 3300, 100, 1125
A. Finch, 334, 666, 5000, 200, 1800
E. Brewer, -, -, -, 20, 214
A. J. Ioone, 100, 200, 2400, 50, 525
R. Burton, 54, 54, 1296, 120, 480
Francis Gregory, 67, 133, 1200, 40, 766
S. M, Gregory, 300, 330, 3300, 50, 563
Wm. Bacon, 618, 1236, 18500, 200, 2760
S. E. Burwell, 284, 566, 6150, 100, 665
J. S. S. R. Burwell, 400, 800, 7200, 100, 1450
W. R. Smith, 500, 500, 6000, 150, 1637
Julia Moore, 96, 40, 1450, 100, 420
Thos. R. Elam, 140, 841, 1686, 60, 310
Wm. H. Lunsford, rented, rented, land, 20, 122
Matilda Perkins, 105, 105, 860, 05, 268
Thomas Kite, 30, 72, 816, 20, 40
Pat. Williams, -, -, -, -, -
G. W. P. Pool, 287, 553, 4150, 300, 1185
Jo. Furloins, 30, 200, 980, 30, 800
Ira Glasscock, rented, rented, land, 20, 75
James Heat, 73, 200, 1092, 75, 475
Richd. Glasscock, -, -, -, 20, 150
Richd. Yancey Sr., 100, 300, 800, 100, -
Aplen Puryear, 123, 200, 2000, 100, -
Erasmus Matthews, -, -, -, 20, -
Clayton Yancey, 50, 100, 750, 30, -
Allen Jones, -, -, -, 20, -

Aeurietta Jones, 28, 70, 294, -, -
Saml. Puryear, 150, 150, 600, 30, -
A. Bowen, -, -, -, 10, -
Wm. Gold, -, -, -, 15, -
J. W. Talley, -, -, -, 20, -
Elia A. Moore, 50, 200, 1250, 30, -
Beverly Talley, 73, 100, 692, 25, -
John Puryear, -, -, -, 30, -
Richd. Phillips, 40, 40, 400, 15, -
Hampton Malone, 45, 45, 450, 20, -
H. Crusius, -, -, -, 15, -
Abram Talley, 10, 52, 310, 15, -
James Overbey, 85, 200, 2280, 20, -
O. G. _. Glasscock, 25, 40, 640, 30, -
Thos. Blanks, 27, 27, 216, 30, -
Wm. P. Pool, 162, 324, 4374, 200, -
Jos. Keaton, 50, 132, 1820, 20, -
Harvey Sizemore, 80, 82, 972, 20, -
John Bowen, -, -, -, 15, -
R. O. Robertson, 213, 213, 5125, 175, 600
C. Y. Richards, 1200, 1200, 20000, 400, 3000
Dr. R. R. Puryear, 75, 300, 5625, 200, 1050
Chas. McCutcheon, 150, 80, 1610, 100, 390
R. A. Crowder, 93, 200, 3416, 60, 300
J. G. Sternbridge, 75, 75, 1500, 50, 270
Wm. Loafman, 237, 238, 3800, 125, 750
H. _. Jeffress, 337, 338, 6940, 100, 890
T. R. Brame, 100, 100, 2000, 50, 430
J. F. Royster, 62, 63, 600, 25, 180
Wm. Sternbridge, 44, 100, 864, 25, 62
W. Townsend, rented, rented, land, 20, 40
W. Davis, rented, rented, land, -, 40
James Moody, 433, 433, 4330, 75, 490
John P. Williams, 80, 135, 3240, 100, 660

Jos. Watson, 125, 100, 3000, 100, 440
Sam. Dedman, 900, 600, 15000, 100, 1245
Wm. Wilson, 87, 88, 1750, 75, 200
H. C. Moss, 35, -, 700, 500, 700
B. R. Royster, 177, 178, 2840, 150, 275
R. Crute, 177, 100, 2116, 100, 330
R. R. Pulliam, 1000, 1000, 20000, 500, 1000
R. _. Hayes, 150, 150, 3000, 50, 380
J. H. Barnes, 750, 750, 12000, 300, 1800
Wm. H. Blanch, 212, 212, 4000, 100, 380
Chas. Bridy, 200, 100, 4000, 50, 700
C. Wood, 662, 663, 12000, 200, 1500
E. Overton, 200, 200, 2000, 100, 450
J. F. Barnes, 125, 200, 3000, 100, 960
J. H. Gregory, No, land, land, 25, 175
John Cosby, No land, land, 25, 330
Wm. Carter, No, land, land, 25, 150
Wm. A. Smith, 100, 100, 2000, 500, 725
J. A. Clivourne, 50, 110, 1600, 50, 200
B. Puryear, -, -, -, 25, 240
J. J. Puryear, 225, 175, 4000, 50, 250
Dr. Tho. Field, 325, 995, 13000, 108, 1000
E.M. Holloway, No, land, land, 50, 375
E. M. Toone, 200, 510, 7000, 100, 500
J. S. Field Sr., 1000, 2000, 45000, 750, 3200
J. W. Love, 772, 773, 45000, 600, 3100
Rhoda Puryear, 160, 400, 4000, 100, 1000
Bentley Wilkinson, 125, 200, 3000, 50, 500

J. T. Wotton, 230, 300, 4000, 100, 1000

John Brewer, 35, -, 80, 25, 50

Edwd. Allgood, 20, 23, 250, 25, 70

R. Y. Overbey, 2320, 3000, 75000, 1400, 7000

J. E. Harris, 140, 200, 2400, 100, 250

S. T. Harris, 59, 60, 600, 50, 175

H. Burton, 75, 75, 1200, 50, 600

C. D. Gregory, 137, 100, 1800, 50, 175

Andrew Gregory, 140, 100, 2000, 60, 150

J. A. Keaton, -, -, -, 50, 110

J. E. Lawson, 50, 50, 1000, 75, 110

Betsy Baplut(Bassleet,Basteet Baptest), 100, 100, 2000, 60, 160

Sally Baplut (Bassleet, Basteet), 250, 300, 4000, 100, 640

Wm. Smith, 300, 617, 9000, 150, 900

E. Keene, 300, 650, 9000, 500, 1550

J. R. Walker, 200, 300, 6000, 200, 800

J. Cunningham, 300, 300, 6000, 500, 1800

P. R. Burwell, 859, 857, 10000, 300, 1750

D. J. Bigger, 295, 200, 35000, 160, 1000

E. B. Burwell, 200, 200, 32000, 150, 560

El__ Gale, 650, 400, 8000, 200, 1200

D. Thompson, 250, 250, 6000, 100, 600

Jno. Oliver, 75, 75, 1300, 100, 500

Wm. Tucker, 650, 735, 9000, 510, 1450

Hy Tucker, 333, 333, 6000, 300, 900

_____ Pettus, 273, 200, 2000, 600

J. Boswell Jr., 166, 167, 2400, 200, 600

J. Boswell Sr., 285, 285, 3000, 75, 500

J. W. Hardy, 185, 200, 3000, 300, 1000

T. H. Pettus, 398, 399, 7000, 300, 1100

T. T. Pettus, 333, 333, 10000, 300, 900

S. A. Crenshaw, 200, 200, 2000, 50, 300

E. Parish, 50, -, 500, 25, 350

W. L. French, -, -, -, 75, 600

W. A. Keaton, -, -, -, 50, 175

J. C. Gregory, 200, 311, 4000, 300, 950

Jno. Hardy, 200, 235, 3000, 200, 600

B. A. Coleman, 175, 200, 2500, 200, 500

J. G. Masson, 37, -, 400, 50, 400

A. W. Garner, 182, 183, 2000, 75, 300

S. E. Dance, 97, 100, 1000, 75, 400

Jno. N. Craddock, 150, 150, 1200, 50, 160

J. C. Wootton, 250, 250, 4000, 100, 420

J. H. Jeffress, 600, 600, 10000, 150, 1800

J. L. Wootton, 100, 300, 4000, 100, 900

N. Carter, 241, 241, 4000, 100, 575

J. W. Wootton, 175, 175, 3000, 100, 1050

R. A. Masson, -, -, -, 25, 280

J. B. Foy, 180, 100, 1680, 50, 180

J. Murray, -, -, -, 20, 100

J. Mason, 400, 400, 6000, 100, 1100

A. Adkins, -, -, -, 20, 150

L. M. Wilson, 403, 403, 8000, 300, 1700

Dr. W. H. Jones, 750, 750, 18000, 1000, 3800

E. Tarry, 75, 2250, 36000, 500, 1750

A. L. Rainey, 400, 400, 9000, 450, 1300

J. H. Taylor, 450, 650, 12000, 350, 1750

Geo. Tarry, 150, 150, 3000, 100, 900

Wm. Shanks, 333, 667, 5000, 100, 600

T. G. Burwell Ex., 900, 900, 10000, 400, 1450

S. D. Booker, 500, 500, 25000, 500, 1750

J. S. Field Jr., 908, 709, 20000, 500, 1700

Jas. Morson, 750, 750, 18000, 500, 1800

Young Averett, 112, 113, 1225, 30, 180

Hy. Averett, 100, 100, 1000, 30, 140

J. Ramsay, 60, 100, 800, 25, 250

J. E. Haskins, 85, 100, 3700, 150, 1200

J. H. Johnson, 220, 220, 8800, 150, 900

J. Y. Chandler, 200, 200, 3200, 100, 400

W. Daniel, 75, 75, 1200, 75, 275

W. A. Porter, -, -, -, 75, 150

Clarke, 50, 50, 500, 50, 40

J. J. Struckmaster, 160, 300, 2700, 100, 250

Thos. Nash, -, -, -, 75, 75

Jas. Clark, 160, 100, 1000, 70, 375

S. Glasscock, -, -, -, 30, 30

R. Daniel, 150, 100, 1200, 50, 150

M. Nelson, 200, 200, 2000, 100, 400

M. Creath, 77, 100, 800, 50, 90

H. Nelson, 160, 200, 2500, 75, 390

W. R. Wilkins, -, -, -, 50, 350

A. Apple, 242, 243, 3000, 100, 800

John W. Ezell, 30, 155, 500, 75, 325

Sterling Smith, 280, 885, 5000, 250, 500

Wm. Cabaness, 100, 239, 1200, 50, 250

Edward Howel, 10, 33, 200, 10, 125

M. Brandson, 10, 33, 200, 10, 50

Thos. Wertman, 250, 325, 2500, 50, 450

Thos. P. Cleator(Cleaton), 50, 100, 450, 10, 150

Saml. J. Vaughan, 10, 5, 75, 5, 20

Elizabeth Walker, 20, 20, 300, 5, 100

C. Dnusnight, 100, 350, 2000, 100, 350

E. G. Denton, 100, 185, 1200, 50, 450

W. W. Connell, 50, 86, 500, 15, 100

W. W. Winkler, 100, 151, 1200, 75, 200

J. E. Walker, 250, 400, 2500, 75, 400

A. J. House, 20, 57, 250, 10, 60

J. B. Northington, 600, 750, 11000, 200, 1000

E. J. Mosley, 70, 169, 2000, 75, 250

W. D. A. Jeeter, 10, 10, 200, 5, 200

E. M. Hite, 400, 330, 6000, 200, 1000

Wm. G. Rook, 60, 200, 1500, 25, 300

George Williams, 60, 200, 1500, 25, 350

John B. Kidd, 200, 860, 15000, 150, 500

Zack Jones, 10, 10, 100, 15, 150

Wm. Elvin, 20, 30, 300, 200, 75

J. Overby, 75, 180, 2000, 15, 350

Wm. Overby, 25, 75, 1000, 75, 100

Wm. J. Mason, 20, 135, 1000, 50, 100

Mary A. Peckinson, 75, 400, 2000, 10, 100

Thos. F. Dnusnight, 20, 100, 500, 10, 30

Nancy Sands, 150, 336, 2500, 25, 300

Wm. H. Taylor, 75, 270, 1500, 25, 300

L. Crutchfield, 20, 10, 200, 25, 200

Wm. D. Crutchfield, 100, 129, 1200, 50, 200

Wm. H. Hightour, 20, 70, 800, 15, 100

Thos. C. R. Linch, 10, 60, 200, 10, -

Wm. W. Thomason, 10, 90, 500, 10, 50

Willus R. Smelly, 50, 100, 1000, 25, 250

Wm. T. Smelley, 30, 150, 1200, 20, 200

Robert W. Davis, 50, 152, 2000, 25, 250

P. H. Hudgins, 50, 140, 1200, 25, 200

Isaac Baerd, 50, 48, 400, 15, 150

W. W. Bowen, 60, 95, 800, 25, 300

B. W. Hines, 83, 100, 1200, 25, 300

John Oslin, 40, 65, 600, 10, 150

J. M. Crutchfield, 75, 120, 1200, 30, 350

Isaac Benford, 100, 97, 3000, 100, 350

John W. Gregory, 100, 180, 3000, 100, 450

Emma F. Walker, 100, 172, 2000, 100, 600

Jessee Q. Gee, 110, 390, 5000, 150, 500

Wesley A. Nash, 50, 105, 800, 25, 250

Wm. G. Thompson, 10, 73, 400, 5, 50

A. G. Nicholson, 75, 232, 2500, 25, 300

B___ Taylor, 25, 275, 2000, 10, 50

July Miller, 15, 86, 500, 10, 50

John J. Waler, 10, 17, 200, 5, 25

Mary W. Singleton, 10, 120, 200, 5, 75

Thos. J. Bauer, 30, 270, 1500, 20, 80

Francis House, 20, 41, 250, 10, 60

John W. Danel, -, -, -, -, 40

Martha Cooper, -, -, -, -, 75

John P. Bower, 20, 22, 200, 10, 25

John C. Davis, 10, 50, 400, 10, 10

Rebecca Chapman, 20, 49, 200, 5, 50

James L. Thomason, 30, 53, 600, 15, 150

C. G. Thomason, 50, 32, 500, 15, 250

Sudy Baum, 10, 60, 300, -, -

Wm. D. Nash, 10, 130, 1500, 25, 300

John W. Nash, 50, 239, 2000, 50, 300

Henry J. Bauer, 50, 153, 1500, 20, 200

W. W. Hudson, 25, 50, 300, 10, 150

Jessee Gee, 10, 211, 1200, 50, 250

John Hawthrone, 150, 280, 3000, 300, 700

Wm. H. Moore, 200, 521, 4000, 75, 350

Wm. G. Hazlewood, 15, 85, 400, 10, 100

T. M. Butterworth, 120, 360, 2500, 300, 350

James McAden, 170, 260, 2500, 200, 550

Alen Pouns(Downs), 50, 105, 1000, 10, 220

Wm. H. Hendley, 25, 31, 300, 10, 50

Jane Whitenarl(Whitemore), 20, 36, 300, 10, 25

Mary O. Singleton, 15, 2, 200, 10, 150

Wm. J. Manness, 25, 30, 400, 20, 150

Richard C. Gregory, 250, 993, 5000, 100, 600

James W. Thomason, 70, 20, 200, 5, 200

Macklin Soward, 10, 27, 100, 5, 100

David Oslin, 10, 40, 200, 10, 150

James W. Bowen, 10, 30, 150, 10, 100

C. P. Mooreing, 6, 48, 400, 5, 50

John H. Hudgins, 100, 175, 300, 25, 500

Benja. Mongomary, 20, 50, 250, 10, 75

P. Burnett, 40, 108, 500, 10, 180

Jos. C. Fanner, 100, 500, 3000, 50, 500

W. L. Merryman, 15, 50, 300, 50, 20

Mark A. Baley, 10, 30, 150, 5, 20

Ralph Hubboard, 100, 200, 1000, 100, 500

John W. McAdler, 100, 640, 3000, 150, 500

Thos. Lambert, 20, 80, 250, 10, 120

A. H. Middagh, 150, 600, 5000, 250, 1000

Jas. S. Gregory, 50, 143, 1200, 25, 175

A. W. Hutcherson, 10, 175, 2000, 50, 250

Edwin Benford, 20, 25, 800, 15, 250

Wm. A. Pace, 85, 185, 2000, 200, 300

Jos. E. Smith, 60, 210, 2000, 50, 200

Thos. Dnusnight, 60, 100, 1200, 20, 200

Thos. D. Crutchfield, 50, 74, 500, 10, 100

Peter Crutchfield, 75, 200, 1200, 15, 250

Spencer Guy, 10, 50, 300, 15, 50

Wilsher Burnett, 15, 50, 300, 10, 75

James L. Wright, 10, 20, 200, 10, 50

George W. Smith, 50, 100, 800, 15, 250

Wm. C. Crutchfield, 50, 100, 1000, 15, 250

Wm. S. Biggs(Brigg, Bugg), 10, 30, 600, 10, 150,

John P. Hudson, 20, 86, 600, 10, 200

B. B. Vaughan, 40, 55, 1000, 200, 250

Wm. N. Puryear, 100, 650, 2500, 150, 600

Benga J. Walker, 100, 281, 2500, 250, 600

Evans Tanner, 300, 700, 4000, 350, 1000

John Merryman, 10, 30, 100, 5, 50

Benja. Vaughan, 5, 30, 150, 5, 50

James Matthews, 50, 100, 1000, 20, 300

John R. Barnes, 75, 125, 1000, 10, 150

Nancy Barnes, 125, 413, 2500, 25, 300

Wm. P. Cook, 100, 420, 2500, 50, 500

F. Floyd, 50, 150, 2000, 10, 150

Stephen G. Vaughan, 35, 23, 700, 100, 200

Wyatt Adams, 200, 65, 1200, 40, 250

Amaida Nash, 150, 197, 1500, 200, 550

Eliza. Crowor, 50, 72, 800, 10, 180

Thos. C. Thompson, 75, 125, 1000, 50, 350

P. Cleaton, 120, 223, 2000, 75, 800

John Mize, 10, 50, 250, -, 25

James Beeid, 50, 86, 500, 25, 150

Isaac Taylor, 100, 242, 1500, 125, 500

James B. Jones, 75, 148, 1200, 125, 400

Charles Jones, 10, 50, 250, 50, 150

Lew Jones, 50, 77, 1200, 10, 25

E. S. Cleaton, 100, 100, 1000, 75, 300

Wm. R. Gregg, 400, 270, 6000, 200, 1500

Teney W. Thomas, 150, 275, 1500, 50, 200

J. J. D. Pecinson, 25, 75, 500, 10, 200

Wm. Cousins, 10, 50, 300, 5, -

E. C. Jones, 20, 50, 300, 5, -

James Burton, 100, 200, 1600, 20, 250

David Duglas, 30, 30, 300, 15, 150

Nancy Thomas, 150, 275, 2000, 50, 200

Robin Thomas, 200, 300, 3000, 100, 350

E. H. Riggin, 225, 175, 3000, 150, 800

Buga(Benja) Chilouss, 1, 2, 100, -, 30

John Gray, 31, 40, 300, 10, 100

James A. Taylor, 30, 70, 500, 10, 100

Judith Barker, 10, 20, 100, 5, 20

B. W. Barker, 30, 70, 400, 10, 100

John R. Cole, 50, 150, 1200, 50, 300

Alice Cole, 60, 20, 500, 75, 100

Beverley Valentine, 30, 100, 600, 50, 250

Edward King, 10, 50, 300, -, 75

Charles G. Turner, 400, 400, 4000, 300, 1000

Richard T. King, 65, 200, 1200, 25, 150

Mark McLen, 10, 20, 150, 10, 75

Thos. McLin, 10, 50, 300, -, 30

Elizabeth W. Rainey, 200, 237, 2500, 50, 600

Elizabeth Bacy (Baly), 200, 230, 2500, 50, 500

John J. Rainey, 150, 250, 2000, 80, 500

Benga. Baskerville, 10, 20, 150, 5, 60

Thos. McLin Sr., 10, 30, 200, 5, 60

Henry T. Keicker, 15, 25, 250, 25, 60

Samuel Bass, 10, 20, 100, 5, 50

Jeremiah R. George, 70, 100, 1800, 25, 300

Mary McLin, 5, 45, 150, -, 30

Wm. H. A. Mayo, 10, 30, 150, -, 15

David T. Rideout, 75, 75, 1200, 70, 150

John Stewart, 62, 50, 500, 10, 200

_. F. Lambert, 100, 150, 1500, 20, 300

Wright King, 400, 800, 12000, 200, 1600

Hugh D. Bacy (Baly), 200, 280, 4500, 100, 400

Solomon Guy, 10, 20, 300, 5, 50

Alex. Shaw, 250, 300, 5500, 700, 1500

Wm. H. Mayo, 10, 20, 300, 5, 50

N. M. Thornton, 130, 40, 1500, 100, 100

P. Ellington, 200, 300, 5000, 100, 800

Green Blanton, 6, 10, 200, -, 10

O. M. Moss, 100, 220, 1500, 25, 250

O. H. P. Tanner, 2550, 303, 5000, 200, 600

Jane Cannon, 150, 246, 2500, 100, 400

Robert Taylor, 30, 100, 800, 25, 100

George G. K. King, 50, 43, 800, 60, 300

John Cypress, 10, 30, 200, 5, 150

Mary Cypress, 10, 20, 100, -, 25

F. J. Burton, 50, 150, 1000, 50, 200

Baxter Lambert, 250, 307, 4000, 100, 1000

Wm. B. Thomas, 40, 50, 500, 10, 125

Green Jones, 20, 41, 300, 10, 50

G. G. Sylaier, 10, 50, 300, 5, 25

Evennet King, 17, 60, 500, 25, 100

John G. Sectar, 200, 237, 3000, 75, 300

James McLin, 10, 10, 200, 5, 25

Wm. Rittenburg, 25, 80, 500, 10, 75

A. G. Boyd, 600, 700, 11000, 250, 2000

Duland Mayo, 10, 20, 150, -, 25

Willis Cannon, 200, 270, 2000, 50, 350

An. E. Caureigton, 700, 900, 20000, 300, 2500

Thos. R. Handock, 500, 865, 11000, 800, 2100

Martha L. Thomas, 160, 150, 1000, 10, 200

L. _. Rose, 182, 181, 2000, 75, 500

Brighteart Hannass, 30, 70, 800, 10, 157

Robert Tannmew, 500, 450, 8000, 250, 400

Wm. J. & T. Brady, 100, 200, 3000, 40, 500

B. Brady, 1000, 1300, 20074, 450, 2500

Joel H. Rainey, 160, 200, 3000, 25, 400

Wm. B. Clanton, 500, 500, 5000, 400, 1200

Fannie _. Cleaton, 300, 400, 4000, 300, 1500

Burke & Taylor, 200, 216, 2000, 100, 1000

Martha Wartmann, 50, 150, 500, 25, 100

John Dayson, 10, 20, 100, 5, 25

Thos. J. & T. Brady, 250, 191, 3000, 150, 1000

Wm. T. Pennington, 300, 400, 7000, 150, 1000

R. D. V. Vaughan, 40, 50, 800, 150, 200

G. P. Cleaton, 20, 30, 250, 10, 100

Martha Gid_mas, 10, 30, 200, 10, 50

S. H. Vaughan, 150, 200, 2000, 50, 150

L. G. Crutchfield, 60, 100, 1000, 30, 200

Saml. Watson, 25, 50, 350, 10, 100

Edward _.Walker, 200, 325, 5000, 200, 400

G. G. Caraway, -, -, -, -, 25

B. W. Barker, 30, 50, 4000, 10, 75

Lewis G. Wright, 75, 50, 800, 50, 150

Joseph Bennett, 350, 380, 6000, 50, 800

Walker Evans, 100, 131, 2000, 150, 500

Saml. T. Moore, 350, 400, 7000, 150, 1000

Archer Jackson, 200, 200, 4000, 150, 1000

David D. Walker, 50, 150, 1600, 150, 300

Saml. G. Harriss, 200, 375, 8000, 200, 1500

Mary Goode, 200, 125, 6000, 100, 800

John E. Forbes, 50, 75, 1000, 25, 300

Wm. E. Walker, 70, 70, 800, 25, 300

Alice E. Hannass, 500, 500, 20000, 300, 1500

Wm. R. Baskerville, 1000, 1700, 50000, 300, 2000

M. Alexander Jr., 450, 375, 20000, 500, 1700

A. W. Harsead, 800, 683, 14000, 500, 3000

Saml. D. Ferguson, -, -, -, -, 100

Wm. D. Pulley, 50, 200, 2000, 25, 250

H. F. Pulley, 30, 120, 1200, 10, 100

Henry Simmons, 75, 250, 2000, 100, 300

N. C. Bugg, 30, 50, 400, 10, 100

Isaac D. Watson, 30, 80, 600, 20, 150

Elizabeth Ezell, 200, 200, 2500, 150, 500

M. W. Jarden, 10, 30, 600, 25, 1200

Wm. H. Simmons, 772, 400, 8000, 150, 1000

Lucy Baskerville, 400, 618, 10000, 150, 1500

Henry Davis, 100, 200, 3000, 200, 150

John H. Walker, 300, 400, 7000, 200, 800

James W. Walker, 100, 115, 1700, 50, 300

Peter Evans, 30, 70, 500, 10, 20

R. D. Baskerville, 500, 375, 20000, 400, 2500

Wm. Baskerville Jr., 800, 1500, 30000, 400, 3500

John F. Oglanne, 150, 250, 3500, 75, 1000

Thos. Winn, 45, 66, 600, 30, 150

Jos. H. Janes, 400, 760, 10000, 250, 1500

Wm. Evans, 50, 160, 800, 25, 200

John E. Lambert, 400, 733, 12000, 250, 1000

Rebecca Cole, 50, 165, 1500, 50, 280

Brien T. Cradle, 50, 800, 4000, 25, 200

Sarah Evans, 300, 500, 5000, 200, 500

Wilson E. Evans, 50, 150, 1200, 10, 300

Benja. Fennell, 50, 110, 1200, 30, 400

Deborah A. Hudson, 50, 75, 500, 20, 200

Wm. N. Wright, 40, 50, 300, 10, 100

John Watson, -, -, -, -, 100

G. W. Cleaton, 25, 75, 500, 75, 150

Robert W. Ezell, 30, 150, 800, 40, 10

John Cook, 500, 700, 10000, 100, 1200

Wm. B. McAden, 300, 300, 6000, 150, 1000

Wm. H. Cassley, 30, 27, 500, 75, 300

Martha E. Byassed, 30, 145, 1000, 20, 150

Ezekiel Crowder, 50, 40, 800, 50, 400

Thos. Thomason, 30, 100, 800, 15, 100

John R. Harriss, 100, 100, 1500, 75, 800

Wm. M. Bugg, -, -, -, -, 150

Lucy Bucks, 20, 39, 290, 10, 75

D. E. Hazlewood, 100, 513, 3500, 10, 100

P. F. Smith, 40, 100, 1000, 50, 200

David Smith, 100, 200, 2500, 25, 250

Sarah H. Rukes, 50, 80, 1000, 10, 100

Philadelphen Stone, 200, 235, 4000, 100, 1000

F. Stone, 50, 189, 1500, 50, 100

Allen T. Anorens, 200, 214, 4000, 150, 700

N. C. Thompson, 50, 189, 1500, 50, 100

John W. Crowder, 30, 100, 600, 10, 50

John W. Waler, 5, 20, 100, 5, 25

Conway D. Whittle, 150, 630, 8000, 200, 1000

Wm. Stone, 80, 240, 3000, 35, 400

Samuel H. Warner, 200, 205, 4000, 200, 1500

John Smith, 200, 284, 5000, 100, 200

John A. Smith, 50, 100, 1000, 30, 150

John A. Harriss, 77, 100, 1500, 30, 300

C. W. Oglanne, 100, 200, 2000, 100, 500

John Piney, 100, 266, 2500, 150, 500

Richard J. Smith, 35, 75, 1000, 10, 100

David Roberts, 50, 100, 1000, 10, 50

Susan Highton, 50, 150, 1000, 25, 100

Thos. Cook, 100, 200, 3000, 20, 300

Thos. W. Smith, 75, 120, 2000, 25, 350

Hartwell Johnson, 30, 160, 1000, 20, 100

Wm. Evans, 100, 135, 1500, 25, 300

Samuel Crame, 30, 170, 1000, 25, 200

John W. Arnold, 100, 270, 1000, 50, 400

Elisha Andrews, 40, 80, 1000, 20, 300

Maval Bowen, 4, 98, 500, -, 40

James Roberts, 20, 10, 150, 20, 75

Michael E. Crutchfield, 25, 75, 800, 20, 150

Mary Recie, 50, 150, 800, 110, 50

B. C. Smithson, 100, 197, 1200, 25, 250

Evy D. Hall, 50, 50, 1000, 25, 75

James Calaham, 20, 83, 500, 25, 100

Richard H. Wilson, 30, 297, 3000, 20, 675

Charles P. Calley, 30, 275, 1200, 30, 100

Richard Nething, 20, 180, 1000, 30, 50

Mary Jones, 40, 180, 1500, 30, 200

Mary W. Johns, 20, 180, 1500, 20, 75

Joseph A. Reese, 20, 300, 1500, 20, 15

Charles Hudson, 10, 600, 4000, 30, 500

Isaac Nanny, 50, 32, 500, 20, 150

Reubin Clark, 30, 150, 1000, 5, 100

Wm. Roberts, 100, 230, 1500, 25, 300

E. R. Edmuson, -, -, -, -, 1500

Wm. H. Hames, 150, 150, 2000, 300, 1000

James T. Walker, 300, 250, 7000, 250, 1500

B. C. Anderson, 50, 100, 2000, 125, 250

Howel Mallet, 30 79, 1000, 25, 100

John J. Daws, 25, 100, 1000, 10, 50

Wm. A. Adams, 50, 200, 2000, 75, 150

Jacob Chavaes, 25, 100, 1000, 150, 300

Edwin Cooper, 20, 30, 250, 10, 50

James M. Hankins, 40, 60, 1000, 100, 150

Alex. Johnson, 100, 50, 1000, 30, 300

Benga. S. Watson, 100, 270, 2500, 250, 800

John S. Meachum, 20, 90, 500, 10, 100

James Johnson, 40, 60, 500, 15, 250

Charles Nanny, 30, 100, 500, 10, 100

Alex. Maning, 60, 144, 1000, 50, 300

Pleasant Jones, 10, 10, 100, 20, 60

Richard Hayse, 5, 5, 100, -, 450

John D. Hughes, 350, 243, 1000, 200, 1000

Marsha E. Daws(Davis), 100, 140, 2500, 130, 1000

John H. Winkler, 250, 759, 10000, 500, 1500

Isaac B. Sallie, 50, 300, 2000, 20, 150

Gillem Chavaus, 8, 4, 200, 10, 100

Thos. Roffe, 20, 30, 250, 10, 50

J. V. Daly, 500, 1200, 16000, 500, 2000

S. P. Thrower, 18, -, 2500, 20, 600

Wm. E. Roffe, 60, 46, 8000, 50, 600

John W. Williamson, -, -, -, -, 600

Alfred Boyd, 500, 900, 13000, 400, 2200

Benga. Lewis, 150, 40, 2000, 100, 600

H. Arnold, 10, 5, 300, 25, 350

O. Poor Mecklenberg VA, 100, 350, 4500, 25, 350

George F. Griffin, 100, 253, 3500, 40, 250

Wm. Hayse, 20, 86, 500, 80, 200

Lucy Clark, 20, 100, 500, 10, 100

Saml. Emery, 80, 113, 200, 200, 500

Thos. L. Jones, 250, 409, 6000, 250, 1000

Wm. Quenechell, 20, 8, 200, 40, 120

H. P. Mallett, 20, 10, 200, 10, 50

Berenly Mallett, 42, 100, 100, 50, 250

Jesse Mallet, 25, 12, 300, 5, 50

Henry E. Janes, 300, 350, 3000, 100, 300

James Folense, 30, 55, 500, 10, 50

James Bevil, 30, 20, 250, 25, 100

Sand Simmons, 100, 243, 5000, 20, 450

Benga. T. Wiland, 75, 200, 1000, 100, 700

Jessee Roffe, 300, 640, 6000, 100, 500

Wm. Evans, 30 79, 500, 15, 200

Wm. A. Moss, 125, 155, 3000, 130, 600

Peter Parish, 30, 30, 300, 25, 100

Robert W. Algood, 25, 175, 1000, 25, 150

Richard C. Pope, 80, 300, 4000, 75, 400

Mary Leuker, 100, 207, 1500, 50, 150

Mary Williams, 200, 400, 5000, 100, 1000

Wm. Gerceltney, 50, 134, 800, 30, 100

Wm. N. Meins, 50, 200, 1500, 30, 200

G. W. Burwell, 250, 350, 4500, 250, 700

Wm. R. Moss, 75, 240, 2500, 60, 350

Mary C. Hutcherson, 100, 200, 3000, 25, 300

Charles S. Hutcherson, 500, 500, 10000, 250, 1200

Ann A. F. Lane, 300, 200, 10000, 50, 600

Jas. W. Simmons, 250, 366, 5000, 150, 800

R. M. Hutcherson, 250, 350, 3000, 150, 700

Mary Hutcherson, 700, 107, 1500, 25, 300

Benga. Octohuson, 60, 60, 800, 25, 200

Elizabeth Dunn, 30, 30, 300, 20, 50

Jane J. Whittemore, 75, 100, 1000, 20, 200

Benga. Cram, 50, 50, 600, 25, 100

Silas Wells, 20, 65, 600, 25, 50

D. P. Evans, 20, 30, 400, 10, 50

M. P. Simmons, 200, 500, 4000, 200, 1000

James McCargo, 150, 150, 3000, 50, 400

James Hayse Jr., 300, 400, 5000, 100, 1500

Peter H. Rainey, 30, 45, 400, 100, 200

Edward M. Patillo, 150, 110, 2500, 30, 350

H. R. Thomas, 200, 150, 4000, 30, 500

M. W. Jones, 350, 650, 10000, 150, 1000

Charles H. Oglanne, 1200, 1800, 30000, 1000, 3000

Dennis R. Fielder, 300, 582, 8000, 350, 800

Martin Phillips, 200, 100, 3000, 60, 600

Benga. D. Hatcher, 500, 500, 10000, 200, 1500

Mathew A. Rainey, 40, 100, 1200, 30, 250

Pettus C. Phillips, 150, 250, 4000, 75, 500

Mary Simmons, 50, 150, 1500, 20, 200

George L. Hayse, 400, 425, 11000, 250, 1500

Peter T. Fergerson, 100, 400, 4000, 100, 500

John W. Simmons, 260, 440, 7000, 250, 1500

Robert B. Chappel, 250, 400, 6400, 250, 1500

Chr. Gale, 350, 300, 7000, 250, 1700

Thos. W. Davis(Daws), 100, 240, 2000, 100, 500

Peter Daws (Davis), 300, 470, 7000, 150, 1000

Isaac H. Jones, 200, 500, 6000, 150, 1000

James S. Moss, 600, 1300, 25000, 500, 2000

Thos. D. Phillips, 30, 40, 7000, 150, 1000

Mark Alexander Sr., 2000, 2382, 100000, 1500, 3500

Jas. T. Alexander, 500, 740, 12000, 500, 2000

Deaketiar M. Young, 700, 1000, 34000, 300, 2500

Dillebery Watson, 500, 500, 20000, 300, 1500

Jacob L. Bugg, 200, 125, 4000, 150, 500

Fenloe Skipwith, 1300, 1200, 40000, 1000, 3000

Richard D. Bugg, 300, 344, 5000, 300, 1100

Wm. Tacams(Yocums) Jr., 500, 1150, 35000, 1000, 3000

Giles Harriss, 300, 330, 10000, 500, 2000

M. T. Winckler, 75, 190, 1500, 50, 700

Francis H. McGuire, 340, 200, 12000, 250, 700

John T. Ball, 600, 1100, 17000, 1000, 2000

Wm. T. Craete, 300, 400, 7000, 200, 1000

David W. Brame, 40, 90, 1000, 30, 200

Saml. Noel, 150, 155, 3000, 50, 500

Daniel M. Smith, 250, 350, 7000, 100, 700

James H. Rudd, 30, 72, 1000, 50, 300

Edwin H. Turpin, 500, 1100, 2000, 500, 3000

George W. House, 15, 85, 600, 75, 200

J. H. Pattillo, 150, 300, 4000, 100, 400

Benga. Towler, 100, 409, 4000, 150, 500

John Adams, 25, 25, 400, 10, 100

Venab. C. Create, 230, 230, 3500, 30, 700

James M. Tunstill, 20, 15, 500, 20, 200

John B. Tunstill, 50, 125, 1500, 60, 200

Pettus Famer, 600, 375, 10000, 50, 1600

Matthew C. Gill, 100, 200, 1500, 50, 500

Saml. Famer, 300, 600, 6000, 150, 1000

Jos. C. Hutcherson, 450, 600, 10000, 300, 1500

Saml. G. Famer, 140, 140, 2000, 30, 500

James Wickeler, 100, 200, 2000, 100, 400

Martha A. Rainey, 10, 63, 350, 10, 25

Ann Vdreidson, 100, 400, 4000, 20, 400

Harriet Curtis, 100, 150, 2000, 25, 500

Nancy Reese, 20, 30, 200, 10, 100

Thos. Ryland, 50, 30, 500, 20, 300

George Canly, 25, 76, 500, 25, 100

Robert A. Walkins, 150, 300, 3000, 75, 600

Rasehal H. Barnes, 150, 377, 3000, 50, 350

George Being, 30, 100, 1000, 20, 300

Farlvian Winn, 10, 40, 200, 5, 100

Green Curtis, 50, 50, 500, 20, 200

Peter E. Lett, 125, 150, 3000, 125, 700

Jos. H. Lett, 100, 200, 3000, 25, 300

Sarah T. Holmes, 100, 282, 2000, 25, 200

Martha G. Jackson, 500, 500, 10000, 500, 1500

John & Robert Faves, 150, 357, 4000, 50, 500

P. L. Hinton, 400, 500, 30000, 500, 2500

Wm. H. Wartham, 40, 80, 500, 10, 100

Est. J.M. Fites, 500, 1060, 30000, 250, 2000

James H. Fittes, 300, 345, 12000, 100, 2000

David E. Jiggetts, 900, 1030, 60000, 1500, 3500

Robert A. Jackson, 300, 500, 8000, 400, 900

Henderson Crander, 50, 30, 1000, 50, 300

Mary A. Hendrick, 1000, 1500, 17000, 1000, 3000

S. E. Finch, 500, 267, 12000, 300, 1500

John M. Wright, 200, 300, 6000, 300, 1000

Est. of A. P. Wright, 300, 294, 6000, 50, 600

George Jefferson, 400, 800, 15000, 400, 200

Jaro E. P. Wright, 160, 173, 3000, 100, 500

Benga. Spruiel, 150, 180, 3000, 150, 1100

John Largley, 230, 200, 4000, 100, 600

Henry Newby, 20, 20, 400, 10, 50

Lewis G. Meachum, 100 170, 2000, 100, 300

Mary A. Davies (Davis,Daws), 600, 400, 30000, 500, 2500

John Davis(Davies, Daws), 300, 300, 15000, 200, 1000

Levy Tally, 250, 300, 6000, 200, 1000

Charles Abenason, 500, 437, 12000, 200, 1500

Neshel Alexander, 700, 850, 20000, 400, 1500

Arthur H. Daies(Davis, Daws), 400, 600, 6000, 300, 800

Wm. H. Steagal, 400, 75, 100, 10, 50

Helen Jones, 800, 500, 40000, 1000, 5000

Hiram Richardson, 25, 100, 1250, 100, 150

Henry E. Colman, 400, 600, 16000, 500, 2000

Willis Steagal, 30, 80, 500, 10, 250

Metilda Baltshaf, 100, 200, 3000, 25, 400

Janco G. Newton, 20, 80, 500, 10, 50

Thos. C. Haskins, 200, 43, 2000, 50, 200

Richd. Nicholson, 50, 50, 1000, 30, 200

Ephraim Johnson, 300, 500, 8000, 100, 600

Wm. H. Reed, 301, 600, 9000, 200, 1000

Thos. Kersey, 35, 20, 200, 10, 100

George Terrey Sr., 1300, 12620, 40000, 2000, 5000

Lewis J. Peeples, 200, 400, 4000, 100, 600

Thos. C. Newton, 25, 100, 1000, 10, 100

Victor M. Epps, 500, 600, 30000, 500, 3000

John Newton, 100, 230, 2500, 75, 250

Edward A. Rawlins, 850, 1000, 40000, 2000, 6000

George Terry Jr., 300, 700, 160000, 50, 500

H. R. Thompson, 60, 100, 1000, 20, 300

Merit Tucker, 35, 40, 300, 10, 50

David H. Crowder, 20, 26, 200, 5, 50

Peter Inscoe, 25, 29, 200, 10, 50

Ira _. Crenshaw, 125, 300, 4000, 150, 800

George Guy, 30, 20, 250, 10, 100

Benga. W. Davis, 400, 600, 1000, 300, 1200

Est. Wm. Jones, 1500, 1500, 20000, 500, 360

Est. Jno. P. Smith, 800, 900, 10000, 300, -

David A. Walker, 75, 155, 1500, 50, 400

Benja. H. Merryman, 10, 50, 200, -, 25

Wm. Harriss, 100, 170, 1500, 200, 500

Middlesex County, Virginia
1860 Agricultural Census

The Agricultural Census for Virginia 1860 was microfilmed by the University of North Carolina Library under a grant from the National Science Foundation from original records at the Virginia Department of Archives and History in 1963.

There are forty-eight columns of information on each individual. Only the head of household is addressed. I have chosen to use only six columns of information because I feel that this information best illustrates the wealth of individuals. The columns are:

1. Name
2. Improved Acres of Land
3. Unimproved Acres of Land
4. Cash Value of Farm
5. Value of Farming Implements and Machinery
13. Value of Livestock

Philomen T. Woodward, 100, 50, 5000, 300, 900
M. B. Gaessitt, 6, -, 1000, 100, 300
Geo. E. Shackleford, -, -, -, -, 200
R. H. Woodward, 15, 18, 2000, 50, 400
Geo. D. Rilee, 1, -, 1000, 100, 85
John H. Figg, ½, -, 500, -, -
John Creswell, -, -, -, -, -
Robt. H. Stiff, 2 ½, -, 800, 30, 150
Thos. Hutchings, 100, 130, 4000, 100, 530
John J. Boss, 30, -, 1250, 25, 145
J. H. Jackson, 90, 90, 4000, 100, 285
Malici Christopher, 5, 9, 240, -, 30
Rurwom Greenwood, 5, 9, 240, -, -
James L. Wilson, 5, 10, 460, -, 80
James T. Marguinoy, 5, -, 100, -, 40
Sarah Wilson, 9, 84, 2000, -, 40
J. B. Jackson, 20, 23, 860, 10, 50
Robt. Dudley, -, -, -, -, 230
Wm. H. Purkins, 250, 111, 5000, 250, 500
D. Van Wagenner, 7, 8, 600, -, 20
Geo. Snider, -, -, -, -, 8

Wm. S. Christian, 25, 79, 600, 10, 350
Mary A. Purkins, -, -, -, -, 20
Thos. Healey, 300, 200, 7000, 250, 1075
Thos. Street, 1000, 948, 30000, 400, 1000
B. L. Robinson, 20, 45, 4000, 10, 175
John D. Gressitt, -, -, -, -, 71
John Norton, -, -, -, 10, 250
E. T. Purkins, -, -, -, 7, 275
P. Skinner, -, -, -, -, -
Thos. R. Sutton,-, -, -, -, 150
Robt. Healey Jr., 180, 520, 10000, 500, 1050
Robt. B. Fauntleroy & Bro., 400, 180, 9000, 200, 1200
N. H. Christopher, 4, 4, 250, -, 75
Jane Dudley, 50, 56, 2000, - ,-
Virginia Norris, 75, 24, 2000, 30, 150
C. Beadon, 3,-, 100, -, -
Wm. M. Harrow, 10, 8, 400, -, 30
W. W. Coleman, 40, 26, 1350, 15, 145

Richd. H. Bristow, 5, 3, 1500, 50, -
John G. Anderton, 30, 3, 800, 200, 330
Henry C. Palmer, 120, 85, 4500, 100, 370
R. T. A. Gresham, 239, 150, 8000, 500, 750
James E. Bourne, 90, 30, 4000, 150, 153
Charles Roane, 150, 160, 6000, 100, 384
John H. Bowden, 9, 3, 1000, 75, 40
John P. Bristow, 200, 300, 15000, 300, 731
K. R. Daniel (manager), 75, 55, 1200, 25, 100
Richd. M. Clements, -, -, -, -, 165
Orinder Clements, 80, 25, 1800, -, 105
Richerson Slaughter, 160, 85, 2200, 150, 380
John S. Cropper, 6, 10, 800, 25, 75
C. J. Marston, -, -, -, -, -
L. Oaks, 220, 330, 15000, -, 800
Benjamin Major, -, -, -, -, 100
Thos. Jessee, 228, 176, 6000, 500, 750
John Hardy (manager), 600, 500, 16500, 300, 1338
John A. Miles, 250, 61, 8000, 400, 800
M. W. Towell, 530, 230, 10000, 500, 1000
John L. Blake, 300, 150, 4500, 400, 903
John H. Saunders, 60, 140, 4000, 250, 360
John K. Lumpkin, 30, 106, 2000, 150, 275
A. C. Crittenden, -, -, -, 50, 150
J. S. Saunders, 23, 20, 1300, 20, 150
A. C. Martin, -, -, -, -, 125
E. P. Carter, 10, -, 500, -, -
J. D. Ailsworth, 12, 3, 1000, 150, 140
G. W. Daniel, 50, 85, 2500, 20, 175

Geo. W. Nelson, 125, 120, 4000, 100, 400
Thos. D. Weston, 45, 22, 1000, 10, 100
Geo. W. Davis, 15, 9, 600, 10, 66
John S. Norton, 5, 4, 450, -, 175
Able Kellum, 20, 16, 650, 30, 35
Robt. J. Boss, 50, 12, 2000, 10, 77
G. A. Carter, 40, 6, 1500, 20, 150
Wm. Powell, 15, -, 300, 15, 91
Joseph Beadles, 5, -, 500, -, -
Henry D. Barrack, 300, 150, 1500, 300, 730
Able Parks, -, -, -, -, 40
Mary Sullivan, 70, 15, 1000, -, 170
John D. Miller, 18, 55, 2500, 200, 310
John D. Robinson, 12, 10, 220, 5, 30
Alfred Palmer, 5, -, 2000, -, 45
C. R. Haines, -, -, -, -, -
R. A. Davis Sr., 216, 82, 6000, 400, 606
R. A. Davis Jr., 100, 56, 2300, 100, 428
Loretta Lee, -, -, -, -, 20
J. T. Healey, -, -, -, -, 50
Robt. Seward, 200, 120, 2500, 250, 612
J. C. Keiningham, -, -, -, -, 50
Wm. H. Armstrong, -, -, - -, 70
Fanny P. Rowen, 500, 15, 14000, 500, 1300
Henry Hackney, 170, 80, 5000, 150, 456
R. A. Christian Sr., 370, 170, 11000, 400, 1000
Wm. P. Woodward, 350, 285, 15000, 200, 1140
Alex. Campbell, 400, 300, 15000, 400, 1500
Joseph Christian, 201, 275, 7500, 300, 950
Wm. R. Legar & Bro., 225, 165, 7500, 150, 522
John W. Callis, 450, 300, 10000, 500, 1414

James H. Callis, 120, 50, 2500, 50, 600

R. H. Mackan, 100, 150, 2500, 100, 250

Thos. M. Bray, 40, 110, 2000, 50, 225

Robt. H. Bray, 250, 250, 5000, 150, 365

Robt. S. Major, 150, 50, 5000, 150, 800

A.B. Evans, 400, 200, 12000, 300, 1556

Thos. B. Evans, 800, 500, 20000, 300, 1350

E. T. Montague, 350, 130, 8000, 200, 410

J. M. Evans, 250, 150, 8000, 150, 550

James Chewning, 170, 530, 8000, 200, 815

Wm. H. Hail, 75, 75, 1000, 20, 146

John R. Faucett, 150, 315, 4000, 75, 411

Usula Street, 600, 400, 10000, 100, 500

Wm. L. Gatewood, 300, 335, 15000, 200, 1350

Wm. H. Daniel, 100, 107, 3000, 100, 425

Thos. Trice, 100, 150, 1500, 100, 360

Jas. Perciful, 125, 68, 5000, 265, 210

Elijah Dungy, 75, 91, 1600, 75, 260

Lewis B. Seward, 300, 500, 9000, 100, 600

Enos Walden, 150, 60, 2500, 50, 360

John Newbill, -, -, -, -, 20

G. L. Seward, -, -, -, -, 40

Wm. F. Newcomb, 75, 98, 2000, -, 216

Carter Williams, 16, 23, 500, 25, 100

Gideon Keiningham, 15, 39, 1000, 25, 90

Andrew South, -, -, -, -, 30

Peter Gregg, -, -, -, -, 25

A. J. Groom, 18, 22, 1000, 50, 185

Robt. Haile, 50, 95, 1500, 10, 150

Charles Dudley, -, -, -, -, 90

D. M. Thurston, -, -, 2500, -, 85

S. H. Roane, -, -, -, -, 430

H. M. Parron, 85, 50, 1350, 50, 370

J. E. Smither, 150, 100, 1800, 50, 180

Ellison Daniel, 250, 400, 4000, 75, 325

John Brown, 75, 75, 750, 25, 145

John A. Beazley, 100, 500, 6000, 100, 465

R. L. Fleet, 300, 200, 5000, 100, 620

Richd. Taylor, 90, 64, 1800, -, 300

Thos. Corn, 100, 340, 4000, 100, 275

G. W. Sadler, -, -, -, 10, 150

A. A. Bass, 90, 117, 2500, 50, 820

Robt. L. Montague, 400, 212, 15000, 500, 1536

J. R. Kidd, 60, 90, 1300, 25, 115

Lee Campbell, 20, 46, 300, 50, 100

Elias Dungy, -, -, -, -, 155

E. E. Dungy, -, -, -, -, 20

B. B. Sibley, 160, 90, 4000, 250, 665

S. W. Fisher, 500, 500, 20000, 250, 360

Jos. Walden, 100, 50, 1500, 100, 150

Lewis W. Gardner, 275, 270, 12000, 100, 680

J. J. Harrow, 10, -, 1200, 10, 170

Wm. J. Daniel, 20, 4 ½, 1000, 25, 165

Job Moore Sr., 150, 136, 6000, 50, 520

Robt. H. Montague, 200, 330, 6000, 100, 500

Wm. T. Miles, 20, 12, 600, 20, 100

Wm. G. Wortham, 20, 14, 1200, 100, 160

Wm. H. Hudgins, 5, -, 800, 100, 165

John R. Taylor (M. R. Gray Pro.), 700, 560, 25000, 1000, 835

Eliza Bailey, 300, 300, 30000, 1000, 1600

Lewis S. Bristow Jr., 50, 50, 2000, 50, 260

A. L. Bristow, 100, 30, 2500, 50, 405
Fanny B. Woodward, 450, 50, 10100, 200, 775
A. J. Harrow, 10, -, 200, -, 45
R. M. Glenn, 50, 30, 1500, 25, 175
Mat. Glenn, -, -, -, 35, 210
Wm. L. Trainyer, 10, 2 ¾, 300, -, 30
Thos. Christopher, 5, -, 75, -, 24
Billy Christopher, 5, -, 75, -, 24
Wm. Clark, -, -, -, -, 20
John T. Craton, 40, 10, 1000, -, 20
Jas. R. Jackson, 16, 10, 1000, 20, 150
John L. Johnston, 35, 73, 1500, 50, 90
Thos. Williams, 4, 12, 500, -, -
John Cundiff, 10, -, 200, -, 20
Thos. Hundley, 3, -, 150, -, 27
Thos. Jones Sr., 300, 400, 7000, 100, 1368
Dudley Horsley, 80, 95, 3500, 50, 355
Andrew J. South, 80, 120, 4000, 50, 300
Wm. R. Didlake, 50, 91, 2200, -, 265
B. F. Howlett, -, -, -, -, 35
Edwin Greenwood, -, -, -, -, 150
Joseph Howlett, -, -, -, -, 15
Sarah Owen, 170, 102, 3500, 100, 400
John A. Owen, -, -, -, 100, 450
Stage L. Sibley, 60, 34, 1000, 50, 362
Lewis S. Bristow Sr., 150, 50, 2500, 200, 460
Geo. L. Nicholson, 500, 400, 18000, 600, 1700
John T. Games, 70, 111, 1800, 35, 400
R. Wilkins, 8, 2, 200, 20, 50
Robt. Healey Sr., 900, 1075, 40000, 1000, 2650
P. A. Blackburn, -, -, -, -, -
Andrew Stiff, 260, 340, 9000, 500, 1000
Sarah Barrack, -, -, -, 100, 400

Hiram Walker, 150, 100, 3000, 150, 375
James B. Barrack, 90, 110, 3000, 125, 375
Wm. H. Barrack, 10, 2, 200, 20, 186
James Revere, 6, 6, 300, -, 50
James Sylva, 60, 165, 1810, 25, 175
Zerabable Wortham, 50, 30, 800, 10, 65
Holland Walker, 200, 45, 2500, 100, 400
A. Hall, -, -, -, -, 110
Mel. M. Walker, 50, 11, 600, 15, 90
Edward Topping, 28, 11, 600, 25, 115
John L. Humphreys, -, -, -, 25, 250
Julia A. Major, 10, 34, 800, 10, 60
W. C. Moody, 70, 70, 2000, 50, 135
G. G. W. Hall, 15, 37, 500, 20, 100
Robt. M. Blake, 40, 30, 700, 20, 135
Jas. Edwards, 40, 40, 800, 5, 170
John P. Chrispan, 100, 89, 1900, 50, 140
Arwmon Didlake, 20, 36, 700, 20, 80
A. S. Rock, -, -, -, -, 15
E. T. Watkins, -, -, -, -, 55
Thos. H. Crittenden, 45, -, 1000, 40, 200
Zack Blake, -, -, -, -, 24
Wm. T. French, 20, 8, 800, -, 150
Jas. Q. Crittenden, 60, 40, 1500, -, 150
E. P. Jones, 700, 1100, 20000, 500, 2143
N. Gordy,-, -, -, -, 12
Wm. N. Chowning, 800, 500, 19000, 500, 1630
Robt. C. Garland, 80, 123, 4000, 50, 250
Wm. M. Major, -, -, -, 75, 205
J. Wooldridge, -, -, -, -, 25
Wm. S. Blake, 28, -, 300, -, 75
Wm. A. Daniel, 20, 5, 800, -, 100
Robt. Prince, 24, 10, 600, 10, 140
Thos. Y. Lawson 75, 25, 2000, 50, 123

Wm. Stiff, 150, 150, 4000, 200, 625
Joel Revere, -, -, -, -, 45
E. B. Stiff, -, -, -, -, 30
Jas. A. Eubank, 400, 400, 12000, 200, 800
Wm. D. Turner, -, -, -, 10, 45
John C. Meras, 8, 3,800, 25, 200
F. M. Walker, 10, 26, 500, -, 45
Wm. B. Dunlavy, 14, 7, 1000, -, 75
James Mercer, 10, 3, 500, -, 35
N. C. Montague, 50, -, 400, 50, -
Uriah Bird, -, -, -, -, 65
Michelboro Daniel, 250, 200, 4500, 200, 500
Robt. Daniel Sr., 300, 475, 8000, 300, 600
R. T. Bland, -, -, -, -, 350
A. M. McTire, -, -, 400, 15, 90
James Kidd, -, -, -, -, 100
Judson Walker, 150, 170, 6000, 50, 350
Ludwell Blake, -, -, -, -, 150
James Gardner, 200, 400, 6000, 15, 350
Joseph Eubank, 650, 400, 20000, 400, 1000
Mortimer Evans, 200, 250, 6000, 150, 670
John Richerson, 60, 69, 1200, 5, 50
Miles F. Mason, 100, 135, 2500, 300, 750
James Watts, 75, 152, 3500, 25, 400
Philip Montague, 80, 25, 1600, 10, 190
Thos. H. Montague, -, -, 1000, -, -
Robt. Daniel Jr., 80, 80, 3000, -, 110
Geo. Ingram (Manager), 1500, 356, 35000, 200, 1550
O. W. Lee, 300, 253, 7500, 250, 800
Edinmed Ailworth, 20, 80, 1000, 25, 300
Wm. C. Gardner, -, -, -, -, 25
Wm. H. Groom, 30, 30, 600, -, 315
G. J. W. Key, -, -, -, -, 50
John W. Mears, 42, 34, 1000, 40, 140

J. R. Butler, -, -, -, -, 310
Thos. W. Fauntleroy, 1200, 1360, 55000, 1000, 3200
Griffin Cundiff, 60, 77, 1200, 15, 100
Geo. W. Smith, 170, 230, 10000, 250, 624
D. R. Sibley Jr., -, -, -, -, 24
John W. Daniel, 30, 10, 1500, 40, 300
Jennette Wood, 250, 80, 4000, 30, 215
Wm. S. Blackburn, 180, 420, 6000, 350, 950
Braxton Clark (Proprietor), 250, 90, 4000, 200, 750
John R. Smith Sr., 80, 75, 1500, 30, 340
William Enos, 30, 60, 1400, 25, 180
Enos Healey, 200, 186, 5800, 250, 550
P. T. Woodward guardian for R. B. Woodward, 40, 145, 1850,-, -
Robt. Parson, 20, 43, 390, 25, 165
Fredk. Williams, -, -, -, -, 35
Thos. H. Parson, -, -, -, -, 165
E. L. Dillard, -, -, -, -, 125
Alfred Healey, 300, 300, 15000, 500, 1250
Douglass Pitt, 140, 63, 4500, 75, 450
Henry L. Layton, 300, 75, 7500, 500, 1100
Z. Blake (Proprietor), 200, 23, 2250, -, 20
Robt. Blake, 175, 77, 5000, 250, 455
Lewis Jones Sr., 225, 260, 7000, 250, 580
Walter Bristow, 4, 23, 400, -, 20
Wm. Kieningham, 75, 125, 1200, 40, 285
Ryborn Smither, 100, 121, 2000, 25, 330
Ed. S. Seward, 200, 160, 3000, 50, 300
Martha C. Claybrook, 420, 330, 7500, 250, 650

John C. Daniel, 450, 350, 10000, 500, 1680
E. W. Beazley, 50, 54, 1800, 5, 270
Thos. N. Roane, 80, 330, 4500, 100, 450
Wm. S. Hackney, 75, 50, 2000, 100, 300
Thos. Key, 60, 53, 1000, 50, 140
John J. Kemp, -, -, -, -, 130
Wm. P. Vaughan, 140, 145, 3500, 100, 530
Wm. J. Blake, 25, 18, 850, 50, 105
James A. Blakey, 450, 150, 10000, 250, 1040
John E. Leger, 50, 10, 5000, 200, 330

Sandy Parker (Prop), 10, -, 300, -, 50
Arma. F. Williams, 40, 20, 600, -, 60
R. C. Blackly, 100, 42, 1500, 100, 265
M. A. Seward, 20, 75, 800, 25, 50
Gabriel Robinson, 20, 12, 350, 10, 50
Joseph Fogg, 100, 50, 2000, 25, 150
Henry Bristow, 15, 45, 600, 10, 50
Geo. S. Davis, 150, 150, 4000, 200, 600
Wm. Roane, 100, 100, 4000, 100, 300
Zachariah W. Bristow, 60, 40, 1000, 25, 200

Index

Abagast, 40
Abenason, 203
Abernethy, 158
Abert, 84
Abington, 33, 35
Able, 121
Abraham, 80, 86
Abrahams, 82
Acree, 69-70, 72, 86
Acres, 19, 67
Acros, 141
Adam, 125-126
Adams, 4-6, 19, 22, 49-50, 61, 81-83, 103, 106, 110, 121, 132, 145, 150, 186, 196, 200, 202
Adkins, 16, 29, 193
Adkinson, 164
Aikin, 15
Ailsworth, 205
Ailworth, 208
Akline, 123
Alberger, 22
Albert, 104
Aldee, 118
Alder, 127, 130
Aldridge, 135
Alen, 102, 104
Alestork 145
Alexander, 5, 62, 72, 80-81, 83, 87, 116, 127, 198, 201, 203
Algood, 200
Allder, 130, 134
Allen, 1, 4, 13, 17-19, 25, 67, 71, 82-83, 129, 134, 158, 177-178
Allenson, 54
Allensworth, 65
Allenworth, 80
Alley, 1, 24
Allgood, 193
Allison, 4
Allman, 87
Allnnard, 52
Allnutt, 115
Almond, 155

Ambler, 114, 150, 168
Ammons, 16
Ancarrow, 82
Anderson, 2, 5, 8, 12, 21, 23, 34,, 70, 75, 82, 92, 106-107, 128, 145, 149, 151, 156, 160, 178, 200
Anderton, 205
Andess, 103
Andis, 104
Andish, 103
Andrew, 21
Andrews, 48, 54, 156-157, 200
Anorens, 199
Anthony, 7, 9, 146
Apperson, 12, 60
Apple, 194
Arbagast, 39
Arbagest, 38
Arbogest, 39
Archer, 7, 50
Arensking, 140
Arington, 34
Armistead, 57, 181-182, 185
Arms, 161
Armstrong, 26, 38, 43, 94, 145-146, 149-150, 205
Arnal, 103
Arnett, 130, 141
Arnold, 5, 29, 32, 63-64, 79-80, 102, 123-124, 199-200
Aroy, 102
Arrial, 98
Arrington, 168, 176-177
Arundle, 115
Arvine, 162, 164, 166
Asberry, 112
Ash, 50
Ashburn, 55
Ashby, 46
Ashnoth, 158
Ashton, 62-63, 79
Ashworth, 153, 162
Athy, 35
Atkins, 48-50, 75, 77, 80-83, 142

Atkinson, 3, 14, 25, 54, 82, 84, 158
Atkison, 145, 151
Atlee, 17
Attkison, 145
Attwell, 117
Atwell, 64, 67, 129, 160-161
Aukers, 113, 115
Aukeson, 145
Auldeer, 122
Ault, 118
Ausburn, 108, 110
Austin, 3, 18, 24, 27-28, 177
Auston, 116
Averett, 158, 190, 194
Avis, 55, 203
Avoy, 102
Aylett, 80-81
Aylor, 170-172, 174-175, 177
Ayre, 127
Ayres, 34-35, 97, 129
Ayrs, 125
Ayseo, 97
Babb, 51, 57
Baber, 64
Bacesue, 46
Bacon, 191
Bacy, 197
Baer, 55
Baerd, 195
Bagby, 77, 88, 144, 151, 161
Bagent, 122
Baglary, 98
Bagley, 137-138, 156, 158, 160, 162
Bagly, 78
Bagnall, 47-48
Bailey, 17, 49, 51, 64, 100-112, 160, 162, 195, 206
Bailor, 112
Baily, 54, 154
Baker, 5, 10-11, 16, 18, 32-33, 50, 64, 66, 119, 122, 140-143
Baldwin, 100, 128-129
Balee, 96
Bales, 108
Baleste, 31
Baley, 56, 98, 102, 157

Balis, 96, 99-100
Ball, 17, 19, 84, 90, 94, 97-99, 109-110, 116-118, 122, 202
Ballard, 23, 34, 53, 56, 141
Ballman, 148
Balton, 52
Baltshaf, 203
Baly, 197
Banister, 32
Banks, 51, 60, 156, 170, 178-179, 182-183
Banner, 109
Baplut, 193
Bapteet, 189
Baptest, 189, 193
Barber, 30
Barch, 29
Barish, 140
Barker, 5, 14, 19, 27, 67, 103, 106, 111, 196, 198
Barley, 15
Barlow, 13, 49, 51-52, 164
Barnage, 138
Barnes, 53, 59, 105, 107, 161, 192, 196, 202
Barnett, 31, 92
Barnhouse, 117
Barns, 154, 161, 163
Barr, 25, 55, 129, 170
Barrack, 205, 207
Barracke, 64
Barrell, 52
Barret, 102, 140, 143
Barrett, 52, 118
Barron, 110
Barrott, 56
Barrow, 29
Bartlett, 115, 122, 126
Bartley, 99, 108
Barton, 55
Barvan, 56
Barvau, 56
Basher, 4
Bashier, 109
Baskerville, 197-198
Basket, 4, 70

Bass, 197, 206
Bassett, 5, 31, 33, 187
Bassleet, 193
Basteet, 189, 193
Baswell, 96
Bateman, 27
Bates, 3, 20, 89, 161, 173
Battin, 49, 56
Batton, 52, 176
Batts, 20
Bauckman, 118
Bauer, 195
Baughan, 11-12, 23, 144
Baum, 195
Baylor, 85, 87, 98
Bayne, 119, 156, 164-165
Bayze, 100
Bazile, 2
Bead, 42
Beadles, 82-83, 143, 148, 205
Beadon, 204
Beal, 5, 30, 55
Beame, 191
Beamer, 124
Beane, 90-93
Beanes, 131
Beans, 120, 130-131, 134
Beard, 129
Bearson, 129
Beasley, 13, 189
Beasly, 103
Beath, 38, 45
Beatly, 126
Beatton, 57
Beaty, 106, 122, 130
Beavers, 113, 116, 127, 132
Beazley, 22, 206, 209
Beckeley, 83
Bedily, 117
Bedine, 116
Beeid, 196
Beeler, 55
Begley, 185
Being, 202
Belamy, 148
Belcher, 20, 28, 190

Bell, 28, 47, 49, 103, 155, 163, 166, 170, 187
Bellamy, 141
Belt, 117
Ben, 74-75
Beney, 173-174
Benford, 195-196
Bennett, 23, 198
Bennith, 119
Benson, 40-41, 44, 129
Bentley, 116, 165
Bently, 124, 136
Benton, 74, 76, 132, 169
Berkeley, 10, 81, 125-126
Bernard, 20
Berny, 178
Berrey, 174
Berrick, 90, 94
Berry, 62, 66
Berwell, 155
Besser, 133
Besson, 52
Best, 118, 131, 133
Betts, 51
Beverage, 38-39
Beveridge, 20
Bevil, 200
Bevins, 102
Bew, 85
Bewey, 173-174, 177-178
Bibb, 139, 142-143
Bickenbough, 7
Bickers, 170-171
Bickley, 102
Bigger, 193
Biggers, 142
Biggs, 196
Billingsley, 63-64
Billups, 181, 183-186
Binford, 3, 16, 52
Binns, 58
Birch, 74, 76, 135-136
Bird, 40-41, 69-70, 138, 140, 208
Birde, 75
Birdsall, 82, 133-134
Birgan, 112

Brabille, 100
Brack, 49
Brackit, 18
Bracy, 48, 51, 54-55
Braden, 120
Bradford, 169
Bradley, 16, 18, 144, 174
Bradshaw, 44, 48, 53, 55-57, 114, 124, 163
Brady, 197-198
Bragg, 18, 149, 155-157, 159, 162-163
Brainha___, 143
Brame, 192, 202
Bramhall, 188
Bramican, 65
Branaugh, 129
Branch, 58
Brancharm, 143
Brander, 163
Brandson, 194
Brant, 100
Braswell, 54
Braune, 60
Braxton, 6, 84
Bray, 27, 75, 99, 206
Breeden, 169
Bazil, 169
Brent, 90-92, 94, 98
Bresck, 30
Breslow, 76
Brewer, 191, 193
Brice, 3, 63
Bridgeforth, 159
Bridges, 53, 114, 187
Bridgewater, 18
Bridy, 192
Brigg, 196
Bright, 40
Brim, 27, 36, 96
Brimly, 139
Brisco, 40
Bristow, 78, 205-209
Briton, 74
Britt, 46, 52, 54
Brittain, 97

Brittan, 97
Broach, 72, 76-77
Brock, 2, 21, 49, 51, 54
Brooke, 71, 146
Brooker, 74
Brookes, 13, 73
Brooking, 171
Brooks, 2, 17, 84, 99, 102, 136, 145-147, 151, 181-186
Broumley, 78
Brown, 5-8, 23-24, 36, 42, 44, 65, 69-70, 72, 75-76, 79-80, 91-92, 98-99, 106, 117-118, 131, 133-135, 155, 163, 170, 174-175, 177, 185, 206
Browne, 183, 186
Browning, 3, 60, 106
Brownley, 182-183, 185
Broyles, 170, 173, 176
Bruce, 154, 161, 165
Bruer, 125
Bruice, 148
Bruns, 22
Brushwood, 76
Bruton, 22
Bryan, 63-64
Bryant, 26
Bryce, 13
Brydie, 163
Bryson, 189
Buchanan, 10
Buck, 60, 177
Buckley, 47
Buckner, 125, 138, 169
Buckon, 94
Bucks, 199
Bucktent, 57
Buckwell, 165
Buffin, 15
Bugg, 196, 198-199, 201
Bukers, 170
Bukley, 147
Bule, 53
Bullard, 65, 67-68
Bullock, 13, 138, 150
Bulman, 73-75
Bumgardner, 105

Bumhouse, 119
Bumpass, 11, 13-14, 128, 138
Bunch, 150
Bunck, 148-149
Bundy, 101
Bungardner, 96
Bunkley, 48
Burch, 28, 82, 84
Burchell, 67, 79
Burchet, 106-107
Burchett, 106-107
Burdon, 17
Burgan, 100-101, 103
Burgess, 27, 34, 77, 90, 118
Burgmer, 107
Burk, 45, 100, 110, 147
Burke, 82, 170, 177, 185, 198
Burket, 70
Burkley, 47-48
Burkley, 72
Burnet, 6
Burnett, 4, 69, 104, 154, 157, 160, 177, 195-196
Burnley, 146
Burr, 113
Burriss, 8, 23
Burroughs, 181-182
Burruss, 82, 87-88, 142
Bursougles, 168
Burten, 104
Burton, 6, 15, 17, 20, 22, 24, 74, 167, 172, 191, 193, 196-197
Burwell, 154, 191, 193-194, 201
Busby, 52
Bush, 19, 58, 89, 98, 109
Bushie, 123
Bussick, 4
Butcher, 128
Butler, 6, 8, 10-11, 50, 52-55, 57, 142-143, 147, 208
Buton, 20
Butrick, 74
Butterworth, 195
Butts, 121
Buzzard, 41
Byassed, 199

Byrd, 44, 52
Byrne, 125
Cabaness, 194
Cabiness, 29
Cabranss, 157
Cahall, 27
Cahill, 30
Cake, 185
Calaham, 160, 199
Caldwell, 82-83, 86
Caleis, 115
Caliham, 108
Call, 175
Callaham, 104
Callahan, 89-90
Callaway, 32
Callcote, 56
Calle, 164
Calley, 199
Callia, 93
Callis, 184, 186-187, 205-206
Caltell, 82
Cam, 120
Cambell, 149
Camden, 18, 140
Came, 123
Camp, 121
Campbell, 2, 11, 33, 40-41, 82, 88, 126, 138, 155, 176, 205-206
Camtell, 122
Canada, 67
Canaday, 60
Canly, 202
Cannon, 1, 197
Canntly, 169
Canter, 103
Caraway, 198
Cardle, 160
Cardwell, 74, 87
Carlisle, 44, 130
Carlton, 4, 70-74, 77-78
Carmack, 98
Carmon, 20
Carney, 186
Carns, 103

Claybrooke, 138
Clayton, 70, 169
Cleaton, 194, 196-199
Cleator, 194
Cleft, 28
Clegg, 78
Clements, 81-82, 86, 205
Clemmings, 125
Clendening, 25, 124
Clendenning, 121-122
Clevely, 77
Clift, 63-64, 67
Clifton, 103
Clinger, 105
Clivourne, 192
Clocke, 6
Cloid, 171
Clopton, 5, 27
Clore, 168, 170, 172-174, 178-179
Close, 171, 173-174, 179
Cloud, 97
Clough, 13, 146
Clouson, 102
Clover, 15
Clower, 57
Clowes, 61
Coakley, 62, 64-65
Coales, 138
Coalter, 87
Coan, 37
Coates, 24
Coatney, 172
Coats, 170
Cobb, 34, 88
Cobbs, 21
Coblen, 109
Cobler, 33, 36, 169
Cobs, 42
Cocke, 11, 81, 83-84, 87, 139, 150
Cockral, 89
Cockrane, 12
Cockrell, 115
Cockrill, 119, 130
Cocks, 56
Coe, 133
Cofer, 51-52

Coffer, 51
Cofield, 48, 50
Coke, 59
Colaw, 38-39
Cole, 30, 34, 105, 112, 149, 156, 161, 177, 196, 198
Coleman, 33, 73, 77, 93, 113, 117, 138, 144, 155, 164, 193, 204
Colins, 107
Collens, 5
Colleus, 5
Collier, 102, 105, 110, 167
Collins, 2, 15, 40, 73, 75, 104, 137, 167-168, 171, 178, 187
Collinsworth, 97, 111
Colly, 74
Colman, 203
Colom, 177
Colson, 98
Colton, 67
Colvin, 172, 180
Combs, 99
Comer, 98
Compher, 118-119, 121, 123, 130
Conard, 121-123
Cones, 53
Congrott, 179
Connell, 194
Connelle, 92
Conrad, 20
Conway, 24, 177-178
Coogan, 35
Cook, 50-51, 72, 96, 171, 196, 199
Cooke, 10, 70-71, 76, 84, 88, 138
Cooksy, 156
Coomer, 102
Cooper, 123-124, 195, 200
Copeland, 20, 121, 131
Coppage, 174
Coppedge, 92
Corbett, 57
Corbin, 79, 174-175
Corby, 4
Cordel, 100
Cordell, 118
Corker, 8, 20, 145

Corley, 150
Corn, 206
Cornasean, 138
Cornel, 116
Cornelias, 94
Corr, 76, 84-85, 148
Cosby, 5, 8, 19, 77, 138-140, 147, 192
Cost, 133
Cottrell, 21-22, 24
Couch, 154, 191
Couk, 112
Couling, 20
Council, 72
Councill, 53, 56
Countress, 110
Courtney, 3, 18-19, 23, 73, 77
Cousins, 196
Coveins, 32
Covey, 103
Covington, 34
Cow, 74
Cowadin, 23
Cowan, 47
Coward, 121, 123
Cowden, 116
Cowherd, 143
Cowles, 58, 60
Cowper, 48
Cox, 15, 17, 26, 34, 36, 66, 69, 71, 92, 94, 98, 106, 111, 126, 130, 158, 161, 164
Coy, 142
Crabtree, 98, 100, 103-104
Crace, 108
Craddock, 18, 193
Crade, 158
Cradle, 198
Craete, 202
Crafton, 16, 24, 164
Crage, 129
Craig, 32-33
Cralle, 164
Cram, 201
Crame, 199
Crander, 202

Crane, 133
Crank, 151
Crannage, 66
Crannoss, 66
Craton, 207
Craven, 116, 119
Crawford, 141, 148
Crawley, 164, 166
Creamer, 39
Create, 202
Creath, 194
Crenshaw, 4, 8, 11, 13, 22, 25, 193, 203
Creswell, 204
Crew, 148
Crider, 57
Crigler, 171, 173
Crim, 123
Crimm, 124
Crince, 122
Crisler, 174, 179
Crismon, 65
Crismond, 63, 80
Cristmas, 146, 151
Crittenden, 18, 75-76, 205, 207
Crocker, 49-51
Crocket, 101
Crockett, 97-98, 106
Croftion, 161
Crofton, 161-162
Crokrill, 116
Cronton, 80
Crony, 43
Cropper, 205
Crosby, 151
Croson, 113
Cross, 1, 3, 7, 12-13, 53, 67, 98, 125, 174
Crouch, 4, 19, 24
Crow, 71, 86, 157, 179
Crowder, 155, 163, 189, 192, 199, 203
Crowley, 156
Crowor, 196
Cruise, 123
Crumbacker, 123

Doyle, 34, 41, 46, 57
Drake, 54, 103, 109
Draper, 31
Drew, 22
Drewry, 86
Drian, 48
Driver, 76
Drury, 30
Duck, 53, 55-56
Duckman, 139
Dudley, 19, 82, 85, 204, 206
Dudly, 73
Duff, 109-110, 186
Duffe, 154
Duggins, 14, 144
Dugins, 50
Duglas, 196
Duke, 8, 9, 11, 18, 24, 75, 138-139, 142, 145
Dukerson, 144
Dulancy, 175
Dulaney, 175-176
Dulany, 2, 127
Dulin, 115, 170
Dunavant, 36
Dunaway, 89, 91-92
Duncan, 146
Dungee, 85
Dungie, 75-76
Dungy, 206
Dunlavy, 188, 208
Dunlop, 64
Dunn, 4, 73, 82, 144, 201
Dunnavant, 23
Dunstan, 81
Dunton, 95
Dupriest, 163
Dupsan, 163
Dur, 172
Durbry, 60
Dustry, 4
Dutton, 188
Duval, 24, 85, 145, 188
Dyer, 28-29, 31
Eacho, 18
Eagle, 43, 45

Ealer, 140
Eans, 27, 31
Earley, 168, 177, 179
Earls, 27
Easley, 168, 179
East, 32-33, 164
Eastham, 148
Eastwood, 87
Eavins, 185
Eawter, 142
Edains, 177
Eddleton, 11, 13
Edds, 96, 99
Edens, 99
Edmonds, 94, 160
Edmonson, 96
Edmund, 84, 162
Edmunds, 158-160, 164
Edmundson, 163
Edmuson, 200
Edsall, 106
Edwards, 30-31, 35, 46-52, 54, 57-58, 64-67, 72, 75, 80, 82, 85, 88, 107, 110, 114-115, 119, 121, 141, 150, 186-187, 207
Eggleton, 27-29, 31
Eggy, 18
Egleston, 161-162
Eheart, 167, 170
Eidson, 126
Elam, 190-191
Elder, 155, 159, 163
Eldridge, 96-97, 99
Eley, 48, 52, 54-57
Elgin, 135-136
Elison, 98
Elkins, 67
Ellett, 5, 11, 13, 21, 84, 86
Ellington, 197
Elliot, 7, 104
Ellis, 1, 24, 51, 64, 153-154, 156, 160, 162
Ellison, 61
Ellmore, 114
Ellyson, 3
Ellzey, 135

221

Gahright, 5
Gaines, 5, 57, 91, 174
Galdsay, 15
Gale, 49-50, 56, 193, 201
Gallaher, 23
Galleher, 129
Games, 207
Gammon, 145
Ganes, 129
Gardner, 3, 25, 40, 53, 55, 86, 139, 146, 206, 208
Garey, 189
Garison, 105
Garisson, 49
Garland, 139, 155, 158, 160, 207
Garlick, 78, 80
Garner, 50, 74, 193
Garnet, 86
Garnett, 8, 19, 22, 82, 71, 78, 168, 183
Garret, 176
Garrett, 19, 57, 59-60, 73-76, 81, 87, 99-100, 104, 128-129, 171
Garrison, 112, 126
Garrot, 35
Garrott, 104
Garth, 172, 179
Garthright, 18
Gartick, 7
Gary, 83, 87-88, 156, 164
Gaskins, 52
Gatewood, 16-17, 71-72, 75, 206
Gathright, 5, 16, 19
Gatwood, 58
Gaulden, 6
Gauldin, 5
Gaulding, 165-166
Gay, 15, 40, 55-57
Gayle, 183, 185-186
Geddings, 119
Geddy, 58
Gee, 156-157, 161, 163-165, 195
Gennett, 23
Gentry, 3, 9, 21, 139, 142, 145-146
George, 88, 90-92, 94, 115, 120, 124, 197

Gerceltney, 200
Gesselor, 140
Gibbons, 62
Gibbs, 47, 178
Gibson, 4, 41, 60, 69, 72, 74, 77, 97, 110, 112, 127-128, 134, 137, 139, 143, 146
Gid_mas, 198
Giddings, 117
Gier, 157
Gilbert, 28, 110, 149
Gill, 156, 202
Gillespie, 142, 189
Gilley, 26-27, 34, 36, 105
Gilliam, 160
Gillians, 162
Gills, 163
Gillum, 141
Gilman, 4, 13
Gilmon, 22
Gilmore, 2, 40, 72, 169
Gilvey Farms, 65
Ginn, 40
Gipson, 93
Givens, 104
Givin, 44-45
Glascock, 91
Glass, 4, 6, 110
Glasscock, 127, 191-192, 194
Glazebrook, 13, 20, 25
Glenn, 12, 207
Goalder, 75
Gobble, 103
Goble, 108-109
Gochnauer, 128
Godwin, 47, 50, 54
Goins, 98
Gold, 192
Goliham, 107
Gooch, 22, 143-144, 148, 151
Good, 30, 168
Goodall, 13, 173, 177
Goode, 198
Goodin, 22
Gooding, 175, 179
Goodman, 10, 14, 16, 142

Goodrich, 46
Goodson, 48, 51, 57
Goodsy, 139
Goodwin, 9, 48, 137-139, 141-142, 144-147, 151, 158, 160
Goolsby, 17
Gordon, 23-24, 148-149, 174, 179
Gordy, 207
Gore, 117, 133
Gott, 104
Gouchman, 132
Gouldin, 19
Gouldman, 75
Gower, 172
Goyne, 25
Grabill, 100
Grace, 52
Grady, 101, 127, 140, 144
Gragg, 40
Graham, 44, 100-101, 105, 107
Grame, 85
Grant, 35
Grason, 128
Gravely, 27-30, 37
Graves, 141, 143, 169-170, 172-173, 177-179
Gravis, 169
Gray, 1, 49-52, 54, 116-117, 122, 127, 141, 179-180, 196, 206
Grayson, 171
Gre__, 169
Greeley, 139
Greely, 139
Green, 3, 23, 26-27, 47, 49, 51, 66-67, 85, 104-105, 115, 154, 184-185, 187
Greene, 140
Greenelease, 114-115
Greenhow, 60
Greenlaw, 64, 66
Greenwood, 85, 204, 207
Greever, 103
Gregery, 29-30
Gregg, 81-82, 121, 131-132, 196, 206

Gregory, 69, 84, 87, 143, 154, 157, 163, 191-193, 195-196
Gresham, 70, 72-74, 77-78, 81, 88-89, 91, 93, 205
Gressitt, 204
Griffen, 190
Griffin, 22-23, 42, 50, 63, 190, 200
Griggs, 27, 29-30, 36
Grigsby, 66
Grinnan, 169
Grinstead, 159
Grissom, 104
Grogan, 35-36
Grogg, 45
Groom, 148, 206, 208
Groves, 40
Grovin, 143
Grubb, 108, 115, 119, 121-123, 131
Grubbs, 3-4, 8, 10, 18, 150
Gruver, 103
Grymes, 62, 79
Guen, 17
Gues, 23
Gulick, 116, 129, 136
Gulie, 90
Gullihugh, 177
Gum, 38-41
Gunn, 16, 153
Gunter, 141, 151
Gurthrow, 2
Guthrie, 69, 74-76
Guthrow, 82
Guy, 196-197, 203
Guyn, 186
Gwalthmey, 11
Gwalthney, 49, 69
Gwaltney, 51, 78
Gwathmey, 80
Gwathney, 49-50
Gwin, 40-41, 44
Gword, 110
Gwyn, 85
Haburn, 102
Hacket, 147
Hackney, 205, 209

Haden, 149
Haggard, 90
Hagood, 32
Hague, 174
Hail, 206
Haile, 206
Haine, 49
Haines, 205
Hair, 7
Hairston, 27, 30, 34-35, 37
Hales, 62
Haleway, 20
Haley, 147
Hall, 1, 9-10, 28, 47, 51-52, 57, 66, 70, 85, 92, 101, 104, 109-110, 135, 138-139, 141-142, 144, 152, 183, 191, 199, 207
Hallerman, 49-50
Halliway, 47-49, 51, 54
Hallomard, 48
Hallsay, 47
Halterman, 38
Hamblen, 103, 106
Hambleton, 139-140
Hamblin, 96, 99, 101, 106
Hamelton, 109-110
Hames, 200
Hamilton, 44, 121, 164
Hamlin, 163
Hammer, 39, 114-116
Hammerly, 130
Hammock, 162-163
Hammond, 59
Hammons, 159
Hampton, 47, 99, 130, 135
Hance, 85
Hancock, 7, 9, 139-140
Hand, 129
Handcock, 49
Handley, 28
Handock, 197
Handy, 153
Haneford, 67
Hanes, 8, 22, 109, 118, 133
Hanfield, 27
Haniford, 67

Hankins, 28, 58-59, 112, 200
Hanlen, 33
Hannah, 138
Hannass, 197
Hansel, 38
Hanson, 148
Harber, 112
Harbour, 36
Harcum, 94
Hardgrave, 12
Hardin, 11, 33, 154
Harding, 121, 161
Hardwick, 19
Hardy, 23, 32, 34, 108, 157, 159, 161-162, 164-165, 190, 193, 205
Hargrave, 48, 54
Hargrove, 85
Harley, 29
Harlin, 175
Harlon, 175
Harlour, 149
Harlow, 1, 25, 147, 151, 168
Harper, 72, 101, 118, 140, 143
Harrasson, 49
Harrell, 60
Harris, 1-2, 9-13, 22, 26-27, 30-32, 35, 60, 70, 73, 84, 94, 96, 109, 137-143, 145-146, 148-149, 178, 193
Harrison, 17, 80-81, 116, 127, 135, 139-140, 169, 171-172, 179
Harriss, 24, 55, 157, 198-199, 201, 203
Harrow, 204, 206-207
Harsead, 198
Hart, 52, 69, 73-74, 78, 137, 140, 148, 173-174
Hartgrove, 99
Hartsock, 100
Haruff, 44
Harvey, 17, 120, 161
Harward, 19
Harwood, 1, 19, 78
Hase, 25
Hasher, 151
Haskins, 194, 203
Hatch, 9

Hatchell, 163-164
Hatcher, 28, 134-135, 201
Hatchet, 158
Hatchett, 156
Hath, 120
Hathaway, 90, 94
Hatt, 7
Hattin, 49
Haunshell, 108
Havener, 136
Havenner, 116
Hawes, 72, 83-84
Hawkins, 59, 142, 160, 162, 172, 179
Hawks, 23
Hawling, 128-129, 136
Haws, 128
Hawthorne, 156, 159-160, 195
Haxall, 23
Hay, 4, 35, 83
Hayden, 94
Haydin, 92
Hayes, 191-192
Haynes, 70, 73, 78, 83, 140, 187
Haynie, 89-90
Hays, 55
Hayse, 200-201
Hayslip, 34
Hazlegrove, 3
Hazlewood, 59, 153, 156, 195, 199
Hazzard, 91
Headley, 89
Healey, 204-5, 207-208
Heard, 114
Heart, 123
Heartsock, 101
Heat, 191
Heaton, 118-119, 130, 134
Heckler, 19
Hedrick, 101
Heefner, 118
Hefflefinger, 34
Heler, 140
Heller, 98
Helms, 48
Helsel, 44

Helstine, 22
Heming, 179
Hempston, 117, 136
Henderson, 67-68, 93, 105, 133, 144, 158, 176
Hendley, 195
Hendrick, 12-13
Hendrick, 191, 202
Heneger, 109
Henkel, 168
Henley, 21, 25, 59, 70, 78
Henry, 11, 88
Henshaw, 139, 170-171, 176-178
Hensley, 31
Henson, 148, 149
Heolland, 151
Hepleburne, 154
Herbert, 17
Herd, 99, 105, 108
Hern, 188
Hernaman, 16
Herndon, 111, 167-169
Herrell, 105
Herrin, 98
Herring, 146
Hesler, 140, 142, 144
Heton, 159
Hewltt, 8
Hicklin, 42, 44
Hickman, 41, 110, 118-119, 123-124, 147
Hicks, 58, 60, 79, 127, 168, 185-186
Hidgen, 133
Higdon, 136
Higgason, 150
Higgenbothan, 23
Higgins, 161
Higginson, 10
Higgs, 37
Highton, 199
Hightour, 194
Hiler, 141
Hill, 3, 7, 9, 23, 31-33, 35, 47, 50, 69, 80, 84, 87, 94, 107, 110, 112, 130, 133, 146, 162, 164, 173, 177, 179, 186

Jinkins, 55, 63
Johns, 44, 158, 165, 199
Johnson, 4, 6, 8-9, 15, 27-28, 31, 48-50, 52-58, 67, 82, 85, 104-105, 108, 112, 122, 129, 137, 144-147, 149, 151-152, 159, 163, 170, 190, 194, 199, 200, 203
Johnston, 20, 22, 79, 116, 123, 207
Jolliff, 51
Jonas, 105
Jonathan, 15
Joneds, 168
Jones, 2, 3-4, 7, 10, 12-13, 17, 24-25, 27, 30, 34-36, 38, 45, 49-51, 60-65, 67-68, 71, 73, 77, 79-80, 82, 86, 89, 91, 93, 98, 100, 102, 104, 111, 115, 118-119, 121, 132, 135, 137-138, 140, 143-144, 149-150, 158, 160, 162-163, 165, 169-170, 172, 179, 186, 190-194, 196-197, 199-201, 203, 207-208
Jonson, 96, 100
Jordan, 17-18, 43, 46, 48-49, 51-52, 158, 164
Jordon, 17
Jorrel, 178
Joyce, 31, 34
Joyner, 48, 57
Jude, 24, 160
Judkins, 49, 53
Jurden, 6
Justice, 161, 163
Kary, 174
Kater, 156
Kaufman, 70
Kay, 83
Keach, 17
Kean, 142, 149, 168
Keater, 156
Keaton, 157, 192-193
Keeble, 186
Keeks, 191
Keene, 114-115, 128-129, 193
Keicker, 197
Keiningham, 205-206
Keller, 159, 179

Kelley, 1-2, 23
Kellum, 205
Kelly, 42
Kemp, 16, 69-71, 209
Kemper, 170, 178
Kenady, 140
Kendall, 127
Kendrick, 36, 155
Kenison, 143
Kenn__, 90, 92
Kennedy, 167
Kennen, 141
Kenner, 94, 187
Kent, 4
Kepheart, 115
Keppler, 20
Kerby, 6, 60
Kerr, 25
Kersey, 2, 139, 203
Kervan, 181
Kesterson, 93
Key, 56, 155, 208-209
Kibert, 98
Kidd, 17, 19, 194, 206, 208
Kidwell, 121
Kieningham, 208
Kilgour, 131
Killian, 99
Kimbell, 56
Kimberlain, 106
Kimbrough, 7, 11, 13, 21, 150
Kincaid, 41, 44
Kindrick, 35
King, 2, 4, 6, 12, 19-22, 26-27, 29-30, 35-37, 66, 80, 82, 84, 87-88, 99-100, 109, 150, 184, 197
Kinkead, 39, 44
Kinker, 19
Kinney, 138
Kinnsy, 140
Kinsey, 175, 178
Kinsor, 108, 110
Kirk, 90, 93, 112, 162-163
Kirtley, 172, 178
Kitchen, 57
Kite, 191

Scruggs, 32, 36, 155
Seal, 83, 168, 170, 174, 179
Seale, 96
Seaton, 21, 127
Seay, 30
Sebolt, 99
Sebree, 89, 90
Sectar, 197
Segar, 70
Segars, 118
Seger, 83
Seilst, 140
Seitz, 119
Selcott, 127
Self, 29-30, 68
Sergeant, 147, 149
Serger, 20
Settenglove, 43
Seward, 22, 75, 205-206, 208-209
Sey, 117
Seyars, 33, 35
Seybert, 38, 42, 45
Shackleford, 26-27, 71, 91, 204
Shackleton, 160
Shakelford, 185
Shanks, 194
Sharkling, 63
Sharp, 42, 139, 146, 155
Sharpe, 19
Sharven, 120
Shaw, 24, 197
Shay, 89, 91
Sheffield, 34
Shelburn, 96, 105
Shelburne, 11
Shelley, 47
Shelton, 4, 13, 33-36, 70, 108, 151, 160, 166, 190
Shemuts, 40
Shenen, 74
Shepard, 101
Shepherd, 47-48, 56, 134, 151, 168
Sheppard, 22, 24
Shepperson, 21, 24
Sheridan, 41
Shield, 23

Shields, 23
Shiffler, 140
Shifleman, 156
Shiflett, 142
Shih, 66
Shine, 20
Shipley, 187
Shipman, 18
Shisler, 142
Shiss, 168
Shivers, 50
Shoemaker, 21, 122-124, 132
Shoneburg, 39
Shop, 112
Short, 3, 99, 112, 119
Shotwell, 172, 176
Shreve, 117
Shrewd, 135
Shriver, 122
Shufts, 169
Shumake, 98
Shumaker, 124, 128, 134
Shumate, 29, 31, 136
Shupe, 04
Shuper, 100
Shurm, 19
Shuttice, 183
Sibley, 185, 206-208
Sigon, 5
Sigps, 151
Sikes, 16
Simeon, 74
Simkins, 109
Simmonds, 91
Simmons, 39, 188, 198, 200-201
Simms, 167, 172, 176
Simon, 167
Simons, 43, 124, 170
Simpkins, 104
Simpson, 54, 59, 129, 177-178
Sims, 9-10, 19, 43, 100, 107-108, 138-139, 143-16, 150-151, 177
Singleton, 163, 181, 195
Sinton, 20
Siples, 43
Sisk, 112, 172, 175

242

Toone, 192
Topping, 207
Towel, 106
Towell, 205
Towler, 202
Towles, 93-94
Townsend, 42, 153-154, 156, 164, 192
Tracy, 132
Trader, 186
Trainer, 40
Trainheour, 142
Trainyer, 207
Trammell, 117
Trant, 80, 87
Trass, 100
Travis, 105
Travus, 114
Treacle, 65
Tremyer, 84
Trent, 19, 33, 35-36
Trett, 103
Trevelian, 143
Tribble, 141
Trice, 72, 75, 81, 137-138, 141-142, 144, 206
Tricken, 68
Tricker, 62-63
Trillipfoe, 119
Trimble, 42, 45
Trimmers, 84
Trimyer, 87
Tromson, 119
Trout, 80
Trumell, 160
Trundixler, 137
Trundle, 124
Truslow, 68
Trussel, 127
Tuck, 4, 81-82, 86, 103
Tucker, 4-5, 156-157, 164, 168, 170-171, 191, 193, 203
Tucks, 82
Tulla, 138
Tunstal, 74
Tunstale, 165

Tunstall, 23
Tunstill, 202
Turne, 63
Turner, 4-5, 10, 15, 18, 20, 26-27, 31, 33, 41-42, 47-53, 55, 64, 70, 73, 75, 84, 87, 105, 126, 143, 145, 148, 150, 159, 182, 197, 208
Turpin, 86, 202
Tush, 36
Tuyman, 168
Twickman, 24
Twopence, 82
Twyman, 169, 174, 178
Tyler, 5-7, 20, 106, 129, 138-139, 179
Tynes, 46, 48-49, 51, 56-57
Umbaugh, 118
Underwood, 50, 52, 57
Updike, 132
Upton, 96
Urquhart, 51, 57
Utley, 14
Utz, 168-174, 176-180
Vaden, 160
Vaeden, 137
Vaiden, 58-59
Vail, 49
Valentine, 20, 143, 197
Vallentine, 57
Van Wagenner, 204
Vance, 44
Vandervanter, 134-135
Vandervenre, 131
Vandevender, 42
Vandeventer, 97, 107, 120
Vanhuss, 103, 109
Vansickler, 129
Varner, 42
Vashnoth, 153
Vass, 2, 23
Vaughan, 3, 7-8, 12-14, 17, 53, 165, 190, 194, 196, 198, 209
Vaughn, 54, 56, 78
Vdreidson, 202
Vea, 23
Veal, 116

Weaver, 79, 167, 170, 173-174, 178-180

Webb, 35, 53, 90-91, 101, 150, 158, 161, 163

Weber, 18

Wedderbrism, 75

Weeb, 123

Weedon, 80

Weimer, 120

Welch, 63, 104, 133, 180

Welcher, 11-12

Weldy, 152

Welkins, 190

Wells, 30-35, 100, 111, 201

Welson, 137

Wennber, 119

Wenner, 119-120

Wernel, 132

Werth, 24

Wertman, 194

West Point Land Company, 85

West, 5, 10, 55, 68, 92, 148

Weston, 183, 205

Westy, 55

Whaley, 113, 116, 125

Wharton, 140

Wheat, 8

Wheeler, 98, 103, 140, 149

Wheeley, 81

Whiller, 138

Whisman, 112

Whistleman, 43

Whitaker, 59

White, 4-5, 7, 9, 13, 39, 42, 49, 51-52, 60, 72, 81-82, 87, 117-118, 121, 124, 131, 135, 145, 147, 151, 153-156, 164-165, 181-185

Whitecer, 134

Whiteford, 18

Whitehead, 47, 50, 54-55, 107

Whitemore, 195

Whitenarl, 195

Whitfield, 46, 53, 55-56

Whitley, 46, 51

Whitlock, 5-6, 87, 141, 147

Whitman, 109

Whitmore, 117

Whitt, 105

Whittemore, 201

Whitticoe, 31

Whittle, 199

Whitton, 112

Wiard, 123

Wiatt, 91, 187

Wickeler, 202

Wickes, 5-6

Wickham, 2, 11, 24

Widner, 99

Wien, 140

Wight, 119

Wightman, 124

Wiland, 200

Wilbourn, 133

Wilburn, 104

Wilder, 94

Wiles, 190

Wiley, 51

Wilfong, 40

Wilhoit, 169, 176

Wilkenson, 65, 191

Wilkerson, 49, 54, 66, 152, 159

Wilkes, 162

Wilkins, 194, 207

Wilkinson, 33, 58, 127, 158, 160, 191-192

Willbourne, 190

Willeford, 54

Willeroy, 80

Williams, 5, 22, 36, 42, 48, 60, 72, 75-77, 91-92, 94, 101, 117-118, 120, 140, 148, 152-156, 158, 161-162, 165, 182-183, 185, 187-189, 191-192, 194, 200, 206-209

Williamson, 190-191, 200

Willis, 21, 75, 97, 108, 172

Wills, 47, 50

Willson, 132-133, 153, 155, 162

Wilson, 26-27, 34, 36, 42-45, 47, 49-50, 57, 60, 71, 75, 102-103, 110-111, 113, 125-126, 153, 156, 190, 192-193, 199, 204

Wiltshire, 87

www.ingramcontent.com/pod-product-compliance
Lightning Source LLC
Chambersburg PA
CBHW080234270326
41926CB00020B/4232